Escherichia coli - Old and New Insights

Edited by Marjanca Starčič Erjavec

Published in London, United Kingdom

IntechOpen

Supporting open minds since 2005

Escherichia coli - Old and New Insights
http://dx.doi.org/10.5772/intechopen.95203
Edited by Marjanca Starčič Erjavec

Contributors
Aura Darabă, Antoine Tamsa Arfao, Moïse Nola, Rodney C. Jariremombe, Lawrence Goodridge, Opeyemi U. Lawal, Valeria R. Parreira, Benyamin Djawadi, Mojtaba Mohseni, Noushin Khazaei, Waheed Ullah, Shandana Ali, Sara Kadkhodaei, Gelareh Poostizadeh, Etefia Etefia, Rana Elshimy, Darja Žgur Bertok, Zdravko Podlesek, Utpal Roy, Rehan Deshmukh, M. Basavaraju, B.S. Gunashree, Piotr Skowron, Agnieszka Zylicz-Stachula, Nareshkumar Gattupalli, Archana Gattupalli, Marjanca Starčič Erjavec, Andrej Kastrin

First published in London, United Kingdom, 2022 by IntechOpen
IntechOpen is the global imprint of INTECHOPEN LIMITED, registered in England and Wales, registration number: 11086078, 5 Princes Gate Court, London, SW7 2QJ, United Kingdom
Printed in Croatia

British Library Cataloguing-in-Publication Data
A catalogue record for this book is available from the British Library

Additional hard and PDF copies can be obtained from orders@intechopen.com

Escherichia coli - Old and New Insights
Edited by Marjanca Starčič Erjavec
p. cm.
Print ISBN 978-1-83969-869-9
Online ISBN 978-1-83969-870-5
eBook (PDF) ISBN 978-1-83969-871-2

For EU product safety concerns:
IN TECH d.o.o., Prolaz Marije Krucifikse Kozulić 3, 51000 Rijeka, Croatia, info@intechopen.com or visit our website at intechopen.com.

Meet the editor

Marjanca Starčič Erjavec, Ph.D., is a Professor of Molecular Biology in the Department of Microbiology, Biotechnical Faculty, University of Ljubljana, Slovenia, where she also studied biology (undergraduate course) and biochemistry and molecular biology (graduate course). She defended her Ph.D. thesis in the field of bacterial molecular genetics at Utrecht University, the Netherlands. She was a visiting professor at several universities in Europe and the United States as well as at the Institute of Ecology and Genetics of Microorganisms, Perm, Russia. At the University of Ljubljana, Dr. Erjavec is involved in teaching various subjects to students of microbiology. She conducts research on horizontal gene transfer, including plasmids, natural and clinical E. coli strains, and potential new antimicrobial agents.

Contents

Preface **XI**

Section 1
Introduction to *Escherichia coli* **1**

Chapter 1 **3**
Escherichia coli: An Overview of Main Characteristics
by M. Basavaraju and B.S. Gunashree

Chapter 2 **25**
Exploring the Knowledge Landscape of *Escherichia coli* Research:
A Scientometric Overview
by Andrej Kastrin and Marjanca Starčič Erjavec

Chapter 3 **43**
The *Escherichia coli* SOS Response: Much More than DNA Damage Repair
by Zdravko Podlesek and Darja Žgur Bertok

Chapter 4 **57**
Molecular Diagnostic Platforms for Specific Detection of *Escherichia coli*
by Rehan Deshmukh and Utpal Roy

Section 2
Escherichia coli and Pathogenicity **75**

Chapter 5 **77**
The Biology and the Evolutionary Dynamics of Diarrheagenic
Escherichia coli Pathotypes
by Opeyemi U. Lawal, Valeria R. Parreira and Lawrence Goodridge

Chapter 6 **115**
Escherichia coli O157:H7 and Its Effect on Human Health
by Mojtaba Mohseni, Benyamin Djawadi and Noushin Khazaei

Chapter 7 **131**
Virulence Factors of Uropathogenic *Escherichia coli*
by Etefia Etefia

Chapter 8 141
Pathogenic *Escherichia coli*: An Overview on Pre-Harvest Factors That Impact the Microbial Safety of Leafy Greens
by Aura Darabă

Section 3
Escherichia coli and Antimicrobial Resistance 173

Chapter 9 175
Antimicrobial Resistance in *Escherichia coli*
by Waheed Ullah and Shandana Ali

Chapter 10 197
Mechanisms of Antimicrobial Resistance of *E. coli*
by Rodney C. Jariremombe

Chapter 11 207
Antibiotic Resistance among *Escherichia coli* Isolates, Antimicrobial Peptides and Cell Membrane Disruption to the Control of *E. coli* Infections
by Sara Kadkhodaei and Gelareh Poostizadeh

Chapter 12 223
Escherichia coli (*E. coli*) Resistance against Last Resort Antibiotics and Novel Approaches to Combat Antibiotic Resistance
by Rana Elshimy

Chapter 13 243
Understanding of Cultivability of *Escherichia coli* in Aquatic Microcosm in the Presence of Some Plant Extracts for Possible Treatment of Bacterio-Contaminated Water
by Antoine Tamsa Arfao and Moïse Nola

Section 4
Useful *Escherichia coli* 255

Chapter 14 257
Potential of *Escherichia coli* Probiotics for Improved Health and Disease Management
by Nareshkumar Gattupalli and Archana Gattupalli

Chapter 15 271
DNA-FACE™ - An *Escherichia coli*-based DNA Amplification-Expression Technology for Automatic Assembly of Concatemeric ORFs and Proteins
by Piotr M. Skowron and Agnieszka Zylicz-Stachula

Preface

Escherichia coli is one of the best known and most extensively studied microorganisms that has been proven to be an important microbe. It is famous for its contributions to unraveling fundamental biological processes in genetics, molecular biology, biochemistry, physiology, ecology, and evolution. In addition, it is known for its mobile genetic elements that can carry many genes involved in either bacterial virulence or resistance to antibiotics, even multi-drug resistance, supplementary metabolic pathways, and colicin production. *E. coli* plays an important role not only in the host, whether as a mutualistic bacterium, an intestinal pathogen, or an extraintestinal pathogen, but also in basic and applied research. Moreover, *E. coli* is widely used in recombinant DNA technology and biotechnology.

This book offers insights into all the different aspects of *E. coli*. It is organized into four sections.

The first section introduces the topic. Chapter 1 provides a general overview of the main characteristics of *E. coli*. Chapter 2 presents the knowledge landscape of *E. coli* and highlights the scientific topics that have been the focus of research over the last four decades, as revealed by scientometric analysis. Chapter 3 highlights the largest, most complex, and best-characterized bacterial network induced by DNA damage, known as the SOS response, and shows its broader importance beyond DNA damage repair. Chapter 4 introduces the modern molecular diagnostic platforms for specific detection of *E. coli*.

The second section discusses *E. coli* pathogenicity. Chapter 5 reviews the biology and evolutionary dynamics of diarrheagenic *E. coli*. Chapter 6 describes the "famous" intestinal pathogen O157:H7 *E. coli* and its impact on human health. Chapter 7 presents the virulence factors of uropathogenic *E. coli*. Chapter 8 discusses preharvest factors affecting the microbial safety of leafy vegetables and management strategies for effective on-farm food safety.

The third section deals with antimicrobial resistance and possible new ways to combat antibiotic resistance, since there has been an increase in antimicrobial-resistant strains of *E. coli* posing risks to human health. Chapter 9 provides a general overview of antimicrobial resistance in *E. coli*. Chapter 10 describes the mechanisms of antimicrobial resistance in *E. coli*. Chapter 11 introduces potential new antimicrobial agents, such as antimicrobial peptides. Chapter 12 discusses new vaccines, probiotics/postbiotics/synbiotics, and the CRISPR-Cas system. Chapter 13 discusses antimicrobial plant extracts.

The final and fourth section examines useful *E. coli*. Chapter 14 discusses *E. coli* that can be used to improve health and disease management, the so-called probiotic *E. coli*, whereas Chapter 15 focuses on *E. coli* that can be used in biotechnology to produce useful products, such as *E. coli*-based DNA amplification-expression technology for automatic assembly of concatemeric open reading frames (ORFs) and proteins.

Overall, this book presents old and new insights into all aspects of *E. coli* and I hope it will help readers recognize the importance of this microorganism.

My sincere thanks to all the contributing authors and to the staff at IntechOpen who made the publication of this book possible.

Dr. Marjanca Starčič Erjavec
Professor,
Biotechnical Faculty,
Department of Microbiology,
University of Ljubljana,
Ljubljana, Slovenia

Section 1

Introduction to *Escherichia coli*

Chapter 1

Escherichia coli: An Overview of Main Characteristics

M. Basavaraju and B.S. Gunashree

Abstract

Escherichia coli is a type of bacteria that lives in many places in the environment, including the gastrointestinal system of humans and warm-blooded animals, where it is part of the gut microbiota. Some strains of *E. coli* can be administered as probiotics and are known to have a positive effect on host health. However, some strains can be pathogenic, causing intestinal and extraintestinal infections in humans as well as animals. *E. coli* is hence a bacterium with a wide range of different natural types of strains, each with its own set of features. Because of its unique qualities, such as simplicity of handling, availability of the entire genome sequence, and capacity to grow in both aerobic and anaerobic conditions, *E. coli* is also a popular bacterium for laboratory research and biotechnology. So, *E. coli* is considered to be the utmost widely utilized microbe in the field of recombinant DNA technology, and it is used in a wide range of industrial and medical applications.

Keywords: *Escherichia coli*, Gram-negative bacteria, growth, infection, model organism, pathogenesis

1. Introduction

The bacteria *Escherichia coli* was discovered by German pediatrician Theodor Escherich (1857–1911), who isolated it from babies' feces in 1885 [1]. *E. coli* is a gram-negative, non-sporulating, rod-shaped, facultative anaerobic, and coliform bacterium pertaining to the genus *Escherichia* that commonly inhabits the environment, foods, and warm-blooded animals' lower gut [2]. In the domains of biotechnology and microbiology, it is the most widely studied prokaryotic model organism. It can live for long periods of time in feces, soil, and water, and is frequently used as a water contamination indicator organism. For 2–3 days, the bacterium multiplies rapidly in fresh feces under aerobic circumstances, but its numbers gradually fall after that. *E. coli* is gram-negative, straight, rod-shaped, non-sporing, non-acid fast, and bacilli that exist in single and pairs. Cells are typically rod-shaped, with 1–3 μm × 0.4–0.7 μm (micrometer) in size around 1 μm long, 0.35 μm wide, and 0.6–0.7 μm in volume [3]. It is motile due to peritrichous flagellar arrangement, and very few strains are non-motile. The optimal growth of *E. coli* occurs at 37°C (98°F) but some laboratory strains can multiply at temperatures of up to 49°C (120.2°F). It takes as little as 20 min to reproduce in favorable conditions [4]. Fimbriated strains exist both as motile and non-motile. A polysaccharide capsule has been discovered in some *E. coli* strains isolated from extraintestinal infections. The *E. coli* capsules can be clearly seen using negative staining procedures, which produce a bright halo over a dark backdrop. They have a thin cell wall with only one or two layers of peptidoglycan [5] as shown in **Figure 1**.

It colonizes a newborn's gastrointestinal (GI) tract within hours after birth and even helps to keep our digestive tract healthy. Several strains of *E. coli* have been identified as good and effective probiotics and are currently employed in pharmaceuticals. It truly is a facultative anaerobic chemoorganotroph capable of both respiratory and fermentative metabolism [7]. Although most strains of *E. coli* are safe, some serotypes can induce diarrhea when consumed through contaminated food or drink, while others might cause urinary tract infections (UTIs), anemia, and respiratory or kidney infections [8]. However, certain strains have developed into pathogenic *E. coli* by using plasmids, transposons, bacteriophages, and/or pathogenicity islands to acquire virulence factors [9]. Serogroups, pathogenicity mechanisms, clinical signs, and virulence factors can all be used to classify the pathogenic strain of *E. coli* [10].

The bacterium can be grown easily and inexpensively in a laboratory setting under appropriate conditions. It takes as little as 20 min to reproduce and has been intensively investigated for over 60 years [11]. *E. coli* is the most widely studied prokaryotic model organism and an important species in the field of biotechnology and microbiology, where it serves as the host organism for recombinant DNA and experimental workhorse for DNA manipulation and protein production [12].

2. Habitat of *E. coli*

Escherichia coli can live on a wide variety of substrates. The availability of nutrients within the intestine of host species determines *E. coli* niche. The (GI) tract of humans and many other warm-blooded animals is the principal niche for *E. coli*. It cycles between two major habitats-warm-blooded animal intestines and the

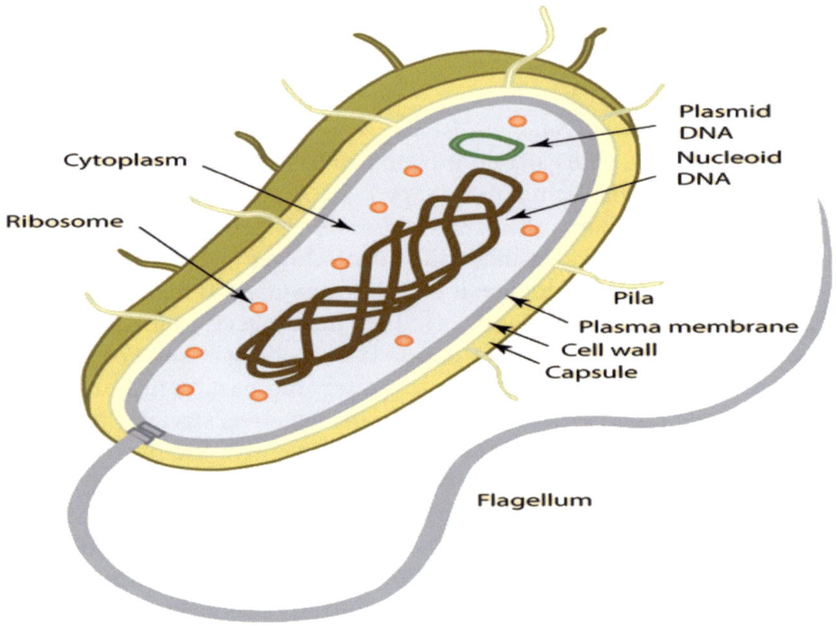

Figure 1.
Structure of E. coli [6].

environment (water, sediment, and soil), which is considerably different in terms of physical conditions, the range, and quantity of nutrients availability. *E. coli* form a mutual relationship with its host. *E. coli* in the colon synthesizes K and B complex vitamins and protects the GI tract against colonization with pathogenic microbes, while the host offers an ecological niche and nutrients. *E. coli* is the most common type of facultative anaerobes in the intestine, accounting for around 0.1% of the gut microbiota [13]. *E. coli* can also be found in hotter conditions, such as on the edge of hot springs and on-ground meats due to slaughterhouse processing [14].

3. Scientific classification

Domain: Bacteria [15]
Kingdom: Bacteria
Phylum: Proteobacteria
Class: Gamma proteobacteria
Order: Enterobacterales
Family: Enterobacteriaceae
Genus: *Escherichia*
Species: *Escherichia coli* (*E. coli*)

4. Antigenic Structure of *E. coli*

E. coli is classified into 150–200 serotypes or serogroups based on 3 antigens, somatic (O) or cell wall antigen, capsular (K) antigen, and flagellar (H) antigen. Seventy five types of the H or flagellar antigen and 173 types of O or somatic antigens 103 types of the K or capsular antigens have been recognized [16] (**Figure 2**).

5. Cultural requirements of *Escherichia coli*

E. coli cells may grow on a solid or in a liquid growth medium under laboratory conditions. It may be grown in a basic minimum of media, which includes glucose as a carbon and energy source, ammonium salts as a nitrogen source, other salts, and trace elements [18]. As *E. coli* have simple nutritional requirements it can be

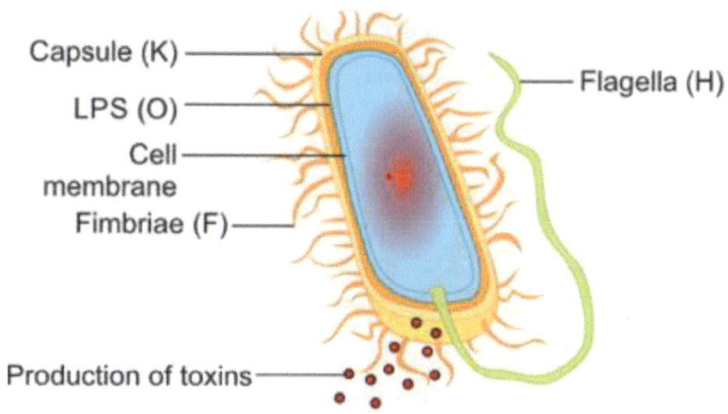

Figure 2.
Antigenic structure of E. coli [17].

easily cultured on a common medium, such as Nutrient agar, Mac Conkey agar, and EMB agar [19].

E. coli can grow at temperatures ranging from 10°C to 40°C, although the optimum temperature for most strains is 37°C (98.6°F), however, some laboratory strains can proliferate at temperatures as high as 49°C (120.2°F) [20]. *E. coli* can survive at 4.5–9.5 pH but the maximum growth is observed at 7.0, i.e., neutral pH. Also, the pH requirements vary with the strains of *E. coli*; [21]. The cultural characteristics of *E. coli* are presented in **Table 1**.

5.1 Nutrient agar

E. coli, on NAM, forms large, thick, greyish white, moist, smooth, opaque, or translucent discs like colonies as shown in **Figure 3**. The smooth forms (S) of colonies seen in fresh isolation are easily emulsifiable in saline. The rough forms (R) of colonies seen in older cultures, with dull surfaces often auto-agglutinable in saline. S-R variation occurs as a result of repeated subcultures and is associated with the loss of surface antigens and usually of virulence [24].

5.2 Blood agar

Some of the strains show beta hemolysis, especially those that are isolated from the pathologic conditions, whereas those which are isolated from normal persons may or may not show hemolysis on blood agar [25, 26] shown in **Figure 4**.

Cultural characteristics	Nutrient agar medium (NAM)	Eosin methylene blue (EMB) agar medium	MacConkey agar medium	Blood agar medium
Shape	Circular	Circular	Circular	Circular
Size	1–3 mm	2–3 mm	2–3 mm	1–3 mm
Elevation	Convex	Convex	Convex	Convex
Surface	Smooth (fresh isolation); rough (repeated subculture, mucoid (capsulated strains)	Smooth fresh isolation rough repeated subculture mucoid (capsulated strains)	Smooth fresh isolation rough repeated subculture mucoid (capsulated strains)	Smooth fresh isolation rough repeated subculture mucoid (capsulated strains)
Color	Grayish white	Green metallic sheen	Pink	Green metallic sheen
Structure	Translucent Opaque	Opaque	Opaque	Opaque
Hemolysis	—	—	—	Beta-hemolysis (in some strains)
Emulsifiability	Smooth from-easily, emulsifiable; roughly forms auto agglutinable hence do not emulsify easily	Smooth from-easily, emulsifiable; roughly forms auto agglutinable hence do not emulsify easily	Smooth from-easily, emulsifiable; roughly forms auto agglutinable hence do not emulsify easily	Smooth from-easily, emulsifiable; roughly forms auto agglutinable hence do not emulsify easily

Table 1.
Cultural characteristics of E. coli [22].

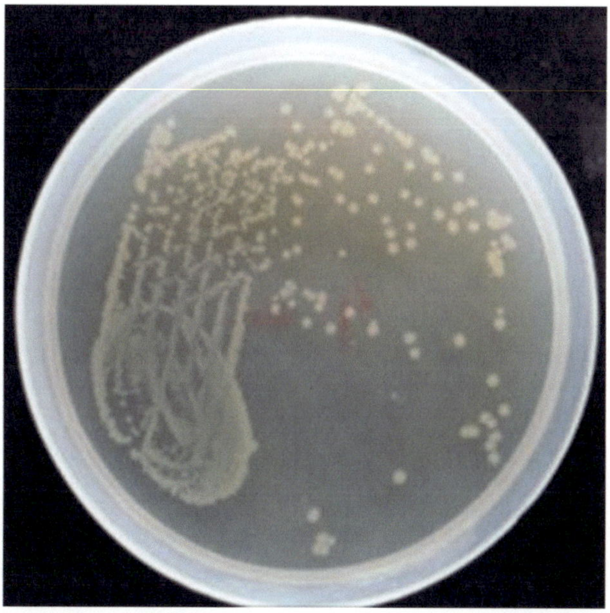

Figure 3.
Growth of E. coli *on nutrient agar [23].*

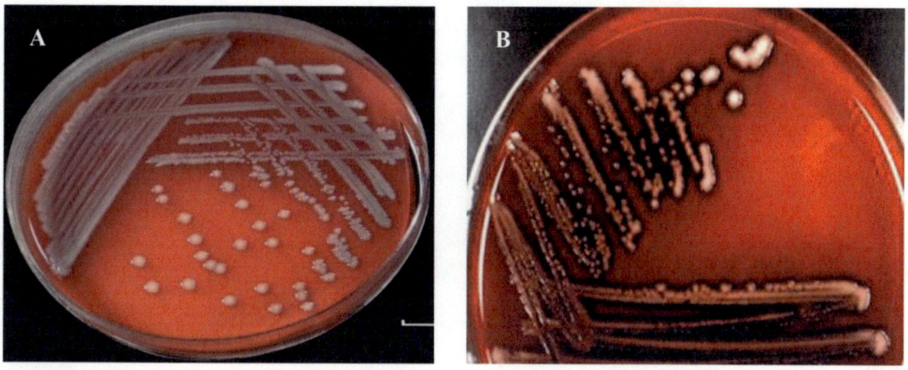

Figure 4.
A. A non-hemolytic E. coli *strain on blood agar [25]. B. A beta-hemolytic E. coli strain on blood agar [26].*

5.3 Mac Conkey agar

The colonies are pink in color due to lactose fermentation, which is important for distinguishing *E. coli* from other bacteria in the specimen, particularly gram-positive bacteria and *Salmonella* species, which are non–lactose fermenters and produce colorless colonies on MacConkey agar media [27] shown in **Figure 5**.

5.4 *E. coli* on Mueller Hinton agar

Starch is added to absorb any toxic metabolites produced and starch hydrolysis yields dextrose, which serves as a source of energy. The use of a suitable medium for testing the susceptibility of microorganisms to sulfonamides and trimethoprim [28] is shown in **Figure 6**.

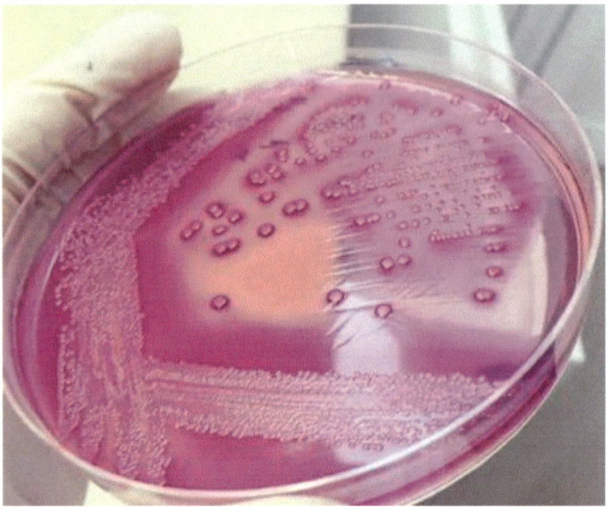

Figure 5.
Colonies of E. coli *on MacConkey agar plate are pink to dark pink, [27].*

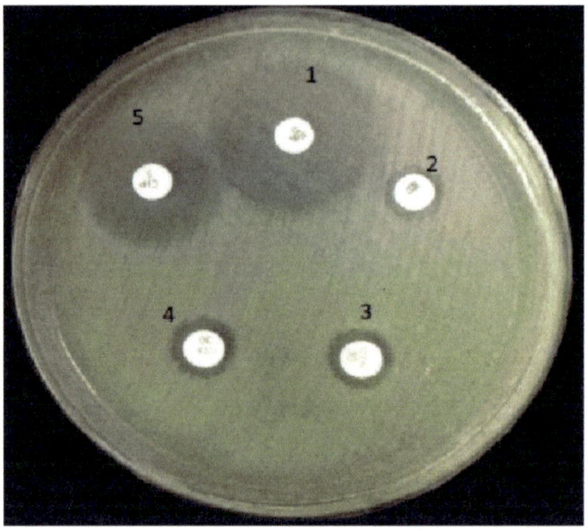

Figure 6.
E. coli *on Mueller Hinton Agar (MHA) tested for susceptibility for five different types of antibiotics [29].*

5.5 Eosin methylene blue agar

The colonies of *E. coli* grow with a green metallic sheen, which is due to the metachromatic property of dyes (eosin and methylene blue in the ratio of 6:1) and the lactose fermenting property of *E. coli*, which changes the pH of the medium to acidic. Hence, making the medium more selective for *E. coli* makes the identification much more easier [30] as shown in **Figure 7**.

5.6 *E. coli* on m-ENDO agar

Coliforms appear as red colonies with a metallic green sheen. In *E. coli*, this reaction is so intense that the fuchsin crystallizes out giving the colonies a metallic green sheen. The selective agents contained in the medium, sodium deoxycholate and sodium lauryl sulfate help to inhibit non-coliforms metabolize lactose with the production of aldehyde and acid [31] shown in **Figure 8**.

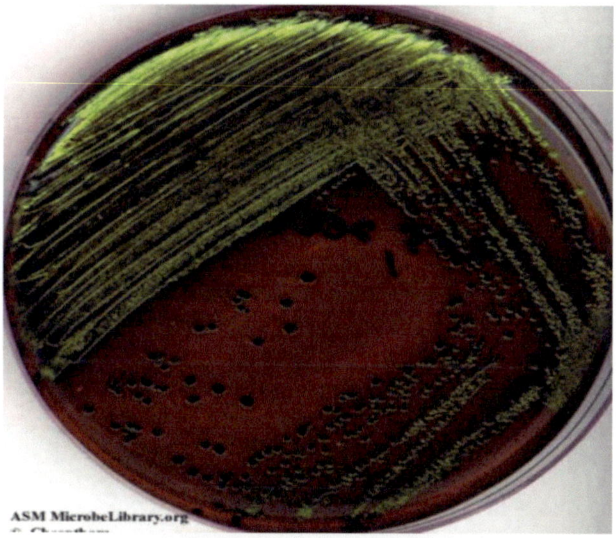

Figure 7.
E. coli *on EMB agar showing green metallic sheen colonies [30].*

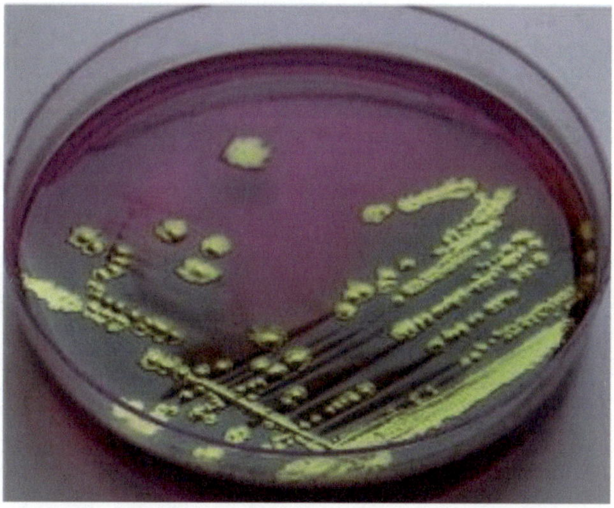

Figure 8.
E. coli *on ENDO agar with green metallic sheen colonies [31].*

5.7 *E. coli* on violet red bile agar

Violet red bile agar (VRBA) is a selective medium used to detect and enumerate lactose-fermenting coliform. Lactose-fermenting microorganisms produce pink to red colonies that are generally surrounded by a reddish zone of precipitated bile. Bluish fluorescence is seen around colonies under UV [32] as shown in **Figure 9**.

5.8 *E. coli* on cystine lactose electrolyte-deficient agar

It promotes the growth and enumeration of UTIs however due to a shortage of electrolytes; it prevents excessive swarming of *Proteus* species. On cystine lactose electrolyte-deficient (CLED) agar, lactose fermenters form yellow colonies, while non-lactose fermenters form blue colonies [33] as shown in **Figure 10**.

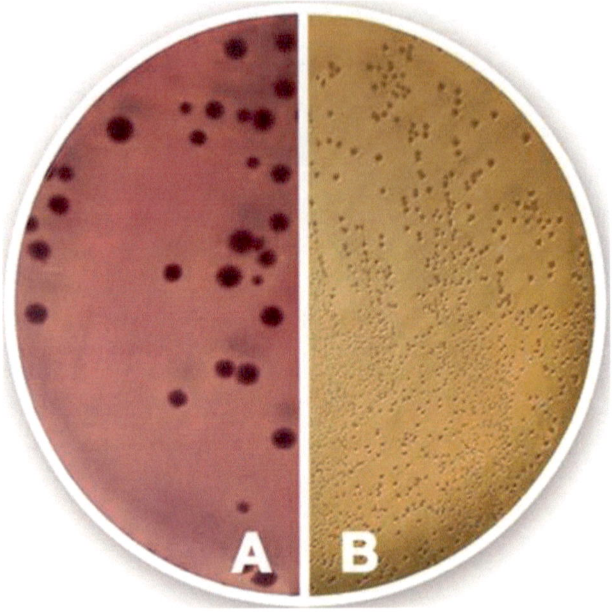

Figure 9.
VRBA agar, A: E. coli, *pinkish red with bile precipitate B: Salmonella gallinarium, fair to good growth; colorless colonies [32].*

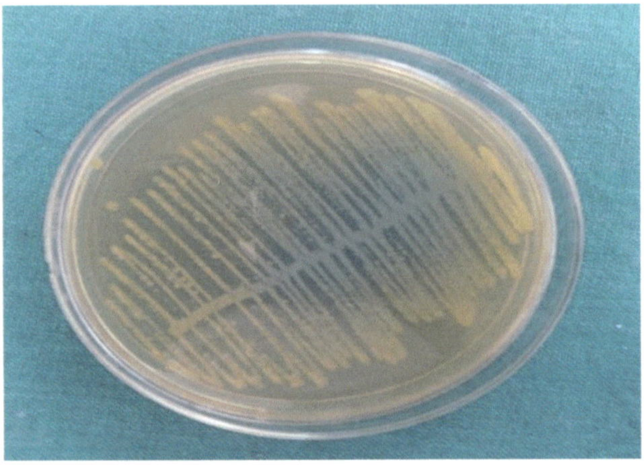

Figure 10.
Growth of E. coli *on cysteine lactose electrolyte-deficient agar, [34].*

6. *E. coli* in liquid media

Within 12–18 h, they demonstrate homogeneous murky development because of the increasing quantity of bacteria, the broth gets hazy. Pellicles grow on the surface of a liquid medium after a long period of incubation (>72 h). Heavy deposits occur, which disperse when shaken [19] as seen in **Figure 11**.

7. Pathogenicity of *E. coli*

The majority of *E. coli* strains in the colon are not harmful, however pathogenic *E. coli* isolates cause intestinal or extraintestinal infections, depending on the array

Figure 11.
Growth of E. coli *on LB liquid medium. (Photo: M. Basavaraju.)*

of virulence-associated genes that they harbor. The intestinal pathogenic *E. coli* (IPEC) strains are divided and classified into several pathotypes (**Table 2**). Diseases associated with various intestinal pathogenic *E. coli* pathotypes in animals are as shown in **Table 3**. *E. coli* is linked also to a number of extraintestinal diseases and is the most prevalent cause of cholecystitis, bacteremia, cholangitis, UTI, traveler's diarrhea, and septicemia as well as neonatal meningitis, etc., [37]. Most infections, with the exception of infant meningitis and gastroenteritis, are endogenous, like *E. coli* from the patient's normal microbiota, cause infection when the patient's defenses are impaired, e.g., through trauma or immune suppression [38]. In order to cause disease *E. coli* to possess several different types of virulence factors: fimbrial and fimbrial adhesins, capsules, toxins (exotoxins, hemolysins, and enterotoxins), iron up-take systems, etc., [39]. Important ExPEC virulence-associated genes, their encoded proteins, function, and connection to different ExPEC pathotypes are given in **Table 4** [41].

8. Antibiotic-resistant *E. coli*

Antibiotic resistance genes have been generated in many gram-negative bacteria and *E. coli* is not an exception. These bacteria evolved different mechanisms that confer resistance to anti-biotics. *E. coli* can produce extended-spectrum beta-lactamase (ESBL) that makes the bacteria resistant to beta lactams (e.g., cephalosporins, monobactams, etc.). Carbapenemase-producing *E. coli* strains, on the other hand, have genes that confer carbapenem resistance (e.g., imipenem, ertapenem, and meropenem). ESBL producing *E. coli* are a rapidly evolving group of β-lactamases, produced by certain types of bacteria where *E. coli* are the major ones. These enzymes can break down the active ingredients by cleaving the beta-lactam ring of penicillin's and cephalosporin antibiotics, resulting in the inactivation of these

Category	Clinical manifestations	Susceptible population	Virulence factors	Diagnostic	Treatment
ETEC (enteropathogenic *E. coli*)	Watery stool (without blood or inflammatory cells) leading to dehydration, headache, fever, nausea and vomiting	Children 0–5 years of age and adults traveling to developing countries.	ST, LT, CFs	Culture, detection of ST (STa, STb) and LT, CFAs using ELISA and PCR-based methods.	Self-limited, responsive to oral rehydration therapy (low response in children <2 years). Antimicrobial therapy on individual cases
EPEC (enteropathogenic *E. coli*)	Secretory and persistent diarrhea, anorexia, low fever, and rapid wasting	Children 0–2 years of age, occasionally adults	pEAF, BFP, LEE and Nle effectors	Culture, adherence patterns (LA LAL, etc.), serotyping, PCR-based methods.	Self-limited, responsive to oral rehydration therapy. Antimicrobial therapy on individual cases.
EAEC (enteroaggregative *E. coli*)	Persistent and acute diarrhea, mucoid stools, abdominal pain, nausea, vomiting, occasionally fever	People of all ages in developing and industrialized countries, HIV-infected adults	EAST Pet Pic ShET-1 Aap AAF/II	Culture, adherence pattern (stacked-brick pattern), pAA DNA probe, multiplex and real-time PCR assays.	Self-limited, responsive to oral rehydration therapy. Antimicrobial therapy on individual cases
STEC/VTE/ EHEC (a hybrid Pathotypes)	Mild uncomplicated diarrhea to hemorrhagic colitis with severe abdominal pain and bloody diarrhea.	During the summer, it's the most common, and the incidence is higher in children under the age of five.	Stx 1 and 2 verotoxins (VT)	Shiga toxins rather and A/E cytopathology.	To drink plenty of fluids to prevent dehydration and blood transfusions and kidney dialysis.
EIEC (enteroinvasive *E. coli*)	Invade and destroy the colonic epithelium, producing a disease characterized initially by watery diarrhea.	A small percentage of patients develop dysenteric illness, which includes fever, stomach pains, and blood and leukocytes in stool specimens.	Inv plasmid, Chromosome, pInv genes	Human stool samples from patients with signs and symptoms of GI infection	Fluoroquinolones, such as ciprofloxacin, macrolides, such as azithromycin, and rifaximin, are antibiotics used to treat non-STEC diarrheagenic *E. coli*.
DAEC (diffusely adherent *E. coli*)	Watery diarrhea	Involved in diarrhea in children but not in adults.	Adhesins	Afa/Dr adhesins	multiplex PCR for DEP genes
AIEC (adherent-invasive *E. coli*)	Type 1 fimbriae, cellular invasion	Associated with Crohn disease.	Persistent intestinal inflammation.	None	Bacteria with antibacterial compounds or with phage therapy, probiotics, or anti-adhesive molecules.

ST: heat-stable toxin, LT: heat-labile toxin, CFA: colonization factors, LEE: locus of enterocyte effacement, HIV: human immunodeficiency virus, pEAF: plasmid enteroadherente factor of EPEC, BFP: bundle forming pilus, EAST: enteroaggregative heat-stable toxin, Pet: plasmid encoded-toxin, Pic: protein involved in colonization, ShET-1: Shigella enterotoxin-1, Aap, dispersin, AAF/II: aggregative adherence factor II.

Table 2.
Pathotypes of human IPEC [35].

Species	Disease (age)	Pathotype	Localization
Poultry	Embryonic mortality	—	Egg
	Swollen head, dermatitis, cellulite (adult)	—	Localized infections
	Diarrhea	—	Intestine
Cattle	New born diarrhea	ETEC	Small intestine
	Hemorrhagic dysentery (1–6 wk)	EPEC STEC	Colon
	Mastitis (adult)	—	Mammary gland
Dog and Cat	Diarrhea (young animal)	ETEC	Small intestine
	Diarrhea (young animal)	EPEC	Small and large intestines
Pig	Newborn diarrhea (0–1 wk)	ETEC	Small intestine
	Young pig diarrhea (2–4 wk)	ETEC	Small intestine
	Post-weaning diarrhea (4–8 wk)	ETEC EPEC	Small intestine
	Edema disease (4–8 wk)	STEC (EDEC)	Small intestine
	Hemorrhagic gastro-enteritis (1–8 wk)	ETEC	Small intestine
Rabbit	Newborn diarrhea	EPEC	Small and large intestine
	Weaning diarrhea	EPEC	Small and large intestine

Source: EcL, APEC: avian pathogenic Escherichia coli, SEPEC: septicemic Escherichia coli, UPEC: uropathogenic Escherichia coli, EDEC: edema disease Escherichia coli.

Table 3.
Diseases associated with various intestinal pathogenic E. coli *pathotypes in animals, [36].*

Virulence gene(s)	Encoded protein(s)	Function	ExPEC pathotype(s)
Adhesions			
fim	Type 1 fimbriae	Factor of colonization in extraintestinal infections, biofilm formation	UPEC, NMEC, SEPEC, APEC
afa	Afimbrial adhesin	The non-fibrous adhesin binds to the DAF receptor on the cell surface epithelium, hemagglutination capacity.	UPEC
dra	Dr fimbriae	Binding to the DAF receptor on the surface epithelial cells and mediation of internalization bacteria to the host cells.	UPEC
pap	P fimbriae	Stimulate the production of cytokines by T lymphocytes, colonization factor in extraintestinal infections.	UPEC, SEPEC, APEC
sfa	S fimbriae	Adhesion to intestinal epithelial cells, kidney, and lower urinary tract cells; facilitate the penetration of bacteria into the tissues.	UPEC, NMEC
foc	F1C fimbriae	Adhesion to renal epithelial cells and endothelial cells of the bladder and kidneys.	UPEC
iha	Iha	Iron-regulated-gene-homologue adhesion.	UPEC

Virulence gene(s)	Encoded protein(s)	Function	ExPEC pathotype(s)
mat	Mat	Meningitis associated and temperature regulated fimbriae.	NMEC
crl, csg	Curli fiber gene	Enable biofilm formation and promote pathogenicity.	UPEC, SEPEC, APEC
agn43(flu)	Antigen43	Protein of autotransporter family, adhesion, and biofilm development.	UPEC
Invasine			
ibeA,B,C	Ibe ABC	Cell invasion into the host tissues	NMEC, SEPEC, APEC
Iron uptake			
iuc,aer	Aerobactin	Siderophore, acquisition of Fe2+/3+ in the host system.	UPEC, APEC
irp	Iron repressible protein	Yersiniabactin synthesis	NMEC
iroN	Salmochelin	Siderophore receptor, use of Fe ions obtained from the body host.	UPEC, NMEC, SEPEC APEC
chu, hma	ChuA, Hma	Enable using of Fe from hemoglobin in the host system.	UPEC, SEPEC
sitA,B,C	SitABC	Transportation of Fe, Mn	UPEC, APEC
Protectins/serum resistance			
traT	Transfer protein	Inhibition of the classical pathway of complement activity.	NMEC, SEPEC APEC
KpsMI-neuA, KpsMII	Capsula antigens	The protection factor against phagocytosis and the spreading factor.	NMEC, SEPEC
omp	Outer membrane protein	Enable intracellular survival, evasion from the body's defense.	UPEC, NMEC
iss	Increased serum survival	The protection factor against phagocytosis.	NMEC, SEPEC, APEC
colV, cvaC	ColV, CvaC	Factor facilitating colonization	NMEC, SEPEC, APEC
Toxins			
pic	Serin protease autotransporter	Degrades mucins, facilitates colonization epithelium, damages the cell membrane.	UPEC
sat	Secreted autotransporter toxin	Proteolytic toxin, effect cytotoxic—influences on cell vacuolization.	UPEC
vat	Vacuolating autotransporter toxin	Proteolytic toxin, induces host cell vacuolization.	UPEC, APEC
hlyA	Hemolysin A	Creating pores in membranes of host cells (cell lysis).	UPEC
cnf	Cytotoxic necrotizing factor	Engaging in cell necrosis	UPEC, SEPEC
cdt	Cytolethal distending toxin	Cytolethal distending factor	SEPEC

Table 4.
Important ExPEC virulence-associated genes, their encoded proteins, function, and association with ExPEC pathotype [40].

drugs, there are at least 200 different types of ESBL enzymes, increasingly isolated as causes of complicated UTIs and remain an important cause of failure of therapy with cephalosporin's and have serious infection control consequences. ESBL producing *Enterobacteriaceae* have been responsible for numerous outbreaks of infection throughout the globe and pose challenging infection control issues [42]. These organisms are associated with multidrug resistance causing a high rate of mortality and treatment failure [43].

9. MUG (beta-glucuronidase) of *E. coli*

MUG is an acronym for 4-methylumbelliferyl-β-D-glucuronide, most strains of *E. coli* (97%) produce the enzyme β-D-glucuronidase hence, the detection of this enzyme is commonly employed in laboratories to identify and differentiate such organisms [44]. β-D-glucuronidase is an enzyme that hydrolyzes the beta-D-glucopyranoside-uronic derivatives to aglycons and D-glucuronic acid. In about 97% of *E. coli* strains, the enzyme-glucuronidase is present [45].

10. Phylogenetic groups of *E. coli*

According to older phylogenetic studies, the *E. coli* strains were classified into four main phylogenetic groups: A, B1, B2, and D. However, recent studies showed that there are more phylogenetic groups seven (A, B1, B2, C, D, E, and F) belong to *E. coli* sensu stricto, whereas the eighth is represented by cryptic Clade I. Apart from clade I, also clades II, III, IV, and V are known to exist [46]. The majority of strains that cause extraintestinal infections belong to the phylogenetic group B2, whereas as strains belonging to the phylogenetic groups A and B1 are known to have low extraintestinal pathogenicity potential but beside commensal strains, strains also cause diarrhea (**Figure 12**). According to Doumith M, et al., *E. coli* strains belonging to various phylogenetic groups displayed diverse phenotypic and genotypic features thought to support fitness in various ecological settings, resulting in niche preference according to scientific findings [48]. To determine *E. coli* phylogroups, several approaches have been described. Polymerase chain reaction (PCR)-based tests, multi-locus sequence typing (MLST), ribotyping, and sequencing of the 16S rRNA gene are among them [49]. For the determination of the original four different phylogroups (A, B1, B2, and D), the Clermont triplex PCR phylogroup method was used [50].

However, research has revealed that this method can only confirm 80–85% of all *E. coli* phylogroups, and in 2013 Clermont et al. [51], proposed a revisited method, the quadruplex PCR, which can be used to classify *E. coli* in the seven phylogenetic groups and clade I [52]. Clermont et al. [53] also proposed a PCR method for the detection of clades II–V.

11. The *E. coli* genome and proteome

The full genome of *E. coli* K12 was published by Science in 1997, making it one of the first species to have its genome completely sequenced. *E. coli* has a circular DNA molecule with 4288 annotated protein-coding genes (arranged into 2584 operons), 7 ribosomal RNA (rRNA) operons, and 86 transfer RNA (tRNA) (data for the *E. coli* laboratory strain K-12 derivative MG1655) [8]. However, *E. coli* core genome (i.e., genes found in all strains) accounts for less than 20% of the pan genome's genes

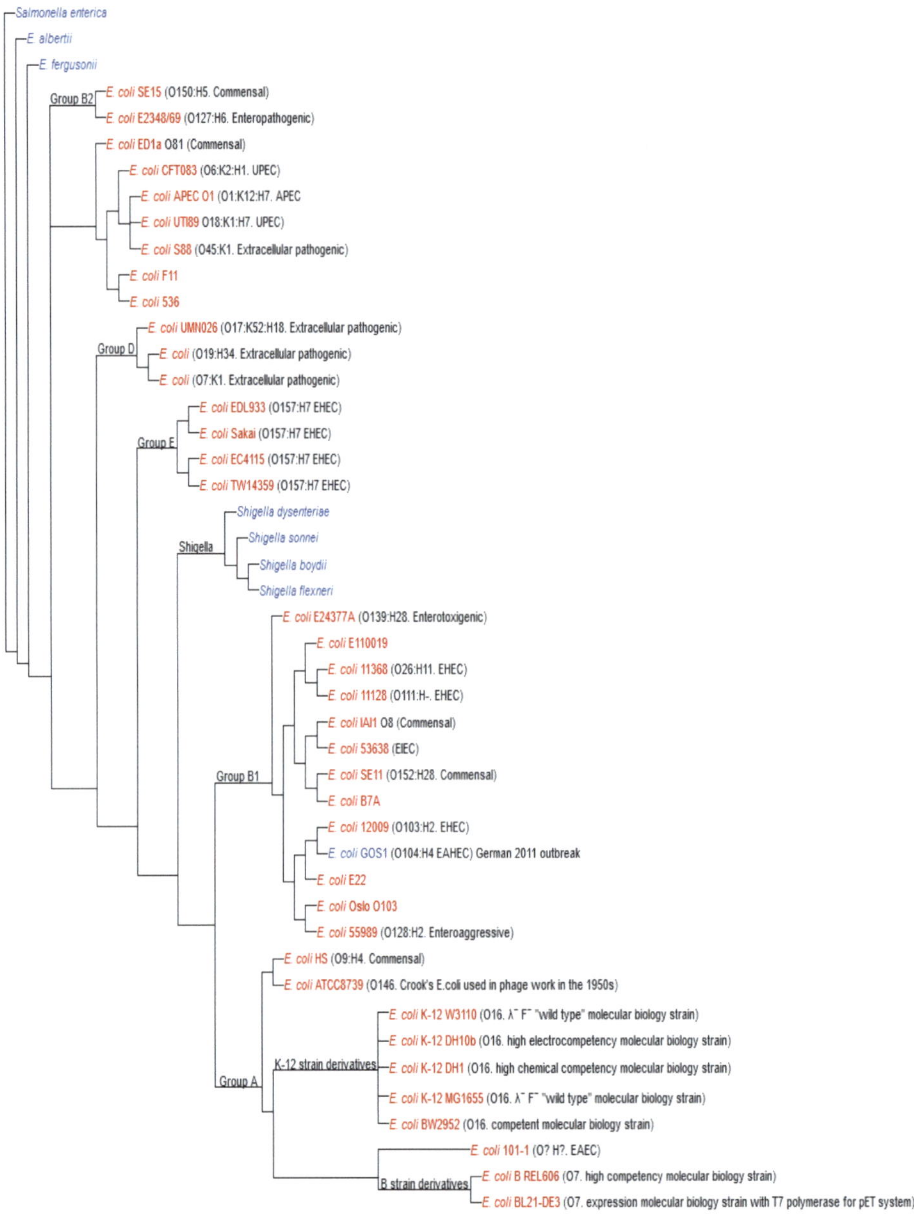

Figure 12.
Phylogenetic tree of E. coli *strains [47].*

or nearly all (90%) of the genomes, leaving only a tiny fraction of genes found in roughly half of the genomes [54]. The *E. coli* core genome is estimated to have less than 1500 genes, while it has a huge pan-genome with more than 22,000 genes [55]. According to genomic analysis many of the genes of the pan-genome could be not yet unidentified but crucial virulence factors [56]. There are 27,621 *E. coli* genome assemblies and annotation sequences available to date and each genome comprises between 4000 and 5500 genes [57]. The *E. coli* genome as a whole is remarkably ordered in terms of local replication direction and oligonucleotides that may be involved in replication and recombination [58].

The diverse behavior of this species is explained by its enormous genetic and phenotypic diversity. With a mean distance between genes of only 118 base pairs, the coding density was found to be extremely high. A multitude of factors contribute to the higher gene density: a. bacterial genes lack introns throughout the genome, and neighboring genes are fairly near together, i.e., there are no many large non-coding DNA sections between genes. There are several transposable genetic elements, repetitive elements, cryptic prophages, and bacteriophage remnants in the genome and a variety of additional patches with unique compositions, showing genome plasticity due to horizontal gene transfer [58, 59].

E. coli is an excellent model for studying the general characteristics of the bacterial proteome, such as its dynamics under different physiological situations, its dynamic range of expression, and its changes. According to the genomic sequence data of the *E. coli* K-12 strain, there are 4364 ORFs or ORF fragments in the *E. coli* K-12 W3110 strain. The *E. coli* proteome has been used as a standard for evaluating and validating new technologies and methodologies in recent years, including sample prefractionation, protein enrichment, two-dimensional gel electrophoresis (2-DE), protein detection, bio-mass spectrometry (MS), combinatorial assays with n-dimensional chromatography and image analysis. In comparison to the proteomes of other organisms such as plants and animals, the *E. coli* proteome is much smaller and with less protein modification and hence provides an excellent model for various research needs. The usage of the *E. coli* proteome as a model is further boosted by the existence of public databases such as SWISS-PROT (http://www.expasy.ch/ch2d/) and NCBI (http://www.ncbi.nlm.nih.gov/), which contain rich information on proteins and corresponding genes of *E. coli* and the existence of the *E. coli* SWISS-2DPAGE maps, which are based on a large amount of biochemical and biological data [60].

12. *E. coli* as a model organism

Escherichia coli is a well-known prokaryotic bacterium that is widely used as a model organism for a variety of research due to its adaptability. *E. coli* is more understood than other living species because of its minimal dietary requirements, rapid growth rate, and, most critically, well-established genetics [61]. *E. coli* cells divide once every 20–30 min on average, allowing them to adapt to their surroundings quickly. It also promotes the growth of numerous bacterial viruses (bacteriophages), allowing researchers to examine the structure and pathogenicity of viruses in greater detail. It is a good model organism for molecular genetics because of its ability to grow quickly on low-cost media and the availability of molecular tools to perform genetic modifications [62].

Recent research on "wild" *E. coli*, for example, has revealed a lot about the bacterial existence in the environment, its variety and genetic development, and its function in the human microbiome and diseases [7]. Vaccine development, bioremediation, biofuel generation, and immobilized enzymes have all exploited modified *E. coli* cells [61]. Furthermore, because *E. coli* reproduce primarily asexually, alterations to the genome are preserved, and the effects exhibited in these mutants are repeatable. Because of these characteristics, *E. coli* is an excellent model organism for molecular genetics and microbiology research, as well as modern biological engineering [62].

13. What discoveries were made using *E. coli* as a model organism

Several key inventions in the field of molecular biology, including molecular genetics, were achieved using *E. coli* as a model organism. This includes an

Year	Nobel-worthy discoveries	Discoverer
1958	Bacterial sex and other methods through which bacteria can transfer DNA	Joshua Lederberg
1959	The process by which life duplicates its genetic code is known as DNA replication	Arthur Kornberg
1965	Gene regulation, how genes are turned on or off	Ellis Englesberg
1968	The genetic code, the language in which our DNA is written.	Nirenberg and Matthaei'
1969	Viral replication is the process by which viruses reproduce within cells.	Max Knoll
1978	Restriction enzymes, also known as "molecular scissors," that enable scientists to cut DNA	Werner, Nathans, and Smith
1980	Recombinant DNA was used to make the first genetically modified DNA	Paul Berg
1982	The first licensed drug produced using recombinant DNA technology was human insulin	Developed by Genentech and licensed as well as marketed by Eli Lilly
1989	Additional uses for RNA such as an enzyme have been revealed	Sidney Altman and Thomas R. Cech
1997	Found ATP, the energy molecule synthesis is the process by which cells keep life going	Paul Boyer and John Walker
1999	Found that protein signal sequences are one way by which cells organize themselves	Günter Blobel
2008	Scientists employed green fluorescent protein as a marker to track cell components	Roger Y. Tsien, Osamu Shimomura, and Martin Chalfie
2009	Bacteria make computers look like pocket calculators; Biologists have created a living computer from *E. coli* bacteria that can solve complex mathematical problems	A team of US scientists DOI: 10.1186/1754-1611-3-11
2015	Mechanistic studies of DNA repair	Tomas Lindahl, Paul Modrich, and Aziz Sancar

Table 5.
Nobel-worthy discoveries of E. coli *organism [8].*

understanding of the genetic code, the mechanisms of DNA replication, the discovery of the genetic operon systems, and the creation of a genetically modified organism. Many proteins previously thought difficult or impossible to be expressed in *E. coli* in folded form have been successfully expressed in *E. coli*. The process of conjugation was discovered in *E. coli* in 1946 by Joshua Lederberg and Edward L. Tatum [63]. The availability of DNA sequence information coupled with vast biochemical and physiological data makes *E. coli* the organism of choice not only for virologists, biochemists, and molecular biologists but for all researchers of biology [8]. The most prominent discoveries made with *E. coli* are presented in **Table 5**.

14. Conclusion

E. coli is a truly resourceful microorganism possessing many facets. It is known for its fast-growing rate in chemically defined media and its adaptability, for ease of handling. So, *E. coli* is the most studied and well-understood organism on the planet. It's been widely used in research, employed as a model organism to investigate biological processing protein engineering, genetic research, and used in biotechnology, its versatility continues to open up new avenues for future investigations.

Author details

M. Basavaraju* and B.S. Gunashree
Department of Studies and Research in Microbiology, PG Centre, Mangalore University, Karnataka, India

*Address all correspondence to: microbasava@gmail.com

References

[1] Feng P, Weagant S, Grant M. Enumeration of *Escherichia coli* and the Coliform Bacteria. Bacteriological Analytical Manual. 8th ed. USA: FDA/ Center for Food Safety & Applied Nutrition; 2007

[2] Campbell NA, Reece JB. Biology. San Francisco: Pearson Education Inc; 2002

[3] Facts about *E. coli*: Dimensions as discussed in bacteria: Diversity of structure of bacteria. Britannica 2015:06-25. Available from: https:// www.britannica.com/science/ bacteria/ The-cell-envelope

[4] Bacteria. Microbiology online. Archived from the original on 27 February 2014. Available from: https:// web.archive.org/web/20140227212658/ DOI: 10.1128/AAMCol.1-2011

[5] Köhler CD, Dobrindt U. What defines extra intestinal pathogenic *Escherichia coli*? International Journal of Medical Microbiology. 2011;**301**:642-664

[6] Enger ED, Ross FC. Concepts in Biology. 10th ed. New York, USA: McGraw-Hill; 2003. Available from: http://hyperphysics.phy-astr.gsu.edu/ hbase/Biology/ecoli.html

[7] Blount ZD. The unexhausted potential of *E. coli*. eLife. 2015;**25**(4):e05826. DOI: 10.7554/eLife.05826

[8] FAQ: *E.coli:* Good, Bad, & Deadly— NCBI Bookshelf; Retrieved 07 June 2021 from https://www.ncbi.nlm.nih.gov/ books/NBK562895. DOI: 10.1128/ AAMCol.1-2011

[9] Kaper J, Nataro J, Mobley H. Pathogenic *Escherichia coli*. Nature Reviews. Microbiology. 2004;**2**:123-140. DOI: org/10.1038/nrmicro818

[10] Sarowska J, Futoma-Koloch B, Jama-Kmiecik A, et al. Virulence factors, prevalence and potential transmission of extra intestinal pathogenic *Escherichia coli* isolated from different. Gut Pathogens. 2019;**11**:10

[11] Bacteria. Microbiologyonline. Archived from the original on 27 February 2014; Retrieved: 27 February 2014. Available from: https://web. archive.org/web/20140227212658/ http://www.microbiologyonline.org.uk/ about-microbiology/introducing-microbes/bacteria

[12] Idalia VN, Bernardo F. *Escherichia coli* as a model organism and its application in biotechnology. In: Samie A, editor. *Escherichia coli*. Recent advances on physiology, pathogenesis and biotechnological applications. London: Intech Open; 2017. DOI: 10.5772/67306

[13] Eckburg PB, Bik EM, Bernstein CN, Purdom E, Dethlefsen L, Sargent M, et al. Diversity of the human intestinal microbial flora. Science. 2005;**308**(5728):1635-1638

[14] Li H, Gänzle M. Some like it hot: Heat resistance of *Escherichia coli* in food. Frontiers in Microbiology. 2016;**7**:1763

[15] *Escherichia coli* (Migula, 1895) Castellani & Chalmers, 1919 in Döring M (2022). English Wikipedia - Species Pages. Wikimedia Foundation. Checklist dataset. DOI: 10.15468/c3kkgh. Available from: https://www.gbif.org/ species/113662963

[16] Van Dijk WC, Verbrugh HA, Tol ME, Peters R, Verhoef J. Role of *Escherichia coli* K capsular antigens during complement activation, C3 fixation, and opsonization. Infection and Immunity. 1979;**25**(2):603-609. DOI: 10.1128/iai.25.2.603-609

[17] https://www.chegg.com/learn/ biology/introduction-to-biology/ cell-structure-*escherichia-coli*

[18] Elbing KL, Brent R. Recipes and tools for culture of *Escherichia coli*. Current Protocols in Molecular Biology. 2019;**125**(1):83. DOI: 10.1002/cpmb.83

[19] Bonnet M, Lagier JC, Raoult D, Khelaifia S. Bacterial culture through selective and non-selective conditions: The evolution of culture media in clinical microbiology. New Microbes and New Infections. 2019;**34**:100622

[20] Fotadar U, Zaveloff P, Terracio L. Growth of *Escherichia coli* at elevated temperatures. Journal of Basic Microbiology. 2005;**45**(5):403-404

[21] Available from: https://paramedicsworld.com/escherichia-coli/morphology-culture-characteristics-of-*E-coli*/medical-paramedical-studynotes

[22] Silva RM, Toledo RF, Trabulsi LR. Biochemical and cultural characteristics of invasive *Escherichia coli*. Journal of Clinical Microbiology. 1980;**11**(5):441-444

[23] Patond A, Narang R. A low cost ingenious approach for ultraviolet decontamination of N95 filtering face-piece respirators to deal with dwindling supply during the COVID-19 pandemic. Journal of Mahatma Gandhi Institute of Medical Sciences. 2020;**25**:80-85

[24] Surinder K. Textbook of Microbiology, Enterobacteriaceae: *Escherichia*, Klebsiella, Proteus and Other Genera. New Delhi, India: Jaypee Brothers Medical Publishers (P) Ltd; 2012. Available from: https://www.jaypeedigital.com/book/9789350255100/chapter/ch36

[25] Association of Public Health Laboratories. STEC Work Group..; Centers for Disease Control and Prevention (U.S.); Guidance for public health laboratories: Isolation and characterization of Shiga toxin-producing *Escherichia coli* (STEC) from clinical specimens. 2012. Available from: https://stacks.cdc.gov/view/cdc/21592

[26] Luppi A. Swine enteric colibacillosis: Diagnosis, therapy and antimicrobial resistance. Porcine Health Management. 2017;**3**:16. DOI: 10.1186/s40813-017-0063-4

[27] Available from: https://microbiologie-clinique.com/macconkey-agar.html

[28] Atlas RM, Snyder JW. Handbook of Media for Clinical and Public Health Microbiology. Boca Raton, FL: CRC Press. Taylor & Francis Group; 2014. pp. 324-325

[29] Neupane A, Parajuli P, Bastola R, Paudel A. Bacterial etiology of diarrhoeal disease in children and antibiogram of the isolates. Clinical Microbiology. 2017;**6**:278. DOI: 10.4172/2327-5073.1000278

[30] Archana L, Naowarat C. Eosin-methylene blue agar plates protocol. American Society for Microbiology. 2007:29

[31] Available from: http://www.dr-moeller-und-schmelz.de/agar-e/Endo-Agar.pdf

[32] Davidson R, Gambrel L. In: Wehr, Frank, editors. Standard Methods for the Microbiological Examination of Dairy Products. 17th ed. Washington, DC: American Public Health Association; 2004

[33] Available from: https://www.sigmaaldrich.com/deepweb/assets/sigmaaldrich/product/documents/331/671/55420dat.pdf

[34] Rahman MS, Garg R, Singh VA, Biswas D. Antibiotic susceptibility profile and extended spectrum β-lactamases production by uropathogenic *Escherichia coli* from tertiary care hospital of rural settings. International Journal of Medical Sciences. 2018;**6**(12):4022-4027

[35] Kaper J, Nataro J. Mobley H, Pathogenic *Escherichia coli*. Nature Reviews. Microbiology. 2004;**2**:123-140. DOI: 10.1038/nrmicro818

[36] Available from: http://www.ecl-lab.ca/en/ecoli/diseases.asp

[37] Clinical and public health microbiology. Clinical Microbiology Procedures Handbook. 4th ed. American Society of Microbiology; 2016. DOI: 10.1128/9781555818814

[38] Hemm MR, Weaver J, Storz G. *Escherichia coli* small proteome. EcoSal Plus. 2020;**9**(1):10

[39] Johnson JR. Virulence factors in *Escherichia coli* urinary tract infection. Clinical Microbiology Reviews. 1991;**4**(1):80-128. DOI: 10.1128/CMR.4.1.80

[40] Sora VM, Meroni G, Martino PA, Soggiu A, Bonizzi L, Zecconi A. Extra-intestinal pathogenic *Escherichia coli*: Virulence factors and antibiotic resistance. Pathogens. 2021;**10**(11):1355

[41] HMM A, Martínez LY, Torres AG. Clinical implications of enteroadherent *Escherichia coli*. Current Gastroenterology Reports. 2012;**14**(5):386-394. DOI: 10.1007/s11894-012-0277-1

[42] Michael LW, Loretta G. Laboratory diagnosis of urinary tract infections in adult patients. Clinical Infectious Diseases. 2004:1150-1158. DOI: 10.1086/383029

[43] Gandra S, Tseng KK, Arora A, et al. The mortality burden of multidrug-resistant pathogens in India: A retrospective, observational study. Clinical Infectious Diseases. 2019;**69**(4): 563-570. DOI: 10.1093/cid/ciy955

[44] Ley AN, Bowers RJ, Wolfe S. Indoxyl-beta-D-glucuronide, a novel chromogenic reagent for the specific detection and enumeration of *Escherichia coli* in environmental samples. Canadian Journal of Microbiology. 1988;**5**:690-693. DOI: 10.1139/m88-115

[45] Ananthanarayan P. Textbook of Microbiology. 11th ed. Bengaluru, India: Prithvi Books; 2020

[46] Ahumada-Santos YP, Báez-Flores ME, Díaz-Camacho SP, Uribe-Beltrán MJ, Eslava-Campos CA, Parra-Unda JR, et al. Association of phylogenetic distribution and presence of integrons with multidrug resistance in *Escherichia coli* clinical isolates from children with diarrhoea. Journal of Infection and Public Health. 2020;**13**:767-772

[47] Sims GE, Kim SH. Whole-genome phylogeny of *Escherichia coli/Shigella* group by feature frequency profiles (FFPs). Proceedings of the National Academy of Sciences of the United States of America. 2011;**108**(20):8329-8334

[48] Katongole P, Kisawuzi DB, Bbosa HK, Kateete DP, Najjuka CF. Phylogenetic groups and antimicrobial susceptibility patterns of uropathogenic *Escherichia coli* clinical isolates from patients at Mulago National Referral Hospital, Kampala, Uganda. F1000 Research. 2019;**8**:1828

[49] Logue CM, Wannemuehler Y, Nicholson BA, Doetkott C, Barbieri NL, Nolan LK. Comparative analysis of phylogenetic assignment of human and avian ExPEC and faecal commensal *E. coli* using the Clermont phylogenetic typing methods and its impact on avian pathogenic *E. coli* classification. Frontiers in Microbiology. 2017;**8**:2-9

[50] Clermont O, Bonacorsi S, Bingen E. Rapid and simple determination of the *Escherichia coli* phylogenetic group. Applied and Environmental Microbiology. 2000;**66**(10):4555-4558

[51] Clermont O, Christenson JK, Denamur E, Gordon DM. The Clermont *Escherichia coli* phylo-typing method revisited: Improvement of specificity and detection of new phylo-groups. Environmental Microbiology Reports. 2013;**5**:58-65

[52] Iranpour D, Hassanpour M, Ansari H, Tajbakhsh S, Khamisipour G, Najafi A.

Phylogenetic groups of *Escherichia coli* strains from patients with urinary tract infection in Iran based on the new Clermont phylotyping method. BioMed Research International. 2015;**2015**:846219. DOI: 10.1155/2015/846219

[53] Clermont O, Olier M, Hoede C, Diancourt L, Brisse S, Keroudean M, et al. Animal and human pathogenic *Escherichia coli* strains share common genetic backgrounds. Infection, Genetics and Evolution. 2011;**3**:654-662. DOI: 10.1016/j.meegid.2011.02.005

[54] Touchon M, Hoede C, Tenaillon O, Barbe V, Baeriswyl S, Bidet P, et al. Organised genome dynamics in the *Escherichia coli* species results in highly diverse adaptive paths. PLoS Genetics. 2009;**5**(1). DOI: 10.1371

[55] Nanda Kafle G, Seale T, Flint T, Nepal M, Venter SN, Brözel VS. Distribution of diverse *Escherichia coli* between cattle and pasture. Microbes and Environments. 2017;**32**:226-233

[56] Robins-Browne RM, Holt KE, Ingle DJ, Hocking DM, Yang J, Tauschek M. Are *Escherichia coli* pathotypes still relevant in the era of whole-genome sequencing. Frontiers in Cellular and Infection Microbiology. 2016;**6**:141

[57] Mueller M, Tainter CR. *Escherichia coli*. In: StatPearls [Internet]. Treasure Island (FL): Stat Pearls Publishing; 2021

[58] Blattner FR, Plunkett G, Bloch CA, Perna NT, Burland V, Riley M, et al. The complete genome sequence of *Escherichia coli* K-12. Science. 1997;**277**(5331):1453-1462. DOI: 10.1126/science.277.5331.1453

[59] Zhaxybayeva O, Doolittle WF. Lateral gene transfer. Current Biology. 2011;**21**(7):242-246. DOI: 10.1016/j.cub.2011.01.045

[60] Weiner JH, Li. Proteome of the *Escherichia coli* envelope and technological challenges in membrane proteome analysis. Biochimica et Biophysica Acta (BBA)—Biomembranes. 2008;**1778-9**:1698-1713

[61] Muhammad K, Zohra S, Ji Xiu L, Taj I, Taj MH, Wei Y. *Escherichia coli* as a model organism. International Journal of Engineering Sciences & Research Technology. 2014;**3**:3-8

[62] Vargas MNI, Franco B. *Escherichia coli* as a model organism and its application in biotechnology. 2017. DOI: 10.5772/67306. Available from: https://www.intechopen.com/chapters/53916

[63] Lederberg J, Tatum E. Gene recombination in *Escherichia coli*. Nature. 1946;**158**(4016):558. DOI: 10.1038/158558a0

Chapter 2

Exploring the Knowledge Landscape of *Escherichia coli* Research: A Scientometric Overview

Andrej Kastrin and Marjanca Starčič Erjavec

Abstract

Escherichia coli (*E. coli*) has the hallmark of being the most extensively studied organism. This is shown by the thousands of articles published since its discovery by T. Escherich in 1885. On the other hand, very little is known about the intellectual landscape in *E. coli* research. For example, how the trend of publications on *E. coli* has evolved over time and which scientific topics have been the focus of interest for researchers. In this chapter, we present the results of a large-scale scientometric analysis of about 100,000 bibliographic records from PubMed over the period 1981–2021. To examine the evolution of research topics over time, we divided the dataset into four intervals of equal width. We created co-occurrence networks from keywords indexed in the Medical Subject Headings vocabulary and systematically examined the structure and evolution of scientific knowledge about *E. coli*. The extracted research topics were visualized in strategic diagrams and qualitatively characterized in terms of their maturity and cohesion.

Keywords: *Escherichia coli*, scientometric analysis, knowledge mapping, keyword analysis, co-word analysis

1. Introduction

Escherichia coli (*E. coli*) is widely known and one of the most studied microorganisms in the life sciences. Since its discovery in 1885 by Theodor Escherich [1], *E. coli* has been the subject of intense research. *E. coli* is believed to be one of, if not the, most important organisms in studies aimed at discovering fundamental biological principles and mechanisms, as well as biology field-specific research methods and techniques. However, very little is known about the knowledge landscape in *E. coli* research. In particular, how published empirical findings on *E. coli* have evolved over time and what scientific questions have been the focus of researchers' interest? Answering these questions motivates the present work.

Scientific achievements are traditionally published in the form of a journal article, a paper in a conference volume, or a book chapter. To illustrate the effect of the accumulation of research results, we show in **Figure 1** the annual number of publications in the period 1940–2021 with *E. coli* as the main research topic. Although it is questionable whether the number of publications is directly related to

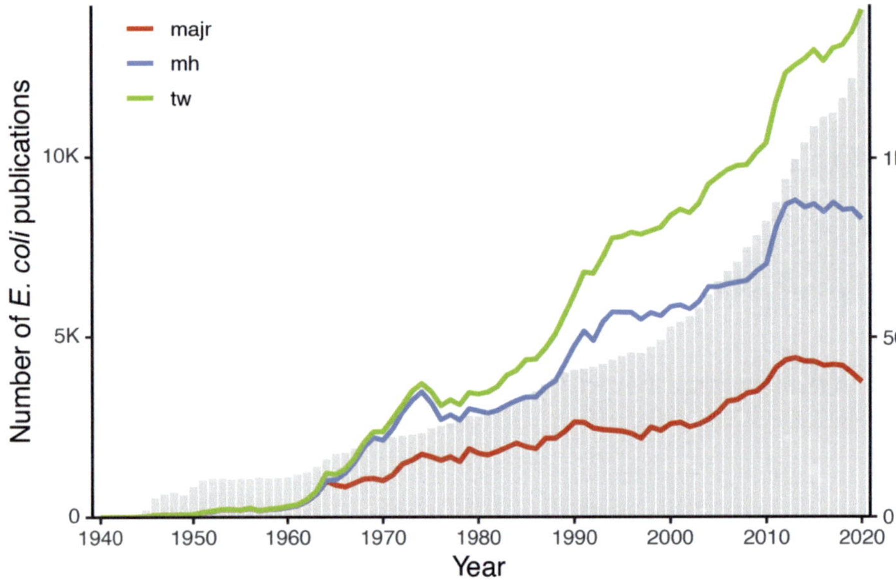

Figure 1.
The annual distribution of PubMed publications on E. coli*. The different colored lines represent the PubMed field tag used to retrieve* E. coli *publications: MeSH heading, which is a major topic of an article (red), MeSH term in general (blue), and all words in the title, abstract, and other relevant fields (green). The gray bars of the histogram, quantified by the second y-axis, show the total number of publications in PubMed in a given year.*

the amount of knowledge in a particular scientific field, we can at least use it as a proxy indicator of research activity in a particular area of interest.

However, the body of scientific literature is growing at an unimaginable pace [2, 3]. PubMed, for example, a leading bibliographic database in the life sciences, has indexed an average of 3800 new articles daily over the past 5 years [4]. Manual review of such a large body of new literature, even in a very specialized field of research, is therefore not only time-consuming, but virtually impossible. Fortunately, we can draw on a rich toolkit of automated methods and techniques offered by a modern scientometric approach to deepen our understanding of science itself [5, 6].

The main objective of this work is to examine the *E. coli* literature from an evolutionary point of view using a data-driven approach. Specifically, the aims are twofold: (i) to provide insights into research topics based on a quantitative text-mining analysis of a large number of *E. coli* papers from 1981 to 2021 indexed in PubMed, and (ii) to highlight the evolution of scientific knowledge in the field from a domain expert's perspective.

2. Background and related work

2.1 Science mapping

If we may paraphrase Ebbinghaus' famous statement, we could say that the field of scientometrics has a long past but a short history [7]. The study of scientific knowledge itself has been the subject of many famous works with great impact on the research community, including contributions by Lotka [8], Zipf [9], Price [10], Merton [11], Garfield [12], and later by Borner [13], Uzzi [14], Wang [15], Clauset [5], and Milojević [16].

Formally, scientometrics is an umbrella term for a set of approaches that aim to describe and understand the (relational) structure between researchers, their institutions, and scientific knowledge—operationalized through ideas, concepts, citations, and keywords—in order to identify and track the driving mechanisms of science [6]. One of these approaches, commonly used in the literature, is also a science map. A science map is a spatial and/or temporal representation of individual authors, their research groups, or the knowledge concepts they have written about [17].

The seminal studies that addressed the organization of scientific knowledge were driven by the study of citation networks, a type of analysis in which we seek to understand common patterns of citation links among articles in a collection of scientific literature [18]. The authors discovered several important structural features, including the famous small-world phenomenon [19, 20], the rich-get-richer mechanism [21], and the hierarchical organization of scientific knowledge [22]. We refer the reader interested in further details to the recently published monograph by Wang and Barabási [23].

Other authors argue that scientific knowledge could be represented more realistically with keywords as basic knowledge elements [24]. Keywords and key phrases are typically extracted from the title and/or abstract of each article using natural language processing tools, or parsed from a list of descriptors already provided by the authors. To overcome the challenges of normalizing keywords, many authors use controlled vocabularies such as Medical Subject Headings (MeSH) in the life sciences [25] or Mathematics Subject Classification in mathematics [26].

2.2 Co-word analysis

Co-word analysis is an improved version of pure keyword-based co-occurrence analysis. By combining various theoretical concepts from graph theory, co-word analysis allows a simple but efficient reduction of a massive network of co-occurring keywords to a higher-level network of clustered keywords. First described by Callon et al. [27] in the 1980s, co-word analysis is a powerful method for mapping the detailed intellectual structure of unstructured text data [17]. The method has been used in a variety of scientific fields, including microbiology [28]. However, to our knowledge, it has not yet been applied to elucidate the intellectual structure of knowledge about *E. coli*.

Technically, the input for co-word analysis is a network of keywords, as described in Section 2.2. In the next step, we use a type of cluster analysis—often referred to as community detection in the language of complex networks—to partition nodes into a smaller number of communities based on the similarity of their wiring patterns. The clustering algorithm is optimized with the objective of maximizing both homogeneity within communities and heterogeneity between communities. Finally, we create a strategic diagram to uncover and explore interesting patterns within the detected community structure based on a set of predefined heuristic rules [29].

3. Methods

In this study, we used a scientometric methodology to capture the structural and dynamic features of the knowledge landscape in *E. coli* research. In Section 3.1, we explain the details of compiling the dataset from the PubMed database. Then, in Sections 3.2 and 3.3, we explain the procedure for extracting keywords and creating co-word networks. Finally, in Section 3.4, we describe a method for identifying broad research topics and interpreting them.

3.1 Data collection

The literature collection was created using an automated procedure from PubMed distribution. We retrieved all PubMed records indexed with the major MeSH descriptor "*Escherichia coli*" and restricted to the English language. Full bibliographic records were downloaded via the PubMed API and stored locally in XML format. To restrict a PubMed search result by the specified date range, we set the "datetype" parameter in an API call to "pdat". The last query update was performed on October 1, 2022.

3.2 Keyword extraction

The co-word analysis presented here is based on MeSH terms to overcome problems with the normalization of plain keywords, as described previously in Section 2.1.

Each PubMed record is manually annotated by human indexers at the National Library of Medicine using the MeSH terms. MeSH is a controlled vocabulary consisting of biomedical terms at different levels of granularity. There are several types of MeSH terms, two of which are important for further understanding of the present work: Main MeSH headings (or descriptors) and MeSH subheadings (or qualifiers). Descriptors are the main elements of the thesaurus and denote the main topic of the paper. For example, the MeSH descriptors for a paper dealing with adherent-invasive *E. coli* pathovar strains in the context of Crohn's disease might be "Bacterial Adhesion", "Crohn's Disease", and "*Escherichia coli*". Qualifiers are optionally assigned to the descriptors to express a particular aspect of the knowledge concept.

For further processing, we extracted all pairs of mesh heading/subheading terms along with the publication date of each bibliographic record.

3.3 Co-word network

In Section 2.2, we introduced the notion of co-word analysis, which aims to detect communities of keywords that frequently occur in conceptually similar articles. Formally, we first created a co-occurrence network based on the MeSH term lists from all retrieved documents. A node in the co-occurrence matrix refers to a particular MeSH heading/MeSH subheading pair, and a relationship between two nodes is established when both headings occur together in a particular document. In the following paragraphs, the phrase "MeSH heading/MeSH subheading" is referred to as "term" or simply "heading".

In the next step, the co-occurrence network was weighted according to the number of observed pairs of MeSH headings. For example, if MeSH heading i and MeSH heading j appear together in 100 papers, the weight of their co-occurrence was set to 100. Finally, the raw edge weights were normalized to account for the unbalanced number of MeSH headings in the papers. For normalization, we used an association measure defined as

$$e_{ij} = \frac{c_{ij}^2}{c_i c_j}, \tag{1}$$

where c_{ij} is the number of co-occurrences of headings i and j [29]. Also, c_i and c_j are the numbers of occurrences of MeSH headings i and j, respectively. The normalized value is zero if the MeSH heading pair is not associated at all, and is equal to one if a given pair occurs together in each paper.

3.4 Identification of research topics

On a prepared co-occurrence network, we ran Louvain's community detection algorithm to identify clusters of homogeneous MeSH headings [30]. Each of the detected clusters groups together several contextually similar MeSH headings and plays the role of a research topic.

The interpretation of the research topics followed the procedure described by Callon et al. [27]. We calculated two measures, centrality and density, to represent a particular research topic in a two-dimensional plot called a strategic diagram. Centrality represents the relatedness of an observed research topic to other topics in a strategic diagram. The stronger this relatedness is, the more central the topic is in the observed network. In practice, we interpret centrality as the strength of a research topic in the entire scientific domain. Formally, the centrality of a topic is defined as

$$c = 10 \times \sum e_{kh}, \tag{2}$$

where k is a MeSH heading from the observed topic, h is a MeSH heading belonging to other topics, and e_{kh} is the normalized co-occurrence frequency of the pair of MeSH headings k and h according to Eq. (1).

Density, on the other hand, represents internal cohesion, i.e., how strongly an observed research topic is conceptually developed. Density is formally defined as

$$d = 100 \times \left(\frac{\sum e_{ij}}{w}\right), \tag{3}$$

where i and j are MeSH headings associated with a cluster, and e_{ij} is the normalized frequency of co-occurrence of the two MeSH headings. The w in denominator represents the total number of MeSH headings in a given research topic.

Finally, considering centrality and density, we created a strategic diagram to represent the structural landscape of knowledge. The diagram is centered by the median of the two axis values and divides the plot area into four quadrants characterized by different types of research topics [29]. A particular topic can be assigned a unique qualitative description based on its position in the diagram as follows:

1. The *motor* research topics in quadrant I are characterized by high centrality and high density. These topics are well defined, mature, and have been worked on over a long period of time by already well-developed research groups.

2. *Niche* topics in quadrant II have low centrality but high density. Such research topics are very homogeneous (i.e., they are characterized by strong internal linkages). However, they have weak external linkages and are therefore not well connected to other research topics.

3. *Emerging or declining* topics in quadrant III are defined by both low centrality and low density and refer to either new (i.e., emerging) or declining research topics.

4. *Basic* research topics in quadrant IV are characterized by high centrality but low density and thus combine transversal and very general research topics.

Although such topics are important to a particular research community, they are not well-developed.

4. Results

A total of 98,085 unique bibliographic records for the period 1981–2021 on the topic of *E. coli* were retrieved and considered for further processing. We identified 13,408 unique MeSH descriptors, which in turn yielded 54,663 unique combinations with MeSH qualifiers.

In the next sections, we present the descriptive results for each subperiod analyzed. In addition to the strategic chart, we have also included a table with a list of MeSH terms that define a particular research topic, as well as a brief qualitative description of the topics.

4.1 Period 1981–1990

For the time period 1981–1990 the performed survey resulted in 20,739 documents and the MeSH-based co-word analysis resulted in 10 different clusters. The strategic diagram is shown in **Figure 2**. The description of the identified research topics is summarized in **Table 1**.

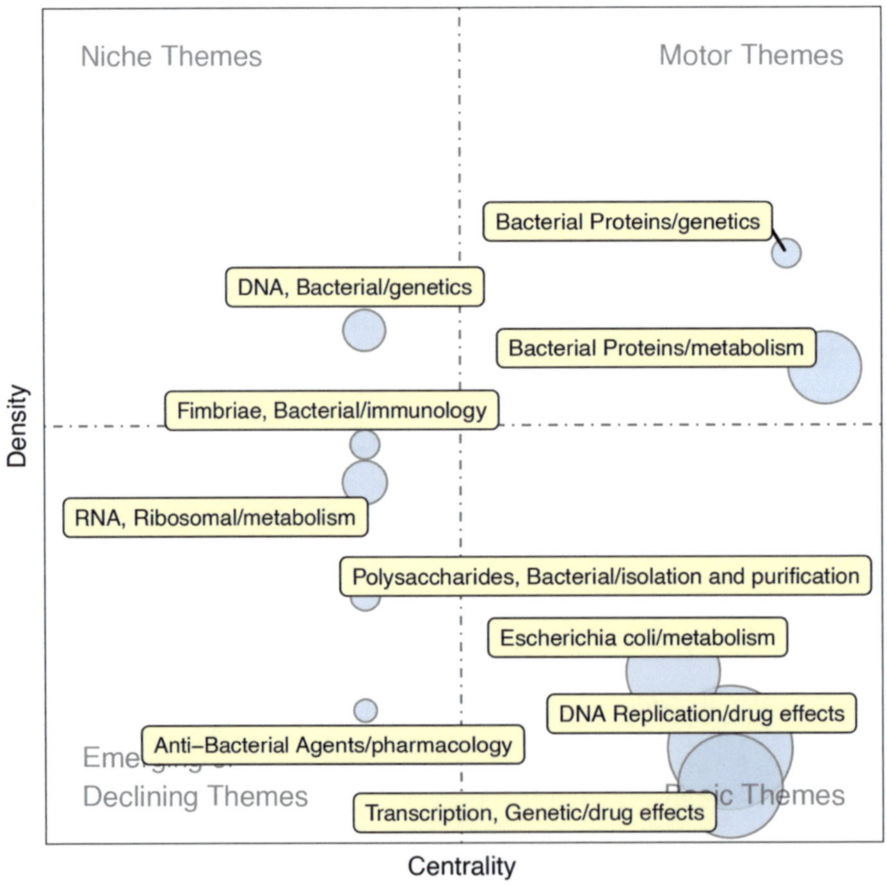

Figure 2.
Strategic diagram of the period 1981–1990. Each research topic is represented by a node and labeled with the most frequent pair MeSH heading/MeSH subheading. The size of the node is proportional to the number of MeSH heading/MeSH subheading pairs in each cluster.

ID	Size	MeSH headings
1	127	DNA Replication/drug effects, DNA Repair/radiation effects, Plasmids/drug effects, DNA Repair/drug effects, Genes, Bacterial/radiation effects, DNA Replication/radiation effects, Genes/radiation effects, Plasmids/radiation effects, Recombination, Genetic/radiation effects, Transformation, Bacterial/drug effects
2	81	Transcription, Genetic/drug effects, Genes, Bacterial/drug effects, Gene Expression Regulation/drug effects, Genes/drug effects, Protein Biosynthesis/drug effects, Operon/drug effects, Promoter Regions, Genetic/drug effects, Gene Expression Regulation, Bacterial/drug effects, Suppression, Genetic/drug effects
3	64	*Escherichia coli*/metabolism, *Escherichia coli*/drug effects, *Escherichia coli*/genetics, RNA, Transfer/genetics, *Escherichia coli*/isolation and purification, *Escherichia coli*/radiation effects, *Escherichia coli*/growth and development, *Escherichia coli*/immunology, Bacteriolysis/drug effects, RNA, Transfer/metabolism
4	40	Bacterial Proteins/metabolism, Bacterial Proteins/isolation and purification, Ribosomal Proteins/isolation and purification, Transcription Factors/metabolism, Repressor Proteins/metabolism, Ribosomal Proteins/analysis, Bacterial Proteins/immunology, DNA-Binding Proteins/isolation and purification, Flagellin/metabolism, Ribosomal Proteins/immunology
5	21	RNA, Ribosomal/metabolism, RNA, Bacterial/metabolism, RNA, Ribosomal/isolation and purification, Ribosomal Proteins/metabolism, RNA, Bacterial/isolation and purification, RNA, Transfer, Amino Acyl/metabolism
6	20	DNA, Bacterial/genetics, DNA, Bacterial/analysis, DNA, Bacterial/metabolism, DNA, Bacterial/isolation and purification, DNA/analysis, DNA/genetics, DNA, Superhelical/radiation effects
7	15	Bacterial Proteins/genetics, DNA-Directed DNA Polymerase/genetics, DNA Polymerase I/genetics, Viral Proteins/genetics
8	15	Fimbriae, Bacterial/immunology, *Escherichia coli*/ultrastructure, Fimbriae, Bacterial/ultrastructure, *Escherichia coli*/analysis
9	15	Polysaccharides, Bacterial/isolation and purification, Antigens, Bacterial/isolation and purification, Lipopolysaccharides/isolation and purification
10	13	Anti-Bacterial Agents/pharmacology, Anti-Bacterial Agents/metabolism

Table 1.
Principal research topics related to E. coli *research in the period 1981–1990.*

In the period 1981–1990 the biggest cluster was the cluster "DNA Replication/drug effects" comprised of the MeSH headings dealing with drug or radiation effects on DNA replication and DNA repair. This cluster was found to be the basic theme in this time period. A typical representative article for this cluster is the article published by Fram et al. with the title "DNA repair mechanisms affecting cytotoxicity by streptozotocin in *E. coli*" [31]. A further important basic research time in this period was also encompassed in the cluster "Transcription, Genetic/drug effects" consisting of the MeSH headings relating to the gene expression and regulation of gene expression and drug effects on the gene expression and regulation of this expression. A typical representative article for this cluster is the article published by Goda and Greenblatt with the title "Efficient modification of *E. coli* RNA polymerase *in vitro* by the N gene transcription antitermination protein of bacteriophage lambda" [32]. The major motor theme in this period was covered by the cluster "Bacterial Proteins/metabolism" involving MeSH headings related to metabolism and isolation and purification of different bacterial proteins, e.g., ribosomal proteins, transcription factors, repressors, and other DNA-binding proteins. A typical representative article for this cluster is the article published by Thomas et al. with the title "Amplification and purification of UvrA, UvrB, and UvrC proteins of *Escherichia coli*" [33]. Only one niche theme was detected—the cluster

"DNA, Bacterial/genetics"—covering MeSH headings dealing mostly with isolation and purification of DNA and with DNA analysis. A typical representative article for this cluster is the article published by Klaer et al. with the title "The sequence of IS4" [34].

4.2 Period 1991–2000

For the period 1991–2000, the retrieval yielded 23,470 documents from PubMed. The top 10 clusters that emerged from the co-word analysis are presented in the form of a strategic diagram in **Figure 3**. The corresponding summary of the identified research topics is given in **Table 2**.

In the observed period, the biggest cluster was the cluster "Gene Expression Regulation, Bacterial/drug effects" comprised of the MeSH headings dealing with drug or radiation effects on the genes, their expression, and regulation. This cluster was found to be the basic theme in this time period. A typical representative article for this cluster is the article published by Lutz and Bujard with the title "Independent and tight regulation of transcriptional units in *Escherichia coli* via the LacR/O, the TetR/O and AraC/I1-I2 regulatory elements" [35].

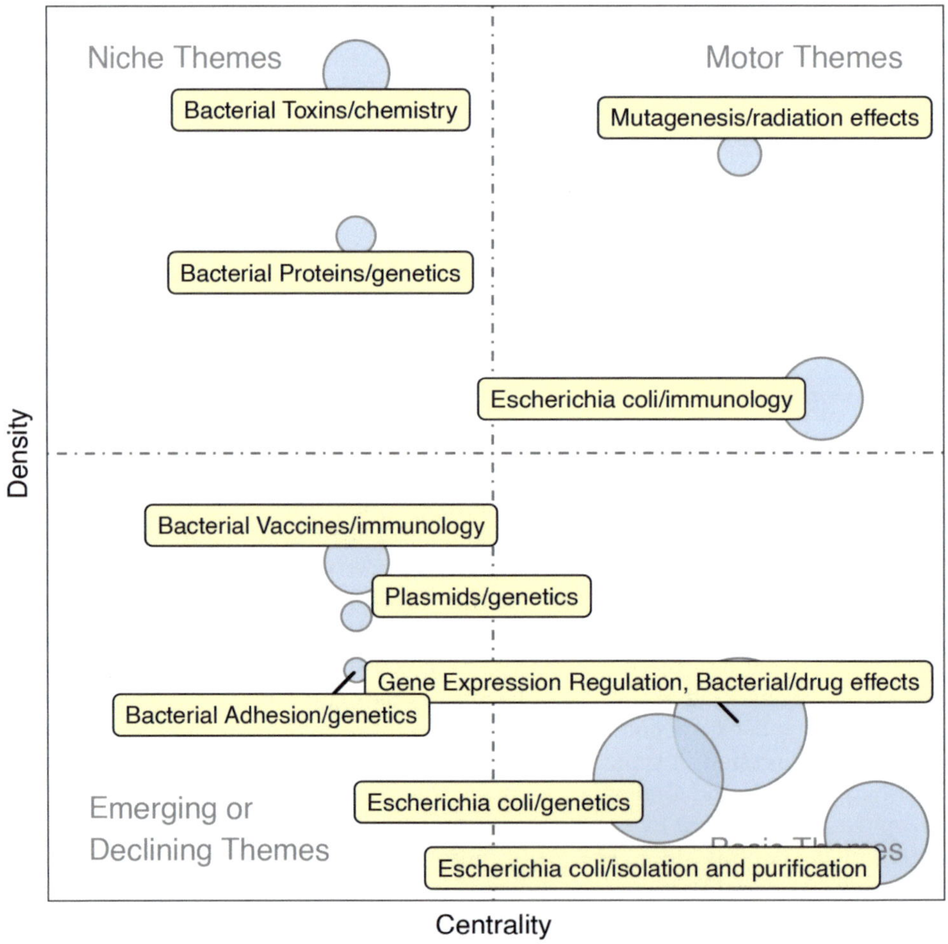

Figure 3.
Strategic diagram of the period 1991–2000. Each research topic is represented by a node and labeled with the most frequent pair MeSH heading/MeSH subheading. The size of the node is proportional to the number of MeSH heading/MeSH subheading pairs in each cluster.

ID	Size	MeSH headings
1	122	Gene Expression Regulation, Bacterial/drug effects, Transcription, Genetic/drug effects, Protein Biosynthesis/drug effects, Gene Expression Regulation, Enzymologic/drug effects, Operon/drug effects, Genes, Bacterial/radiation effects, Protein Processing, Post-Translational/drug effects, SOS Response, Genetics/drug effects, SOS Response, Genetics/radiation effects, Gene Expression Regulation, Bacterial/radiation effects
2	110	*Escherichia coli*/genetics, *Escherichia coli*/metabolism, *Escherichia coli*/drug effects, *Escherichia coli*/chemistry, *Escherichia coli*/growth and development, *Escherichia coli*/ultrastructure, *Escherichia coli*/cytology, Bacterial Outer Membrane Proteins/genetics, *Escherichia coli*/enzymology, *Escherichia coli*/radiation effects
3	53	*Escherichia coli*/isolation and purification, *Escherichia coli*/classification, *Escherichia coli*/pathogenicity, *Escherichia coli*/physiology, Germ-Free Life/immunology
4	30	*Escherichia coli*/immunology, *Escherichia coli* Infections/diagnosis, *Escherichia coli* Infections/epidemiology, *Escherichia coli* Infections/microbiology, *Escherichia coli* Infections/therapy, *Escherichia coli* Infections/drug therapy, *Escherichia coli* Infections/physiopathology
5	19	Bacterial Toxins/chemistry, Bacterial Toxins/metabolism, Bacterial Toxins/toxicity, Bacterial Toxins/genetics, Enterotoxins/genetics, Enterotoxins/chemistry, Enterotoxins/metabolism, Enterotoxins/toxicity
6	18	Bacterial Vaccines/immunology, Bacterial Vaccines/administration and dosage, Bacterial Vaccines/toxicity, Bacterial Vaccines/standards
7	10	Mutagenesis/radiation effects, DNA, Bacterial/radiation effects, Frameshift Mutation/drug effects, Mutagenesis/drug effects
8	9	Bacterial Proteins/genetics, Bacterial Proteins/chemistry, Bacterial Proteins/metabolism, Bacterial Proteins/physiology
9	7	Plasmids/genetics, Plasmids/chemistry
10	6	Bacterial Adhesion/genetics, Bacterial Adhesion/immunology

Table 2.
Principal research topics related to E. coli *research in the period 1991–2000.*

A further important basic research time in this period was also encompassed in the cluster "*Escherichia coli*/genetics" including the MeSH headings relating also to *E. coli* metabolism and enzymology, and growth and development. A typical representative article for this cluster is the article published by Hiraga with the title "Chromosome partition in *Escherichia coli*" [36]. The major motor theme in this period was covered by the cluster "*Escherichia coli*/immunology" involving also MeSH related to epidemiology, diagnosis, pathogenesis, and drug therapy of *E. coli* infections. A typical representative article for this cluster is the article published by Johnson with the title "Virulence factors in *Escherichia coli* urinary tract infection" [37]. The major niche theme was the cluster "Bacterial Toxins/chemistry", covering different MeSH headings dealing with genetics, chemistry, metabolism, and toxicity of bacterial toxins and enterotoxins. A typical representative article for this cluster is the article published by Gyles with the title "*Escherichia coli* cytotoxins and enterotoxins" [38].

4.3 Period 2001–2010

For the period 2001–2010, the search strategy retrieved 24,266 documents. Selected research topics are shown in **Figure 4** (here we point out to the reader that we have included the 11 most important research topics, while the ranks of clusters 7–9 are tied). The contextual meaning of the clusters is summarized in **Table 3**.

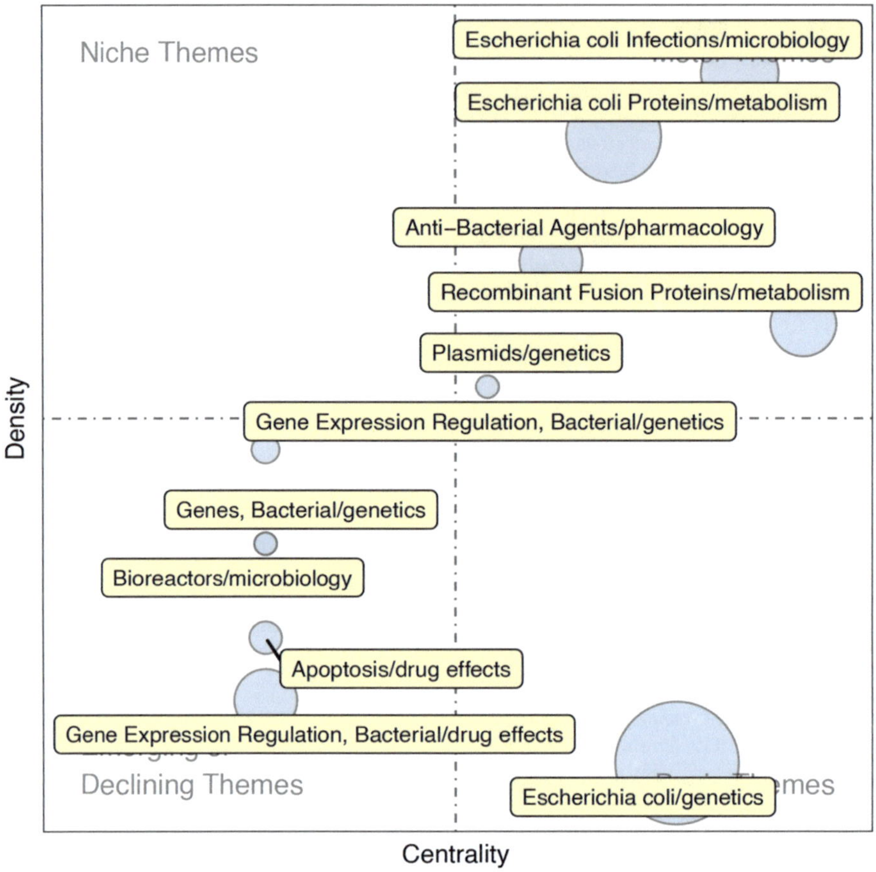

Figure 4.
Strategic diagram of the period 2001–2010. Each research topic is represented by a node and labeled with the most frequent pair MeSH heading/MeSH subheading. The size of the node is proportional to the number of MeSH heading/MeSH subheading pairs in each cluster.

In the period 2001–2010, the biggest cluster was the cluster "*Escherichia coli*/genetics" comprised of very diverse MeSH headings dealing with genetics and classification, but also with enzymology, metabolism and drug effects, isolation, purification and chemistry, cytology, growth, and development and also pathogenicity. This cluster was found to be the only basic theme in this time period. A typical representative article for this cluster is the article published by Tenaillon et al. with the title "The population genetics of commensal *Escherichia coli*" [39].

The major motor theme in this period was covered by the cluster "*Escherichia coli* Proteins/metabolism" involving MeSH headings related to metabolism and isolation and purification of different native, but also recombinant bacterial proteins. A typical representative article for this cluster is the article published by Bell with the title "Structure and mechanism of *Escherichia coli* RecA ATPase" [40]. Among motor themes, another cluster of very similar size was revealed—the cluster "*Escherichia coli* Infections/microbiology" covering the MeSH heading relating to important *E. coli* infection topics associated with pathogenic *E. coli*, e.g., urinary tract infections and bacteremia, but also antimicrobial agents with special emphasis on the topic of beta-lactamases. A typical representative article for this cluster is the article published by Croxen and Finlay with the title "Molecular mechanisms of *Escherichia coli* pathogenicity" [41]. In this time period no niche themes were found.

ID	Size	MeSH headings
1	666	*Escherichia coli*/genetics, *Escherichia coli*/metabolism, *Escherichia coli*/drug effects, *Escherichia coli*/enzymology, *Escherichia coli*/isolation and purification, *Escherichia coli*/chemistry, *Escherichia coli*/growth and development, *Escherichia coli*/pathogenicity, *Escherichia coli*/ classification, *Escherichia coli*/cytology
2	185	*Escherichia coli* Proteins/metabolism, *Escherichia coli* Proteins/genetics, Drug Resistance, Bacterial/genetics, *Escherichia coli* Proteins/chemistry, Bacterial Proteins/chemistry, Bacterial Proteins/metabolism, Bacterial Proteins/genetics, Membrane Proteins/metabolism, *Escherichia coli* Proteins/isolation, and purification, Recombinant Proteins/genetics
3	83	*Escherichia coli* Infections/microbiology, *Escherichia coli* Infections/epidemiology, beta-Lactamases/genetics, Urinary Tract Infections/microbiology, *Escherichia coli* Infections/drug therapy, Anti-Infective Agents/pharmacology, beta-Lactamases/biosynthesis, Urinary Tract Infections/drug therapy, beta-Lactamases/metabolism, Bacteremia/microbiology
4	49	Recombinant Fusion Proteins/metabolism, Recombinant Fusion Proteins/genetics, Recombinant Fusion Proteins/isolation and purification, Antimicrobial Cationic Peptides/ genetics, Antimicrobial Cationic Peptides/metabolism, Antimicrobial Cationic Peptides/ pharmacology, Anti-Infective Agents/chemistry, Anti-Infective Agents/metabolism
5	44	Gene Expression Regulation, Bacterial/drug effects, Protein Biosynthesis/drug effects, Nucleic Acid Conformation/drug effects, Transcription, Genetic/drug effects
6	43	Anti-Bacterial Agents/pharmacology, Anti-Bacterial Agents/chemistry, Anti-Bacterial Agents/chemical synthesis, Anti-Bacterial Agents/pharmacokinetics, Anti-Bacterial Agents/ therapeutic use, Anti-Bacterial Agents/administration and dosage, Fluoroquinolones/ pharmacology, Oligopeptides/pharmacology
7	11	Apoptosis/drug effects, Phagocytosis/drug effects
8	9	Gene Expression Regulation, Bacterial/genetics, Gene Expression Regulation, Bacterial/ physiology
9	7	Plasmids/genetics, Plasmids/metabolism
10	7	Genes, Bacterial/genetics, Mutation/drug effects
11	7	Bioreactors/microbiology, Industrial Microbiology/methods

Table 3.
Principal research topics related to E. coli *research in the period 2001–2010.*

4.4 Period 2011–2021

In the last observed period, we analyzed 30,114 bibliographic records from PubMed. The co-word analysis revealed six thematic clusters, as shown in **Figure 5**. The corresponding details are presented in **Table 4**.

In the period 2011–2022 was again the biggest cluster the cluster "*Escherichia coli*/genetics" comprised of very diverse MeSH headings dealing with genetics and genetically modified microorganisms, but also with enzymology, metabolism, biosynthesis, purification and chemistry, and growth and development. This cluster was again the basic theme in this time period. A typical representative article for this cluster is the article published by Yang et al. with the title "*Escherichia coli* as a platform microbial host for systems metabolic engineering" [42]. A further important basic research time in this period was also encompassed in the cluster "*Escherichia coli*/drug effects" consisting of the MeSH headings relating to bacterial drug resistance. A typical representative article for this cluster is the article published by Da Silva and Mendonça with the title "Association between antimicrobial resistance and virulence in *Escherichia coli*" [43]. Only one major motor theme in this period was revealed, covered by the cluster "Anti-Bacterial Agents/ pharmacology" covering MeSH headings related to different aspects, e.g., pharmacology, chemistry of antibacterial agents, including silver and antimicrobial cationic

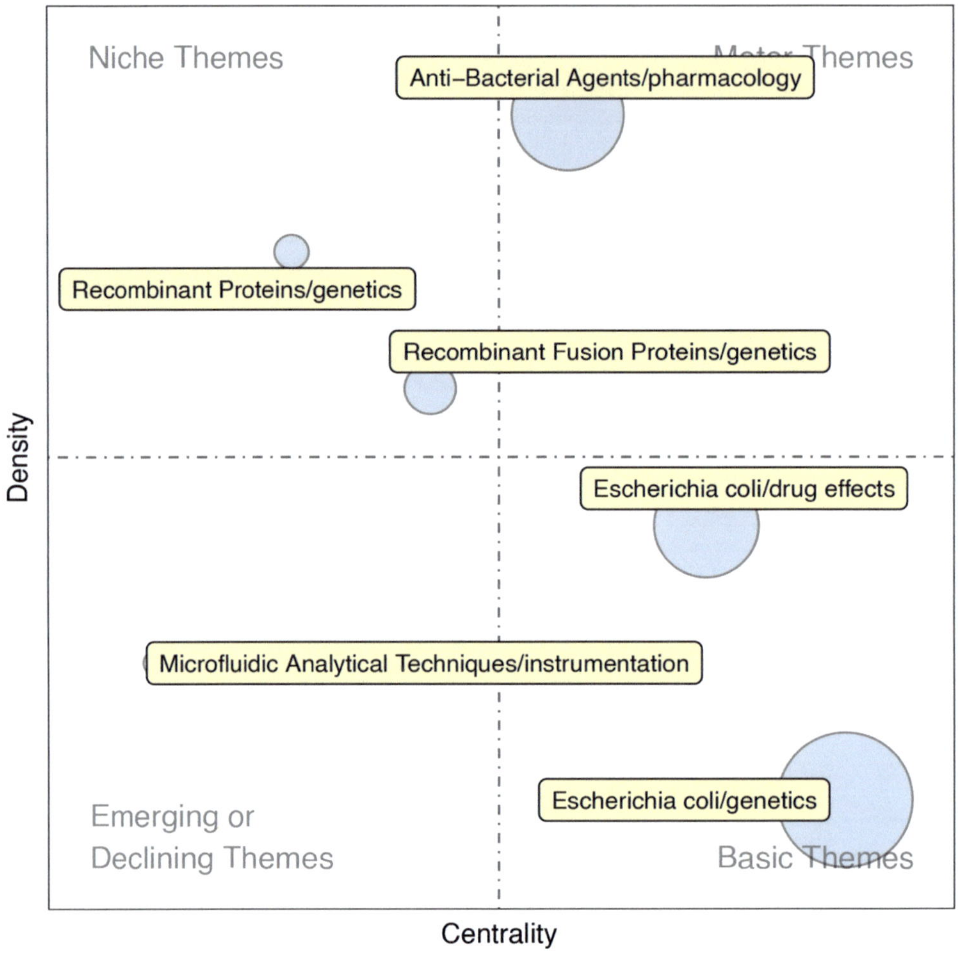

Figure 5.
Strategic diagram of the period 2011–2021. Each research topic is represented by a node and labeled with the most frequent pair MeSH heading/MeSH subheading. The size of the node is proportional to the number of MeSH heading/MeSH subheading pairs in each cluster.

ID	Size	MeSH headings
1	2329	*Escherichia coli*/genetics, *Escherichia coli*/metabolism, *Escherichia coli* Proteins/genetics, *Escherichia coli* Proteins/metabolism, *Escherichia coli*/enzymology, Bacterial Proteins/genetics, *Escherichia coli*/growth and development, *Escherichia coli*/chemistry, Microorganisms, Genetically-Modified/genetics, Bacterial Proteins/biosynthesis
2	921	Anti-Bacterial Agents/pharmacology, Anti-Bacterial Agents/chemistry, Silver/chemistry, Silver/pharmacology, Anti-Infective Agents/pharmacology, Anti-Bacterial Agents/chemical synthesis, Anti-Infective Agents/chemistry, Antimicrobial Cationic Peptides/pharmacology, Antimicrobial Cationic Peptides/chemistry, Titanium/chemistry
3	673	*Escherichia coli*/drug effects, *Escherichia coli*/isolation and purification, *Escherichia coli*/physiology, Drug Resistance, Bacterial/genetics, *Escherichia coli*/pathogenicity, *Escherichia coli* Infections/epidemiology, *Escherichia coli* Infections/microbiology, *Escherichia coli*/classification, Drug Resistance, Bacterial/drug effects, *Escherichia coli* Infections/drug therapy
4	81	Recombinant Fusion Proteins/genetics, Recombinant Fusion Proteins/biosynthesis, Recombinant Fusion Proteins/isolation and purification, Recombinant Fusion Proteins/chemistry

ID	Size	MeSH headings
5	41	Recombinant Proteins/genetics, Recombinant Proteins/metabolism, Recombinant Proteins/chemistry
6	28	Microfluidic Analytical Techniques/instrumentation, Microfluidic Analytical Techniques/methods

Table 4.
Principal research topics related to E. coli *research in the period 2011–2021.*

peptides. A typical representative article for this cluster is the article published by Zhao et al. with the title "Synthesis of Ag/AgCl modified anhydrous basic bismuth nitrate from BiOCl and the antibacterial activity" [44]. The major niche theme in this period was covered by the cluster "Recombinant Fusion Proteins/genetics" wrapping the topics of genetics, biosynthesis, chemistry, and isolation and purification of recombinant fusion proteins. A typical representative article for this cluster is the article published by Jeffery with the title "Expression, solubilization, and purification of bacterial membrane proteins" [45]. The second cluster among the niche themes in this time period is the cluster "Recombinant Proteins/genetics" covering the topics of genetics, metabolism, and chemistry of recombinant proteins. A typical representative article for this cluster is the article published by Gopal and Kumar with the title "Strategies for the production of recombinant protein in *Escherichia coli*" [46].

5. Discussion

E. coli is known to be a versatile microorganism—it is a commensal in the gut microbiota of healthy hosts, but can be found also as a pathogen instigating intestinal but also extraintestinal infections [47]. *E. coli* is also a well-known probiotic bacterium, as some important probiotic drugs including *E. coli* are on the market [48–50]. Further, it is a very well-known model microorganism for Gram-negative bacteria, which was and still is used as a laboratory "workhorse" on which many basic topics of molecular biology, physiology, genetics, evolution, genetic engineering, and biotechnology were and still are studied [51–53].

So there is no surprise in finding many papers published on *E. coli*. Bibliometric co-word analysis has the potential to reveal the topic trends in *E. coli* research. The result of this kind of analysis is two-dimensional plots in which circles (i.e., nodes), whose size corresponds to the number of including MeSH terms, are partitioned into different quadrants. The top right quadrant depicts motor themes with strong centrality as well as high density. The upper left quadrant shows specialized themes, which refer to themes having a high density, but also having inadequate external interactions. The bottom-right quadrant shows the basic themes—these are themes that have a strong centrality, but low density. In the bottom, left quadrant themes are shown that are emerging or declining, as they have in general low density and centrality. For just one strategic diagram, it is usually not possible to determine whether a theme is emerging or declining, however, when data from several graphs, each from a certain period, are compared, for some themes that are found in more graphs a trend can be established. From our analysis, it can be assumed that the cluster "Anti-Bacterial Agents/pharmacology" which appeared in the 1981–1990 period in the quadrant of Emerging or declining themes was in that time period an emerging theme, as the same cluster can be also found in the graphs of the periods 2001–2010 and 2011–2022, namely in the quadrant of motor themes. A similar can be stated for the cluster "Plasmids/genetics", which appeared on the strategic diagram of the period 1991–2000 was an emerging theme, as the same cluster can be found also in the 2001–2010 diagram, namely in the quadrant of motor themes. An example of a declining cluster theme is the cluster "Gene Expression Regulation, Bacterial/drug effects" which is the major basic theme in the diagram of the time period 1991–2000, but moved to the emerging or declining themes quadrant in the plot of the time period 2001–2010. In the strategic diagram of the period 2011–2021 in the emerging or declining themes quadrant the cluster "Microfluidic Analytical Techniques/instrumentation" appeared, which is for sure an emerging cluster as much of the *E. coli* research is now moving into the area of single cell analysis which is enabled by the microfluidic techniques.

6. Conclusions

In the present study, we retrieved nearly 100,000 scientific articles on *E. coli* from the PubMed bibliographic database and investigated the intellectual structure and evolution using co-word analysis. To our knowledge, this is the first systematic knowledge mapping in the field of *E. coli* research. The analysis performed clearly revealed the main research topics in *E. coli* research over the last decades. Based on this analysis, major, niche, and basic topics in *E. coli* research were identified in each decade studied, and new topics are expected to emerge. The future in the field of *E. coli* research lies in single-cell analysis.

Acknowledgements

This work was supported by the Slovenian Research Agency (Grant No. J5-2552 (AK) and P1-0198 (MSE)).

Conflict of interest

The authors declare no conflict of interest.

Author details

Andrej Kastrin[1] and Marjanca Starčič Erjavec[2]*

1 Institute for Biostatistics and Medical Informatics, Faculty of Medicine, University of Ljubljana, Ljubljana, Slovenia

2 Biotechnical Faculty, Department of Microbiology, University of Ljubljana, Ljubljana, Slovenia

*Address all correspondence to: marjanca.starcic.erjavec@bf.uni-lj.si

References

[1] Escherich T. The intestinal bacteria of the neonate and breast-fed infant. Reviews of Infectious Diseases. 1988; **10**(6):1220-1225

[2] Larsen PO, von Ins M. The rate of growth in scientific publication and the decline in coverage provided by science citation index. Scientometrics. 2010; **84**(3):575-603

[3] Bornmann L, Mutz R. Growth rates of modern science: A bibliometric analysis based on the number of publications and cited references. Journal of the Association for Information Science and Technology. 2015;**66**(11):2215-2222

[4] National Library of Medicine (US). MEDLINE PubMed Production Statistics. 2021. Available from: https://www.nlm.nih.gov/bsd/medline_pubmed_production_stats.html [Accessed: November 4, 2022]

[5] Clauset A, Larremore DB, Sinatra R. Data-driven predictions in the science of science. Science. 2017;**355**(6324): 477-480

[6] Fortunato S, Bergstrom CT, Börner K, Evans JA, Helbing D, Milojević S, et al. Science of science. Science. 2018;**359**(6379):eaao0185

[7] Ebbinghaus H. Psychology: An Elementary Text-Book. Boston, MA: D. C. Heath and Company; 1908

[8] Lotka AJ. The frequency distribution of scientific productivity. Journal of the Washington Academy of Sciences. 1926; **16**(12):317-323

[9] Zipf GK. Human Behavior and the Principle of Least Effort. Oxford: Addison-Wesley Press; 1949

[10] Price DJ. Networks of scientific papers. Science. 1965;**149**(3683):510-515

[11] Merton RK. The Matthew effect in science: The reward and communication systems of science are considered. Science. 1968;**159**(3810):56-63

[12] Garfield E. Citation indexing for studying science. Nature. 1970; **227**(5259):669-671

[13] Börner K, Chen C, Boyack KW. Visualizing knowledge domains. Annual Review of Information Science and Technology. 2003;**37**(1):179-255

[14] Uzzi B, Mukherjee S, Stringer M, Jones B. Atypical combinations and scientific impact. Science. 2013; **342**(6157):468-472

[15] Wang D, Song C, Barabási AL. Quantifying long-term scientific impact. Science. 2013;**342**(6154):127-132

[16] Wu L, Kittur A, Youn H, Milojević S, Leahey E, Fiore SM, et al. Metrics and mechanisms: Measuring the unmeasurable in the science of science. Journal of Informetrics. 2022;**16**(2): 101290

[17] Cobo MJ, López-Herrera AG, Herrera-Viedma E, Herrera F. An approach for detecting, quantifying, and visualizing the evolution of a research field: A practical application to the fuzzy sets theory field. Journal of Informetrics. 2011;**5**(1):146-166

[18] Klavans R, Boyack KW. Which type of citation analysis generates the most accurate taxonomy of scientific and technical knowledge? Journal of the Association for Information Science and Technology. 2017;**68**(4):984-998

[19] Watts DJ. Small Worlds: The Dynamics of Networks between Order and Randomness. Princeton, NJ: Princeton University Press; 2003

[20] Newman M. Power laws, Pareto distributions and Zipf's law. Contemporary Physics. 2005;**46**(5): 323-351

[21] Barabási AL, Albert R. Emergence of scaling in random networks. Science. 1999;**286**(5439):509-512

[22] Ravasz E, Barabási AL. Hierarchical organization in complex networks. Physical Review E. 2003;**67**(2):026112

[23] Wang D, Barabási AL. The Science of Science. Cambridge, MA: Cambridge University Press; 2021

[24] Yi S, Choi J. The organization of scientific knowledge: The structural characteristics of keyword networks. Scientometrics. 2012;**90**(3):1015-1026

[25] Lipscomb CE. Medical subject headings (MeSH). Bulletin of the Medical Library Association. 2000; **88**(3):265-266

[26] Dunne E, Hulek K. Mathematics subject classification 2020. European Mathematical Society Magazine. 2020; **115**:5-6

[27] Callon M, Courtial JP, Turner WA, Bauin S. From translations to problematic networks: An introduction to co-word analysis. Information (International Social Science Council). 1983;**22**(2):191-235

[28] Moral-Munoz JA, Lucena-Antón D, Perez-Cabezas V, Carmona-Barrientos I, González-Medina G, Ruiz-Molinero C. Highly cited papers in microbiology: Identification and conceptual analysis. FEMS Microbiology Letters. 2018; **365**(20):fny230

[29] Callon M, Courtial JP, Laville F. Co-word analysis as a tool for describing the network of interactions between basic and technological research: The case of polymer chemsitry. Scientometrics. 1991;**22**(1):155-205

[30] Blondel VD, Guillaume JL, Lambiotte R, Lefebvre E. Fast unfolding of communities in large networks. Journal of Statistical Mechanics: Theory and Experiment. 2008;**2008**(10): P10008

[31] Fram RJ, Mack SL, George M, Marinus MG. DNA repair mechanisms affecting cytotoxicity by streptozotocin in *E. coli*. Mutation Research. 1989; **218**(2):125-133

[32] Goda Y, Greenblatt J. Efficient modification of *E. coli* RNA polymerase *in vitro* by the N gene transcription antitermination protein of bacteriophage lambda. Nucleic Acids Research. 1985;**13**(7):2569-2582

[33] Thomas DC, Levy M, Sancar A. Amplification and purification of UvrA, UvrB, and UvrC proteins of *Escherichia coli*. The Journal of Biological Chemistry. 1985;**260**(17):9875-9883

[34] Klaer R, Kühn S, Tillmann E, Fritz HJ, Starlinger P. The sequence of IS4. Molecular and General Genetics. 1981;**181**(2):169-175

[35] Lutz R, Bujard H. Independent and tight regulation of transcriptional units in *Escherichia coli* via the LacR/O, the TetR/O and AraC/I1-I2 regulatory elements. Nucleic Acids Research. 1997; **25**(6):1203-1210

[36] Hiraga S. Chromosome partition in *Escherichia coli*. Current Opinion in Genetics & Development. 1993;**3**(5): 789-801

[37] Johnson JR. Virulence factors in *Escherichia coli* urinary tract infection. Clinical Microbiology Reviews. 1991; **4**(1):80-128

[38] Gyles CL. *Escherichia coli* cytotoxins and enterotoxins. Canadian Journal of Microbiology. 1992;**38**(7): 734-746

[39] Tenaillon O, Skurnik D, Picard B, Denamur E. The population genetics of commensal *Escherichia coli*. Nature Reviews Microbiology. 2010;**8**(3):207-217

[40] Bell CE. Structure and mechanism of *Escherichia coli* RecA ATPase. Molecular Microbiology. 2005;**58**(2):358-366

[41] Croxen MA, Finlay BB. Molecular mechanisms of *Escherichia coli* pathogenicity. Nature Reviews Microbiology. 2010;**8**(1):26-38

[42] Yang D, Prabowo CPS, Eun H, Park SY, Cho IJ, Jiao S, et al. *Escherichia coli* as a platform microbial host for systems metabolic engineering. Essays in Biochemistry. 2021;**65**(2):225-246

[43] Da Silva GJ, Mendonça N. Association between antimicrobial resistance and virulence in *Escherichia coli*. Virulence. 2012;**3**(1):18-28

[44] Zhao M, Hou X, Lv L, Wang Y, Li C, Meng A. Synthesis of Ag/AgCl modified anhydrous basic bismuth nitrate from BiOCl and the antibacterial activity. Materials Science & Engineering: C. 2019;**98**:83-88

[45] Jeffery CJ. Expression, solubilization, and purification of bacterial membrane proteins. Current Protocols in Protein Science. 2016;**83**:29.15.1-29.15.15

[46] Gopal GJ, Kumar A. Strategies for the production of recombinant protein in *Escherichia coli*. The Protein Journal. 2013;**32**(6):419-425

[47] Kaper JB, Nataro JP, Mobley HL. Pathogenic *Escherichia coli*. Nature Reviews Microbiology. 2004;**2**(2):123-140

[48] Sonnenborn U. *Escherichia coli* strain Nissle 1917—From bench to bedside and back: History of a special *Escherichia coli* strain with probiotic properties. FEMS Microbiology Letters. 2016;**363**(19):fnw212

[49] Wassenaar TM, Zschüttig A, Beimfohr C, Geske T, Auerbach C, Cook H, et al. Virulence genes in a probiotic *E. coli* product with a recorded long history of safe use. European Journal of Microbiology & Immunology. 2015;**5**(1):81-93

[50] Wassenaar TM. Insights from 100 years of research with probiotic *E. coli*. European Journal of Microbiology & Immunology. 2016;**6**(3):147-161

[51] Blount ZD. The unexhausted potential of *E. coli*. eLife. 2015;**4**:e05826

[52] Foster PL. Adaptive mutation: Implications for evolution. BioEssays: News and Reviews in Molecular, Cellular and Developmental Biology. 2000;**22**(12):1067-1074

[53] Tatum EL, Lederberg J. Gene recombination in the bacterium *Escherichia coli*. Journal of Bacteriology. 1947;**53**(6):673-684

Chapter 3

The *Escherichia coli* SOS Response: Much More than DNA Damage Repair

Zdravko Podlesek and Darja Žgur Bertok

Abstract

The *Escherichia coli* SOS response is an inducible DNA damage repair pathway controlled by two key regulators, LexA, a repressor and RecA, an inducer. Upon DNA damage RecA is activated and stimulates self cleavage of LexA, leading to, in *E. coli,* derepresion of approximately 50 SOS genes. The response is triggered by exogenous and endogenous signals that bacteria encounter at a number of sites within the host. Nevertheless, besides regulating DNA damage repair the SOS response plays a much broader role. Thus, SOS error prone polymerases promote elevated mutation rates significant for genetic adaptation and diversity, including antibiotic resistance. Here we review the *E. coli* SOS response in relation to recalcitrance to antimicrobials, including persister and biofilm formation, horizontal gene tranfer, gene mobility, bacterial pathogenicity, as well SOS induced bacteriocins that drive diversification. Phenotypic heterogeneity in expression of the SOS regulator genes, *recA* and *lexA* as well as colicin activity genes is also discussed.

Keywords: SOS response, *Escherichia coli*, DNA damage, antibiotic resistance, persisters, horizontal gene transfer, virulence, biofilms, bacteriocins, phentypic heterogeneity

1. Introduction

Bacteria are constantly exposed to a changing and stressful environment. Coordinated responses by bacterial global regulatory systems enable their survival and adaptation [1].

In all organisms genome integrity is constantly threatened by endogenous and exogenous agents. Exogenous DNA damaging agents are physical (UV and ionizing irradiation, oxidants, drugs) and chemical (oxidizing, crosslinking, alkylating). Endogenous triggers are the result of cellular metabolism such as reactive oxygen species (ROS), stalled replication forks and defects following recombination or chromosome segregation [2, 3]. To cope with DNA damage organisms possess a number of error free and error prone mechanisms [4, 5]. Most bacteria seem to have evolved a coordinated response to DNA damage. In *Escherichia coli* the inducible DNA repair pathway is designated the SOS response and is controlled by two regulators, LexA, a repressor and RecA, an inducer. DNA damage generates an increase in single stranded DNA (ssDNA) as DNA polymerase stalls at a lesion while helicase continues unwinding DNA. RecA is activated (RecA*) by binding to single

IntechOpen

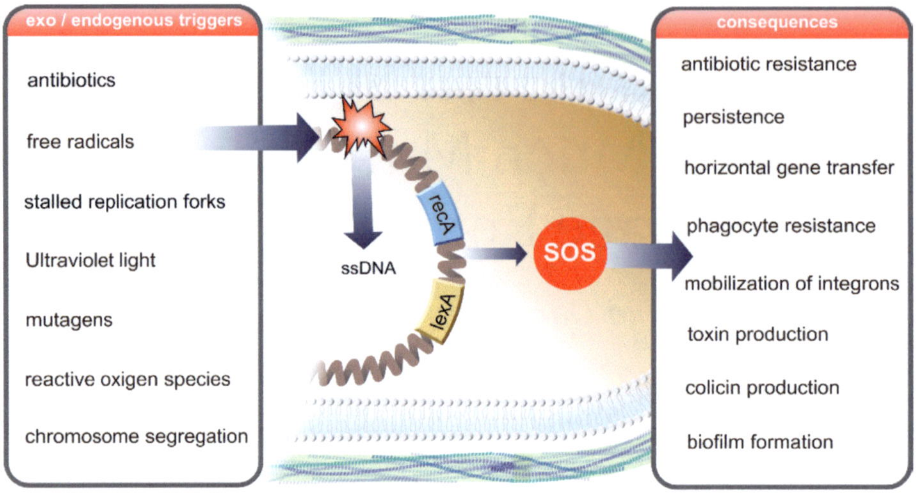

Figure 1.
Exo and endogenous triggers induce the E. coli SOS response leading to antibiotic resistance, persistence, horizontal gene transfer, expression of virulence factors, intraspecies competition and biofilms.

stranded DNA forming a nucleoprotein filament that stimulates self cleavage of LexA and in *E. coli* de-repression of more than 50 SOS genes. A hallmark of the SOS response is its temporal control. High-fidelity repair mechanisms are induced first followed by low fidelity, damage tolerance pathways involving error prone translesion DNA polymerases PolII (*polB*), PolIV (*dinB*) and PolV (*umuC, umuD*). These are active only following extensive and persistent DNA damage. While the error prone/last resort polymerases enable repair of lesions that block DNA replication by the primary replicative DNA polymerase PolIII [4, 5], they also promote an increase in mutation rate.

Even though RecA and LexA are the key SOS regulators, induction/SOS factors may also be governed by other stress response pathways namely, alternative sigma factors RpoS and RpoH of the general stress responses, the stringent response, cAMP and reactive oxgen species (ROS) [6–10].

Whilst the SOS response was initially recognized as regulating DNA damage repair, it is now well established that it plays' a much broader role. Thus, SOS error prone polymerases by promoting elevated mutation rates generate genetic diversity and adaptation. The SOS response is also involved in horizontal gene transfer, virulence factor expression, biofilms, persistence, sustained colonization of the mammalian gut, controls toxin-antitoxin systems as well as intraspecies competition and phenotypic variation (**Figure 1**) [5, 11, 12].

2. Mutagenesis and antibiotic resistance

Antibiotic resistance is one of the most serious global health threats. Resistance occurs by mutation of resident genes or/and by uptake of resistance genes. Antibiotic resistance mechanisms belong to one of several classes with resistance genes encoded on chromosomes and mobile genetic elements such as plasmids, transposons and integrons. As stated above, the error prone translesion DNA polymerases PolII, PolIV and PolV allow translesion DNA replication but also promote an increased level of mutation, significant for evolution of antibiotic resistance.

Exposure of bacteria to antibiotics, even at subinhibitory concentrations, has been shown to increase mutation and recombination frequencies via the SOS response [13, 14]. In *E. coli* as well as a number of other clinically significant bacterial species, some of the most common antibiotics have been shown to induce the SOS response and mutagenesis [15, 16]. Exposure of environmental bacteria to antibiotics, even subinhibitory levels, could thus generate variants with higher rates of genetic modifications and select for resistance.

2.1 Persisters

In addition to antibiotic resistance other mechanisms allow bacterial growth in the presence of antibiotics; (i) population wide tolerance, (ii) persisters, subpopulations characterized by a transient dormant state and transient tolerance [17] and (iii) shielding that protects and enables survival in the presence of antibiotics [18].

Persisters and antimicrobial tolerance have been extensively studied in *E. coli*. One of the first and most thoroughly investigated examples of persister cell formation involving the SOS system, is activation via the toxin-antitoxin TisB/IstR module. TisB is a small membrane-acting peptide that decreases the proton motive force and ATP levels, shutting down cell metabolism and inducing dormancy [19]. The *tisB* gene is repressed by the SOS repressor LexA, while the IstR-1 antitoxin is constitutively expressed. Following DNA damage and SOS induction, *tisB* transcription strongly increases and exceeds that of the antitoxin IstR-1 [20].

Nevertheless, in *E. coli,* the SOS response in persisters also accelerates antibiotic resistance [21, 22]. Thus, from fluoroquinolone (FQ) persisters, the SOS response promotes resistance to unrelated antibiotics following a single FQ exposure [23].

Recently, sub-inhibitory concentrations of ciprofloxacin were shown to, in *E. coli*, induce transient differentiation of a small gambler subpopulation that, generates cross-resistant mutants. Gamblers are characterized by high levels of ROS and a σ^S general stress-response. In gamblers, ROS activate the σ^S response, which allows mutagenic repair of antibiotic-triggered DNA double strand breaks. Further required is SOS induced inhibition of cell division, provoking the presence of multiple chromosomes. Thus, in gamblers, a highly regulated, transient differentiation process with within-cell chromosome cooperation drives evolution of resistance to new antibiotics [24].

3. Mobile genetic elements

Horizontal gene transfer is a significant driving force of bacterial genome evolution, including the emergence and dissemination of antibiotic resistance and virulence genes. The SOS response has been shown to play an important role in gene transfer in a number of bacterial species.

One of the first reports of the involvement of the SOS response in horizontal gene transfer was SOS induction of transfer of antibiotic resistances encoded by the *Vibrio cholerae* integrating conjugative element, SXT [13]. The SOS response has also been shown to induce lambdoid prophages due to SOS induced self cleavage of the CI phage repressor [25].

In turn, as conjugative plasmid DNA transfer and transformation, involve uptake of ssDNA, these mechanisms induce the SOS response [26, 27].

An important class of SOS controlled mobile genetic elements are the wide spread integrons. They are associated particularly with transposons and conjugative plasmids and have played an important role in the evolution of antibiotic resistance among pathogenic bacteria [28]. Integrons encode a site specific recombination system that promotes integration and expression of gene cassettes with antibiotic resistance and metabolism associated functions.

On the basis of integrase sequences, five classes of integrons are distinguished. Class 1 integrons are by far the most prevalent and clinically relevant. Recently, the class 1 integrons were found in a considerable fraction of *E. coli* isolates [29].

Integrons are composed of an *int*I gene encoding an integrase, followed by a recombination site, attI and a variable array of gene cassettes each ending in a recombination attC site [30]. Integron cassette expression is driven by the Pc promoter situated upstream of the array. Cassettes closest to the promoter are expressed at highest levels [31]. The integron integrases are frequently controlled by the LexA protein [11].

Integrons enable bacteria to evolve in response to new antibiotic challenges via rapid optimization of cassette expression. Activity of the integrase allows: (i) modulation of cassette expression, (ii) rapid gain of additional copies of selected cassettes and (iii) elimination of redundant cassettes. Integrase activity does not compromise genome integrity due to the high specificity of integrase-mediated recombination [32–34]. Thus, integrase-mediated cassette re-shuffling in stressful environments could accelerate bacterial evolution allowing bacteria to optimize cassette expression and maximize fitness. Relavant cassettes could be positioned near the Pc promoter for maximal expression, while unnecessary cassettes could be kept at the end of the array and be moved forward when required [33].

DNA acquired by HGT, including pathogenicity islands (PAIs) with virulence factor genes, must either replicate autonomously or be integrated into the bacterial chromose or plasmid. Integration is mediated by recombinases/integrases that are often encoded on PAIs and perform either integration or excision from the chromosome. It was recently shown that SOS inducing antibiotics, including clinically relevant for treatment of UTI, led to in a subpopulation, increased promoter activity as well as increased loss of PAIs [35].

4. *E. coli* virulence

The species *E. coli* encompasses commensals of the gut, pathogens and probiotics. Conditions conducive to SOS induction are encountered by *E. coli* at various host anatomical sites. Recently, the SOS response has been shown to play a vital role in maintaining colonization of the murine gut by commensal *E. coli*. Competing commensal organisms could be a source of genotoxic stress [12].

Pathogenic *E. coli* strains producing virulence factors such as adhesins, iron uptake systems, capsules, toxins and invasins, can provoke infections [36]. Pathogenic strains are broadly classified into two major groups, with regard to their virulence factors and diseases they provoke, the nondiarrheagenic and diarrheagenic. The nondiarrheagenic are designated extraintestinal pathogenic *E. coli* (ExPEC) while diarrheagenic provoke diarrhea and include the Shiga toxin (Stx) producing *E. coli* (STEC) as well as enterohemorrhagic *E. coli* (EHEC) [37].

Among ExPEC infections, the most common are urinary tract infections (UTI) followed by septicaemia and meningitis [36, 38].

UTIs represent a serious worldwide health problem [39] with uropathogenic *E. coli* strains (UPEC) responsible for 75–95% of community-acquired UTIs [40].

To provoke UTI, UPEC undergo a complex intracellular cycle [41] and the SOS response plays an important role in bacterial dissemination and persistence within the urinary tract. UPEC enter the urinary tract through an ascending route and travel up the urethra to colonize the bladder via internalization by the umbrella cells. Infected cells produce nitric oxide that attacks bacterial DNA, inducing the SOS response with inhibition of cell division and UPEC filamentation. Filamentous UPEC successfully resist phagocyte killing, allowing dissemination and persistence within the urinary tract. UTI frequently lead to chronic infection and a persister subpopulation could be responsible for generating relapsing infections [42].

In the intestinal tract, DNA damage and subsequent SOS induction, can be provoked by host factors, e.g. bile salts, and by competing microbes. Intestinal inflammation triggered by infection or the gut immune system involving ROS, also induces the SOS response.

All EHEC strains, including the notorius serotype O157;H7, produce Stx the main virulence factor associated with hemorrhagic colitis [37].

Production of Stx, by O157:H7 is mediated by quorum sensing [43] however, it is also well documented that the SOS response amplifies Shiga toxin production in enterohemorrhagic *E. coli* (EHEC). Stx is encoded on a lambdoid prophage. Induction of the prophage, via repressor autocleavage, and the subsequent upregulation of *stx* expression are controlled by the SOS response [25]. Therefore, DNA-damaging agents, including certain antibiotics, increase Stx synthesis and are counterindicated during treatment of infection [44]. In addition to SOS inducing antibiotics, bacteriocins and microcins secreted by members of the gut microbiota have been shown to amplify Stx synthesis. In the complex intestinal environment, survival involves competition for space and nutrients [45, 46]. Bacteria have therefore evolved mechanisms to counteract competitors [47] such as, production of bacteriocins, proteinaceous toxins, that inhibit growth and survival of usually closely related bacteria competing for similar resources [48]. A subtype of bacteriocins, known as colicins, are produced by *Enterobacteriaceae* while microcins are bacteriocins that are generally smaller than 10 kDa [49].

Thus, a strain producing the nuclease colicin E9 (ColE9) as well as extracted DNase colicins were shown to induce Stx [50]. Recently, microcin B17 (MccB17), a DNA gyrase inhibitor, as well as a putative microcin, were also shown to amplify Stx2a production [51, 52]. Thus in the gut, nonpathogenic *E. coli* strains could, via secretion of DNA damaging colicins and microcins, increase Stx production by O157:H7.

In addition to DNA damage induced by host factors, e.g. bile salts, and by competing microbes, intestinal inflammation triggered by infection or the gut immune system involving ROS, also provokes the SOS response and dysbiosis, suppressing anaerobes and inciting *Enterobacteriaceae* overgrowth with competition for nutrients [53].

5. Biofilms

Biofilms are surface attached structured bacterial communities that create a protective environment for bacterial cells [54]. Biofilm formation is a highly regulated process and is controlled by a number of environmental and genetic factors [55–57]. Biofilms are also induced by antimicrobial stress/SOS response. While biofilm formation is an integral part of the prokaryotic life cycle, biofilms also cause biofilm associated diseases that are difficult to treat, e.g. urinary tract infections (UTI), chronic infections in cystic fibrosis (CF) patients, colonization of medical devices and periodontal diseases [58].

A number of factors allow bacteria in biofilms to survive high dose antibiotic treatment [58, 59]. Antibiotic diffusion is prevented by a mechanical barrier formed by the extracellular matrix. Further, low oxygen and nutrient concentrations within biofilms create niches with low bacterial metabolic activity. In addtion, up to 1% of bacterial cells in biofilms may be dormant persister cells not affected by antimicrobials [60]. Furthermore, high cell density within biofilms enhances horizontal gene transfer and competition, that together with accumulation of metabolic products, microaerobic areas and oxidative stress, incite DNA damage and provoke the SOS response. Starvation stress in biofilm bacteria was shown to increase the level of tolerance to the fluoroquinolone ofloxacin in *E. coli* biofilms and was dependent on the presence of a functional bacterial SOS response [59].

In biofilms, phenotypic variants e.g. small colony variants (SCV), that are slow growing and very tolerant to host defenses and antimicrobials have also been described. SCVs exhibit increased production of exopolysaccharides, can autoaggregate and attach stronly to surfaces [61–63]. They are potentially responsible for difficult to treat persistent infections, wherein bacteria persist in the host for prolonged periods of time despite antimicrobial therapy. Thus, recalcitrance of biofilms to antimicrobials can be due to tolerance, when dispersed biofilm cells exhibit antibiotic sensitivity and low MIC, as well as resistance, characterized by increased MICs and a resistant phenotype of dispersed biofilm bacteria. The SOS response plays a significant role in biofilm formation but in turn, in the dynamic biofilm environment, SOS inducing factors are generated that promote mutagenesis and diversification.

6. Bacteriocins and phenotypic heterogeneity

Colicins are bacteriocins, toxic proteins that are produced by and act against *E. coli* and its close relatives. Sensitive cells are killed by targeting DNA, RNA, cell membranes or by inhibition of peptidoglycan and lipopolysaccharide (LPS) O-antigen. Colicin genes are found within genomic clusters on colicinogenic plasmids. These clusters typically contain the colicin activity gene for the toxin, an immunity gene for a protein that confers self-resistance by binding to and inactivating the toxin protein, and a lysis gene for a protein that aids in colicin release by lysis of the producer cell [64]. Colicin production is found with high frequency among natural *E. coli* isolates [65].

Colicins are expressed from strong promoters whose activity is tightly repressed by the LexA protein. Nutrient limitation and DNA damage are major signals that control colicin production [64, 66]. Nevertheless, additional regulators, in conjunction with LexA, have been found to regulate/delay colicin expression. Thus, the global transcriptional factor, IscR, in response to the nutritional status of the cell and, co-dependently with LexA, delays induction of pore-forming colicin genes following SOS induction [67]. On the other hand, temporal induction of DNA and RNA targeting colicins is co-regulated by the AsnC repressor. At the colicin E8, *cea8* promoter, AsnC repression reflects L-asparagine levels and presumably serves as an indicator of general amino acid abundance and availability [68]. Thus, promoters of nuclease and pore-forming colicins have adopted different transcription regulators and specific metabolic inputs to regulate transcription in conjunction with the LexA repressor.

Colicins have an *in vivo* antagonistic role promoting microbial diversity within *E. coli* populations in the mammalian colon [69] and the potential to promote microbial genetic diversity [70]. Sublethal concentrations of ciprofloxacin have been shown to induce colicin expression in an SOS-dependent manner and imply that SOS-inducing antibiotics could thus affect microbial strain diversification, as well as promote the acquisition and dissemination of antibiotic resistance [71].

Furthermore, monitoring the transcriptional response of *E. coli* to colicins E9, an endonuclease, and E3, an RNase, has shown that the former induces the SOS response while the latter upregulates expression of DNA integrases, invertases, and recombinases. Colicins thus also have the potential to, through the induction of error-prone DNA polymerases, promote microbial diversity, gene transfer, DNA rearrangements affect horizontal gene transfer as well expression of virulence factor genes.

Colicin production has also been found to be a specialized function within a population of genetically identical cells, an example of phenotypic heterogeneity. The colicin K activity gene was shown to be expressed in only a small fraction of a population, while the immunity gene is expressed in the large majority of the cells [72]. A number of colicins are released semispecifically, by cell lysis. Differential expression of the activity and lysis genes prevents excessive lysis. Alternatively, upon DNA damage and induction of the SOS response, all cells express the activity gene. Lysis of the producer releasing colicin as well as lysis of the sensitive target cell, provides material for bacterial shielding or biofilm matrix as well as resources for growth for nonexpressing insensitive cells. A recent study showed ampicillin induced bacterial cell lysis provides a matrix of cell debris that shields viable cells from antimicrobial activity [18]. Further, lysed cells release molecules that could sequester antibiotics.

Subsequently, additional LexA regulated genes, including *lexA* and recA, were also shown to exhibit phenotypic heterogeneity with high level expression, in the absence of DNA damaging agents, in a small subpopulation of cells [73]. Heterogenous expression was found to be established primarily by stochastic factors and the binding affinity of LexA to SOS boxes. Heterogenous expression of *recA* and *lexA* genes could affect a number of phenomenon e. g., subpopulations with higher proficiency in recombination, antibiotic tolerance/persistence, horizontal gene transfer, prophage induction and virulence among pathogenic *E. coli* strains.

7. Conclusions

Given the mounting threat posed by antibiotic resistance, a better understanding of the mechanisms bacteria employ to evolve resistance, persistence as well as pathogenesis is urgently needed. Conditions conducive to SOS induction are encountered by *E. coli* at various host anatomical sites and drive bacterial adaptation to stress, including antibiotic resistance and amplified toxin production. Numerous interdependent mechanisms involving the SOS response are evident, including amplification of the inducing signal in the bacterial population, e.g. (i) SOS induction of horizontal gene transfer which in turn, via ssDNA transfer, induces the SOS response in recipients, (ii) promotion of biofilm formation that generates a dynamic environment with DNA damaging agents and high cell density, conducive to HGT, all in turn inducing the SOS response, (iii) induction of bacteriocins targeting DNA which induce the SOS response in sensitve cells. Nevertheless, our understnding of the modes and the levels of the SOS response, including its connections with other stress response pathways is still lacking. Novel antimicrobial treatment approaches should seek to target the SOS response, possibly the inducer RecA.

Acknowledgements

This work was funded by grant P1-0198 from the Slovene Research Agency.

Conflict of interest

The authors declare no conflict of interest.

Author details

Zdravko Podlesek and Darja Žgur Bertok*
Department of Biology, Biotechnical Faculty, University of Ljubljana, Ljubljana, Slovenia

*Address all correspondence to: darja.zgur.bertok@bf.uni-lj.si

References

[1] Foster PL. Stress-induced mutagenesis in bacteria. Critical Reviews in Biochemistry and Molecular Biology. 2007;42:373-397.

[2] Pennington JM, Rosenberg SM. Spontaneous DNA breakage in single living Escherichia coli cells. Nature Genetics. 2007;39:797-802. DOI: 10.1038/ng2051

[3] McGlynn P, Savery NJ, Dillingham MS. The conflict between DNA replication and transcription. Molecular Microbiology. 2012;85:12-20. DOI: 10.1111/j.1365-2958.2012.08102.x

[4] Butala M, Žgur-Bertok D, Busby SJ. The bacterial LexA transcriptional repressor. Cellular and Molecular Life Sciences. 2009;66:82-93. DOI: 10.1007/s00018-008- 8378-8376

[5] Maslowska KH, Makiela-Dzbenska K, Fijalkowska IJ. The SOS system: A complex and tightly regulated response to DNA damage. Environmental and Molecular Mutagenesis. 2019; 60:368-384. DOI: 10.1002/em.22267.

[6] Layton JC, Foster PL. Error-prone DNA polymerase IV is controlled by the stress-response sigma factor. RpoS, in Escherichia coli. Molecular Microbiology. 2003;50:549-561. DOI: 10.1046/j.1365-2958.2003.03704.x

[7] Dapa T, Fleurier S, Bredeche MF, Matic I. The SOS and RpoS regulons contribute to bacterial cell robustness to genotoxic stress by synergistically regulating DNA polymerase Pol II. Genetics. 2017; 1349-1360. DOI: 10.1534/genetics.116.199471.

[8] Zhao X, Drlica K. (). Reactive oxygen species and the bacterial response to lethal stress. Current Opinion in Microbiology. 2014;21:1-6. DOI: 10.1016/j.mib

[9] Durfee T, Hansen AM, Zhi H, Blattner FR, Jin DJ. Transcription profiling of the stringent response in Escherichia coli. Journal of Bacteriology. 2008;190:1084-1096. DOI: 10.1128/jb.01092-07

[10] Molina-Quiroz RC, Silva-Valenzuela C, Brewster J, Castro-Nallar E, Levy SB, Camilli A. Cyclic AMP regulates bacterial persistence through repression of the oxidative stress response and SOS-dependent DNA repair in uropathogenic *Escherichia coli*. mBio. 2018;9:e02144-17. DOI: 10.1128/mBio.02144-17

[11] Guerin E, Cambray G, Sanchez-Alberola N, Campoy S, Erill I, Da Re S. The SOS response controls integron recombination. Science. 2009;324:1034. DOI: 10.1126/science.1172914

[12] Samuels AN, Roggiani M, Zhu J, Goulian M, Kohli RM. The SOS Response Mediates Sustained Colonization of the Mammalian Gut. Infection and Immunology. 2019;87:e00711-18. DOI: 10.1128/IAI.00711-18.

[13] Beaber J, Hochhut W, Waldor MK. SOS promotes horizontal dissemination of antibiotic resistance genes. Nature. 2004;427:72-74.

[14] Blázquez J, Couce A, Rodríguez-Beltrán J, Rodriguez-Rojas A. Antimicrobials as promoters of genetic variation. Current Opinion in Microbiology. 2012;15:561-569.

[15] Baharoglu Z, Mazel D. Vibrio cholerae triggers SOS and mutagenesis in response to a wide range of antibiotics: a route towards multiresistance. Antimicrobial Chemotherapy. 2011;55:2438-2441.

[16] Andersson DI, Hughes D. Antibiotic resistance and its cost: is it possible to reverse resistance? Nature Reviews Microbiology. 2010;8:260-271.

[17] Wilmaerts D, Windels EM, Verstraeten N, Michiels J. General mechanisms leading to persister formation and awakening. Trends in Genetics. 2019;35:401-411. DOI :10.1016/j.tig.2019.03.007

[18] Podlesek Z, Butala M, Šakanovic A, Žgur-Bertok D. Antibiotic induced bacterial lysis provides a reservoir of persisters. Antonie Van Leeuwenhoek 2016;109:523-528. DOI :10.1007/s10482-016-0657-x

[19] Dörr T, Vulic M, Lewis K. Ciprofloxacin causes persister formation by inducing the TisB toxin in Escherichia coli. PLoS Biology. 2010;8:e1000317. DOI: 10.1371/journal.pbio.1000317

[20] Berghoff BA, Hoekzema M, Aulbach L, Wagner EG. Two regulatory RNA elements affect TisB-dependent depolarization and persister formation. Molecular Microbiology. 2016;103:1020-1033. DOI: 10.1111/mmi.13607

[21] Levin-Reisman I, Ronin I, Gefen O, Braniss I, Shoresh N, Balaban NQ. Antibiotic tolerance facilitates the evolution of resistance. Science. 2017;355:826-830. DOI: 10.1126/science.aaj2191

[22] Windels EM, Michiels JE, Fauvart M, Wenseleers T, Van den Bergh B, Michiels J. Bacterial persistence promotes the evolution of antibiotic resistance by increasing survival and mutation rates. ISME Journal. 2019;13:1239-1251. DOI: 10.1038/s41396-019-03 44-349

[23] Barrett TC, Mok WWK, Murawski AM, Brynildsen MP. Enhanced antibiotic resistance development from fluoroquinolone persisters after a single exposure to antibiotic. Nature Communications. 2019;10:1177. DOI: 10.1038/s41467-019-09058-9054

[24] Pribis JP, García-Villada L, Zhai Y, Lewin-Epstein O, Wang A, Liu J, Xia J, Mei Q, Fitzgerald DM, Bos J, Austin R, Herman C, Bates D, Hadany L, Hastings PJ, Rosenberg SM. Gamblers: an antibiotic-induced evolvable cell subpopulation differentiated by reactive-oxygen-induced general stress response. Molecular Cell. 2019;74:785-800.e7. DOI: 10.1016/j.molcel.2019.02.037

[25] Waldor MK, Friedman DI. Phage regulatory circuits and virulence gene expression. Current Opinion in Microbiology. 2005;8:459 –465. https://doi.org/10.1016/j.mib.2005.06.001.

[26] Baharoglu Z, Bikard D, Mazel D. Conjugative DNA transfer induces the bacterial SOS response and promotes antibiotic resistance development through integron activation. PLoS Genetics. 2010;6: e1001165. DOI:10.1371/journal.pgen.1001165 10.

[27] Baharoglu Z, Krin E, Mazel D. Connecting environment and genome plasticity in the characterization of transformation-induced SOS regulation and carbon catabolite control of the Vibrio cholerae integron integrase. Journal of Bacteriology. 2012;194:1659-1667.

[28] Partridge SR, Kwong SM, Firth N, Jensen SO. Mobile genetic elements associated with antimicrobial resistance. Clinical Microbiology Reviews. 2018;31:e00088-17. DOI: https://doi.org/10.1128/CMR.00088-17.

[29] Halaji M, Feizi A, Mirzaei A, Sedigh Ebrahim-Saraie H, Fayyazi A, Ashraf A, Havaei SA. The global prevalence of class 1 integron and associated antibiotic resistance in *Escherichia coli* from patients with urinary tract infections,

a systematic review and meta-analysis. Microbial Drug Resistance. 2020;26:1208-1218. DOI :https://doi.org/10.1089/mdr.2019.0467, PMID: 32282274

[30] Hall RM, Brookes DE, Stokes HW. Site-specific insertion of genes into integrons: role of the 59-base element and determination of the recombination cross-over point. Molecular Microbiology. 1991; 5:1941-1959.

[31] Collis CM, Hall RM. Expression of antibiotic resistance genes in the integrated cassettes of integrons. Antimicrobial Agents and Chemotherapy. 1995;39: 155-162. DOI: 10.1128/aac.39.1.155

[32] Escudero JA, Loot C, Nivina A, Mazel D. The integron: adaptation on demand. *Microbiology Spectrum.* 2015;3:MDNA3-0019-2014. DOI: 10.1128/microbiolspec. MDNA3-0019-2014.

[33] Gillings MR. Integrons: Past, present, and future. Microbiology and Molecular Biology Reviews. 2014;78:257-277. DOI: 10.1128/MMBR.00056-13

[34] Souque C, Escudero JA, MacLean RC. Integron activity accelerates the evolution of antibiotic resistance. eLife. 2021;10:e62474. DOI: 10.7554/eLife.62474

[35] Chittò M, Berger M, Klotz L, Dobrindt U. Sub-inhibitory concentrations of SOS-response inducing antibiotics stimulate integrase expression and excision of pathogenicity islands in uropathogenic *Escherichia coli* strain 536. *International Journal of Medical Microbiology*. 2019;310: 151361-151361. DOI: 10.1016/j.ijmm.2019.151361

[36] Kaper JB, Nataro JP, Mobley HL. Pathogenic Escherichia coli. Nature Reviews Microbiology 2004;2, 123-140. DOI:10.1038/nrmicro818

[37] Nawrocki EM, Mosso HM., Dudley EG. A Toxic Environment: a Growing Understanding of How Microbial Communities Affect *Escherichia coli* O157:H7 Shiga Toxin Expression. Applied and Environmental Microbiology. 2020;86:e00509-20.

[38] Johnson JR, Russo TA. Molecular epidemiology of extraintestinal pathogenic (uropathogenic) *Escherichia coli*. International Journal of Medical Microbiology. 2005;295:383– 404. https://doi.org/10.1016/j.ijmm.2005.07.005.

[39] Foxman B. Urinary tract infection syndromes: occurrence, recurrence, bacteriology, risk factors, and disease burden. Infectious Disease Clinics of North America 2014;28:1-13. DOI: 10.1016/j.idc.2013.09.003

[40] George DB, Manges AR. (). A systematic review of outbreak and non-outbreak studies of extraintestinal pathogenic Escherichia coli causing community-acquired infections. Epidemiology and Infection. 2010;138:1679-1690. DOI: 10.1017/S0950268810001639

[41] Lewis AJ, Richards AC, Mulvey MA. Invasion of host cells and tissues by uropathogenic bacteria. Microbiology Spectrum. 2016;4. DOI:10.1128/microbiolspec.UTI-0026-2016.

[42] Blango MG, Mulvey MA. Persistence of uropathogenic *Escherichia coli* in the face of multiple antibiotics. Antimicrobial Agents and Chemotherapy. 2010;54:1855-1863. DOI: 10.1128/AAC.00014-10.

[43] Rasko DA, Moreira CG, Li DR, Reading NC, Ritchie JM, Waldor MK, Williams N, Taussig R, Wei S, Roth M, Hughes DT, Huntley JF, Fina MW, Falck JR, Sperandio V. Targeting QseC signaling and virulence for antibiotic development. Science. 2008;321: 1078-1080. https://doi.org/10.1126/science.1160354.

[44] Zhang X, McDaniel AD, Wolf LE, Keusch GT, Waldor MK, Acheson DW. Quinolone antibiotics induce Shiga toxin-encoding bacteriophages, toxin production, and death in mice. Journal of Infectious Diseases. 2000;181:664 – 670. https://doi.org/10.1086/315239.

[45] Hibbing ME, Fuqua C, Parsek MR, Peterson SB. Bacterial competition: Surviving and thriving in a microbial jungle. Nature Reviews Microbiology. 2010;8:15-25. DOI: 10.1038/nrmicro2259.

[46] Stubbendieck RM, Straight D. Multifaceted interfaces of bacterial competition. Journal of Bacteriology 2016;198:2145-2155. DOI: 10.1128/ JB.00275-16.

[47] Conway T, Krogfelt KA, Cohen S. The life of commensal *Escherichia coli* in the mammalian intestine. EcoSal Plus. 2004;1 DOI: 10.1128/ecosalplus.8.3.1.2.

[48] Riley MA, Wertz JE. Bacteriocins: Evolution, ecology, and application. Annual Review of Microbiology. 2002;56:117-137. DOI: 10.1146/annurev. micro.56.012302.161024.

[49] Rebuffat S. Microcins in action: amazing defence strategies of enterobacteria. Biochemical Society Transactions. 2012;40:1456-1462. DOI:10.1042/BST20120183.

[50] Toshima H, Yoshimura A, Arikawa K, Hidaka A, Ogasawara J, Hase A, Masaki H, Nishikawa Y. Enhancement of Shiga toxin production in enterohemorrhagic Escherichia coli serotype O157:H7 by DNase colicins. Applied and Environmental Microbiology. 2007;73:7582-7588. https://doi.org/10.1128/AEM.01326-07.

[51] Xiaoli L, Figler HM, Goswami K, Dudley EG. Nonpathogenic E. coli enhance Stx2a production of E. coli O157:H7 through bamA-dependent and independent mechanisms. Frontiers in Microbiology. 2018;9:1325. https://doi. org/ 10.3389/fmicb.2018.01325.

[52] Mosso HM, Xiaoli L, Banerjee K, Hoffmann M, Yao K, Dudley EG. A putative microcin amplifies shiga toxin 2a production of Escherichia coli O157:H7. Journal of Bacteriology. 2020;202:e00353-19. DOI: 10.1128/ JB.00353-19

[53] Stecher B, Denzler R, Maier L, Bernet F, Sanders MJ, et al. Gut inflammation can boost horizontal gene transfer between pathogenic and commensal Enterobacteriaceae. Proceedings of the National Academy of Sciences U S A. 2012;109:1269-1274. 11.

[54] Hall-Stoodley, L., Costerton, J. W., and Stoodley, P. Bacterial biofilms: from the natural environment to infectious diseases. Nature Reviews Microbiology. 2004;2:95-108. DOI: 10.1038/nrmicro821

[55] López D, Vlamakis H, Kolter R. Biofilms. Cold Spring Harbor Perspectives in Biology. 2010;2:a000398. DOI: 10.1101/cshperspect.a000398

[56] van Gestel J, Vlamakis H, Kolter R. Division of labor in biofilms: the ecology of cell differentiation. Microbiology Spectrum. 2015;3:MB-0002-2014. DOI: 10.1128/ microbiolspec.MB-0002-2014

[57] Qin B, Fei C, Bridges AA, Mashruwala AA, Stone HA, Wingreen NS, et al. Cell position fates and collective fountain flow in bacterial biofilms revealed by light-sheet microscopy. Science. 2020;11:eabb8501. doi: 10.1126/science. abb8501

[58] Costerton JW, Stewart PS, Greenberg EP. Bacterial biofilms: a common cause of persistent infections. Science. 1999;284:1318-1322. doi: 10.1126/ science.284.5418.131

[59] Bernier SP, Lebeaux D, DeFrancesco AS., Valomon A, Soubigou G, Coppée JY, Ghigo JM,

Beloin C. Starvation, together with the SOS response, mediates high biofilm-specific tolerance to the fluoroquinolone ofloxacin. PLoS Genetics. 2013;9: e1003144. DOI: 10.1371/journal.pgen.1003144

[60] Hall CW, Mah TF. Molecular mechanisms of biofilm-based antibiotic resistance and tolerance in pathogenic bacteria. FEMS Microbiology Reviews. 2017;41:276-301. DOI: 10.1093/femsre/fux010

[61] Tashiro Y, Eida H, Ishii S, Futamata H, Okabe S. Generation of small colony variants in biofilms by *Escherichia coli* harboring a conjugative F plasmid. Microbes and Environments. 2017; 32:40-46. DOI: 10.1264/jsme2.ME16121

[62] Soares A, Caron F, Etienne M. Commentary: tolerance and resistance of *Pseudomonas aeruginosa* Biofilms to antimicrobial agents-how *P. aeruginosa* can escape antibiotics. Frontiers in Microbiology. 2019;10:2164. DOI: 10.3389/ fmicb.2019.02164

[63] Greninger AL, Addetia A, Tao Y, Adler A, Qin X. Inactivation of genes in oxidative respiration and iron acquisition pathways in pediatric clinical isolates of small colony variant *Enterobacteriaceae*. Science Reports. 2021; 11:7457. DOI: 10.1038/s41598-021-86764-4

[64] Cascales E, Buchanan SK, Duche D, Kleanthous C, Lloubes R, Postle K, Riley MA, Slatin S, Cavard D. Colicin Biology. Microbiology and Molecular Biology Reviews. 2002;71:158-229. DOI: 10.1128/MMBR.00036-06.

[65] Gordon DM, Riley M A, Pinou T. Temporal changes in the frequency of colicinogeny in *Escherichia coli* from house mice. Microbiology. 1998;144:2233-2240.

[66] Kuhar I, Žgur-Bertok D. 1999. Transcription regulation of the colicin K cka gene reveals induction of colicin synthesis by differential responses to environmental signals. Journal of Bacteriology. 181:7373-7380.

[67] Butala M, Sonjak S, Kamenšek S, Hodoščček M, Browning DF, Žgur-Bertok D, Busby SJW. Double locking of an *Escherichia coli* promoter by two repressors prevents premature colicin expression and cell lysis. Molecular Microbiology. 2012;86:129-139. 10.1111/j.1365-2958.2012.08179.x

[68] Kamenšek S, Browning DF, Podlesek Z, Busby SJW, Žgur-Bertok D, Butala M. Silencing of DNase colicin E8 Gene expression by a complex nucleoprotein assembly ensures timely colicin induction. PLoS Genetics. 2015;11:e1005354. Doi: 10.1371/journal.pgen.1005354

[69] Kirkup, BC, Riley MA. Antibiotic-mediated antagonism leads to a bacterial game of rock-paper-scissors in vivo. Nature. 2004;428:412-414.

[70] Walker D, Rolfe M, Thompson A, Moore GR, James R, Hinton JCD, Kleanthous C. Transcriptional profiling of colicin-induced cell death of Escherichia coli MG1655 identifies potential mechanisms by which bacteriocins promote bacterial diversity. Journal of Bacteriology. 2004;186: 866-869.

[71] Jerman B, Butala M, Žgur-Bertok D. Sublethal concentrations of ciprofloxacin induce bacteriocin synthesis in Escherichia coli. Antimicrobial Agents and Chemotherapy. 2005;49:3087-3090 DOI: 10.1128/aac.49.7.3087-3090.2005

[72] Mulec J, Podlesek Z, Mrak P, Kopitar A, Ihan A, Žgur-Bertok D. A *cka-gfp* transcriptional fusion reveals that the colicin K activity gene is induced in only 3 percent of the population. Journal of Bacteriology. 2003;185:654-659. doi: 10.1128/JB.185.2.654-659.2003

[73] Kamenšek S, Podlesek Z, Gillor OS, Žgur-Bertok D. Genes regulated by the *Escherichia coli* SOS repressor LexA exhibit heterogenous expression BMC Microbiol. 2010; 10: 283. 2010. DOI: 10.1186/1471-2180-10-283

Chapter 4

Molecular Diagnostic Platforms for Specific Detection of *Escherichia coli*

Rehan Deshmukh and Utpal Roy

Abstract

Developing countries due to socio-economic conditions are more prone to frequent pathogenic outbreaks; inadequate sanitation and water quality monitoring are also responsible for such conditions. Therefore, it is of paramount importance to provide microbiologically safe food/water in order to protect public health. Several flaws in traditional culturing methods have sparked a surge in interest in molecular techniques as a means of improving the efficiency and sensitivity of microbiological food/water quality monitoring. Molecular identification of water contaminants, mainly *Escherichia coli*, has been extensively used. Several of the molecular-based techniques are based on amplification and detection of nucleic acids. The advantages offered by these PCR-based methods over culture-based techniques are a higher level of specificity, sensitivity, and rapidity. Of late, the development of a biosensor device that is easy to perform, highly sensitive, and selective has the potential to become indispensable in detecting low CFU of pathogenic *E. coli* in environmental samples. This review seeks to provide a vista of the progress made in the detection of *E. coli* using nucleic acid-based approaches as part of the microbiological food/water quality monitoring.

Keywords: molecular diagnostics, *E. coli*, PCR, LAMP, CRISPR

1. Introduction

Public health protection is of paramount importance that demands the rapid and accurate detection and quantitation of microorganisms in potable water and in various raw and processed foods to prevent undesirable outbreaks of microbial contamination. Water quality has been assessed for potable and recreational activities using culture-dependent quantification and sensing of fecal indicator bacteria (FIB), such as total coliforms, *Escherichia coli*, or Enterococci, an approach that is used as a reference standard in the evaluation of microbial safety of water [1]. The presence of FIBs in large numbers in freshwater, particularly *E. coli* and *Enterococcus*, has been associated with the emergence of waterborne illnesses [2, 3]. Children as young as five years are particularly susceptible to diarrheal infections, with over 800 children dying every day [4, 5]. Amongst coliform bacteria, *E. coli* is commonly regarded as an indicator of fecal pollution of water supplies [6, 7].

Waterborne diseases have been one of the major causes due to the consumption of contaminated water affecting seriously the public health of a humongous number of people in quick succession. In the 2014–2016 survey, the detection rate of pathogenic bacteria was 79.3%, followed by pathogenic *E. coli* (5009 cases, 90%), *Vibrio* spp. (264 cases, 5%), *Shigella* spp. (67 cases, 1%), and *Salmonella* spp. (48 cases, 1%) [8]. The distribution of *E. coli* amongst Korean children suffering from diarrheagenic *E. coli* showed that enteropathogenic *E. coli* (EPEC) was the most common, followed by ETEC and enterohemorrhagic *E. coli* (EHEC) [8]. Of the pathogenic *E. coli*, enteropathogenic *E. coli* (EPEC) was the most common (39%), followed by enterotoxigenic *E. coli* (ETEC) (36%). In a separate study, children suffering from diarrhea were reported [9] in Utah, USA wherein the most commonly detected pathogens included toxigenic *Clostridium difficile* (16%) and diarrheagenic *E. coli* (15%) whereas Shiga toxin-producing *E. coli* were detected in 4% samples [9].

Between 2013 and 2016, a monocentric hospital-based investigation showed that *E. coli* was responsible for about 15% of child infection cases of severe enteritis and EPEC (54%) was the most dominant *E. coli* pathotype, followed by other pathogenic *E. coli* including Shiga toxin-producing *E. coli* [10]. And on the heels of that, in another waterborne outbreak of *E. coli* infection associated with the drinking of contaminated potable water at three different school premises in Korea was reported [11]. As a result of this outbreak, a total of 188 patients with severe gastrointestinal symptoms were reported. The EHEC and EPEC strains isolated from clinical fecal specimens and water samples from water purifiers and water basins respectively were confirmed by the pulsed field gel electrophoresis method [11]. It is warranted therefore to develop rapid and sensitive methods for the detection and quantitation of waterborne bacteria.

Coliforms, particularly *E. coli* is regarded as a primary fecal indicator [12] and indicate the contaminating presence of enteropathogenic bacteria in water and foods supplies [13]. Though these enteric bacteria are abundant in human and warm-blooded animal feces, an umpteen number of the *E. coli* strains have been reported as pathogens [14]. Despite the fact that the wild type of *E. coli* strain is not pathogenic, it could emerge as an infectious agent in immunologically vulnerable people. Furthermore, several *E. coli* O157:H7 outbreaks have been documented in both industrialized and developing economies, resulting in human mortality, notably amongst children under the age of five [15]. *E. coli* serogroup O157:H7 is the most common cause of hemorrhagic colitis in foodborne illness. *E. coli* serogroup O104:H4 was first discovered as an emerging strain in the 2011 German pandemic and was designated a microorganism of serious concern [16]. Perna et al. [17] reported that *E. coli* O157:H7 caused 75,000 cases of foodborne infections per year, of which 85 percent incidences were related to *E. coli* O157 infections [18, 19] with contaminated fruits, vegetables, and water is the principal sources of *E. coli* O157:H7 outbreaks [19].

Traditional microbiological detection techniques consume time as *E. coli* cells require to be isolated, cultivated, and identified using a sequence of biochemical tests [20]. For example, for identification and quantification of *E. coli* in water, the water samples are filtered using the membrane filtration method, followed by the counting of *E. coli* colonies using the plate count method [21]. Furthermore, such processes necessitate 24 to 48 hours to generate observable results and frequently require water samples to be transported to a central laboratory and trained employees to conduct the testing [22].

It is necessary to develop new approaches for detecting *E. coli* in contaminating food and water samples. Optical or impedimetric biosensor systems have evolved as an alternative to the traditional tools for *E. coli* detection, enabling selective, specific, and cost-effective solutions. DNA-based sensing approaches have played an essential role in the development of sensing for the detection of *E. coli*. Due to their rapidity and accuracy, sensing technologies such as the polymerase chain reaction

(PCR), loop associated isothermal method of amplification (LAMP), DNA-based biosensors, and CRISPR/Cas platforms have evolved over time for *E. coli* detection and have been applied in numerous applications in various industries, agriculture, and health care sectors.

2. Polymerase chain reaction (PCR) method

PCR being a mighty and handy tool with molecular biologists showed enormous potential in various forms including multiplex PCR and quantitative real-time PCR. The advantage of PCR is that despite its inability to distinguish between live and dead cells, nonculturable cells may be detected rapidly. In the recent two decades, various PCR-based strategies have been introduced to improve the detection of indicator organisms [23, 24]. Genetic markers such as 23S rRNA and lacZ are often used to establish PCR tests for detecting *E. coli* in environmental samples [25, 26]. The uidA and tuf genes have been identified as potential targets for *E. coli*/Shigella detection using PCR [27, 28]. Most of the PCR assays were reported to amplify the virulence genes, such as *eaeA*, and *stx1, stx2* [29–33] or phenotypic genes, such as *rfbE* (O antigen), and *fliC* (H antigen), *uidA* and *lacZ* which are commonly shared [26, 28, 32]. The ability to generate these lesions is restricted to 43-kb loci of the *E. coli* O157:H7 chromosome [17]. Intimin encoded by eae locus is necessary for early bacterial cell attachment to host cells and the creation of A/E lesions [34, 35]. In a couple of studies, virulence genes like *stx1* and *stx2* were unable to accurately identify a species, owing to the fact that they are widely shared by different species or strains [33]. *Shigella dysenteriae* and *Aeromonas* spp. have been described as the two outliers as non-*E. coli* bacteria bearing Shiga toxin genes [36, 37]. Real-time PCR techniques targeting *Shigella* spp. in food or water utilizing *ipaH* as a target have also been developed to detect enteroinvasive *E. coli* (EIEC) that carries ipaH [36]. Therefore, phenotypic genes such as *rfbE* and *fliC* have been utilized as targets for confirmed identification of *E. coli* in PCR [30].

The *E. coli* genes such as uidA and tuf were used for the detection of *E. coli* and Shigella strains [27, 38, 39]. However, the uidA gene used as a marker was not reported in 3.4% of 116 *E. coli* strains [37]. In another work, Maheux et al. [27] detected Escherichia fergusonii in a PCR targeting the tuf gene. Albeit, it has been extensively reported, neither β-D-glucuronidase activity nor *uidA* gene amplification is the full proof for the accurate molecular detection *E. coli* in the presence of this enzyme or gene has been reported in Flavobacteria and to a great extent in *Shigella, Salmonella* and *Yersinia* [38, 40, 41]. Contrarily, Fricker & Fricker [42] using uidA primer pair detected five non-*E. coli* coliforms in water samples. Recently, Molina et al. [40] designed a set of primers targeting the *E. coli* orphan gene *yaiO* that encodes an outer membrane protein and succeeded in obtaining the *yaiO* amplicon of 115 bp size from unfermented and fermented dairy samples. These workers in terms of specificity claimed superiority of *yaiO* gene-based primers to uidA primers though the study was limited by small sample size. In another recent study, the xanQ-PCR using novel primer set for amplification of *xanQ* gene was demonstrated for specific detection of a large number of *E. coli* strains [41].

Li et al. [43] established a multiplex real-time PCR test that targets the *z3276* and Shiga toxin genes to specifically detect *E. coli* O157:H7 and screen for non-O157 STEC (*stx1* and *stx2*). The reaction mixture contained a primer set; four probes (*z3276, stx1, stx2,* and IAC), and the template DNA of appropriate concentrations. The optimized multiplex assay achieved the limit of detection (LOD) as low as 200 femto grams of bacterial DNA from beef and fresh spinach samples (40 CFU/reaction). In a separate study, a multiplex fluorogenic PCR assay was developed to quantify *E. coli* O157:H7 in manure, soil, dairy wastewater, and cow and calf feces in an artificial wetland. Oligonucleotides were designed to amplify the *stx1* and *stx2* and the *eae* genes of *E. coli* O157:H7 in a simplex reaction [44].

Being a rapid, sensitive, and specific method enabling the detection of multiple pathogens simultaneously this method finds applications in different types of foods and poultry industries. Nguyen et al. [45] developed a multiplex PCR for the rapid and simultaneous detection of three epidemic food-borne pathogens: *E. coli* O157:H7, *Salmonella* spp., and *Listeria monocytogenes* in food samples.

In developing countries, the identification of enteric pathogens in food and other edible items are time-consuming process and often results in wrong and delayed diagnosis. Enteropathogenic *E. coli* (EPEC) has been reported to be frequently associated with outbreaks of infantile diarrhea and recognized as a causative agent for diarrheagenic ailments [46]. In order to detect and identify the Shiga toxin producing *E. coli*, enterohemorrhagic *E. coli* (EHEC), and EPEC primers were designed to amplify *eae* gene and long polar fimbriae (*lpfA*) variants, the bundle-forming pilus gene *bfpA*, and the Shiga toxin-encoding genes *stx1* and *stx2* [47]. This group demonstrated consistent amplification of genes specific to the prototype EHEC O157:H7 EDL933 (*lpfA1–3, lpfA2–2, stx1, stx2*, and *eae-γ*) and EPEC O127:H6 E2348/69 (*eae-α, lpfA1–1*, and *bfpA*) strains using the optimized mPCR protocol with purified genomic DNA (gDNA). A screen of gDNA from isolates in a diarrheagenic *E. coli* collection revealed that the mPCR assay was successful in predicting the correct pathotype of EPEC and EHEC clones grouped in the distinctive phylogenetic disease clusters EPEC1 and EHEC1, and was able to differentiate EHEC1 from EHEC2 clusters. The mPCR assay detection threshold was 2×10^4 CFU per PCR reaction for EHEC and EPEC. Thus, mPCR methodology permitted differentiation of EPEC, STEC, and EHEC strains from other pathogenic *E. coli* and the developed assay has the potential tool for rapid diagnosis of these pathogens. Wang et al. [48] demonstrated the ability of the mPCR assay to detect six bacterial pathogens viz., *E. coli*, *Pasteurella multocida, Proteus mirabilis, Pseudomonas aeruginosa, Salmonella* spp. and *Staphylococcus aureus* in liver, spleen, and blood samples from experimentally infected chicks without cross-amplification with viruses or parasites. In the mPCR assay, gene targets were *phoA, KMT1, ureR, toxA, invA*, and *nuc* of these six pathogens, and six sets of specific primers were designed.

Toma et al. [49] used a single-tube mPCR for the identification of enteropathogenic *E. coli* (EPEC), enteroinvasive *E. coli* (EIEC), enterotoxigenic *E. coli* (ETEC), enteroaggregative *E. coli* (EAEC), and Shiga toxin-producing *E. coli* (STEC). In total six targets were chosen for (*eae*) enteropathogenic *E. coli*, (*stx*) Shiga toxin-producing *E. coli* enterotoxigenic *E. coli*, elt, and est. for enterotoxigenic *E. coli*, (*ipaH*) for enteroinvasive *E. coli* for, and *aggR* for enteroaggregative *E. coli*.

Chen et al. [50] developed a multiplex rtPCR assay for the identification of diarrheagenic *E. coli* (DEC) and claimed it to be a highly sensitive and specific and suggested the rapid identification of DEC in clinical and public health laboratories. Specific virulence genes were selected to identify specific pathogens: *ipaH* for EIEC, *stp/sth/lt* for ETEC, *eaeA/escV* for EPEC, *stx1/stx2* for EHEC, *aggR* for EAEC. The 5′ end of primers were added with a homo tail sequence to reduce the primer dimer formation and the addition of homo tail to 5′ end of primer sequences allowed proper annealing temperature that would fall into broad range in each individual PCR reaction. Molecular beacons were modified and designed using DNA folding form website (http://mfold.rit.albany.edu/?q=mfold/DNA-Folding-Form) [50]. Five categories of DEC were split into two tubes. For tube number one, *stp/sth/lt* for ETEC, *aggR* for EAEC and IAC were included, while *ipaH* for EIEC, *eaeA/escV* for EPEC, *stx1/stx2* for EHEC and IAC were included in tube number two. Carboxy fluorescein (FAM), Hexachloro fluorescein (HEX), Carboxy-X-rhodamine (ROX),

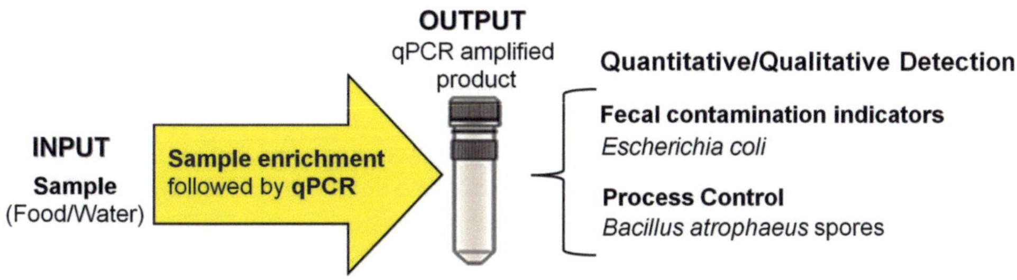

Figure 1.
Schematic depicting the steps in culture-independent detection of E. coli *in a sample using qPCR method. Bacillus atrophaeus Spores are used as an internal control for monitoring of possible PCR inhibition [52].*

Quasar 705, and indodicarbocyanine5 (Cy5) fluorescence were collected and recorded at the end of the annealing step during the third stage.

Detection of harmful bacteria with higher specificity, sensitivity, and reliability is the focus of nucleic acid-based approaches. The desired nucleic acid sequence is hybridized to a synthetic oligonucleotide for specific detection of the pathogen [51]. Nucleic acid-based approaches are routinely used to detect bacterial infections and their toxin-producing genes [51]. Nucleic acid-based methods are rapid and easy to use, and they do not require the pathogens to be cultured (**Figure 1**).

Even a decade ago, the identification and measurement of specific target genes with absolute accuracy and as little as a few copies in a matter of hours was a dream. In the area of water quality assessment, however, qPCR technology has proven to be a powerful technique [53]. Unlike the classical PCR, which needs agarose-gel electrophoresis to identify the end-point PCR products, the qPCR enables assessing PCR product amplification by measuring fluorescence signals released by specialized dual-labeled probes or the intercalating dyes. The fluorescence intensity generated during the qPCR is directly related to the quantity of PCR products produced [12, 54, 55]. The most often used fluorescent systems for qPCR include SYBR green, TaqMan probes, and molecular beacons [56]. The qPCR techniques, which have higher specificity, sensitivity, and reliability than classic culture methods and mPCR [57], allow for the time-efficient detection of harmful bacteria with higher specificity, sensitivity, and reliability [12, 56, 58]. Although the qPCR has been used to detect and quantify *E. coli* O157:H7 in food and clinical samples, it has not been thoroughly evaluated with environmental samples [57, 59, 60].

Utilizing TaqMan probes labeled with different fluorophores, microfluidic qPCR was shown to identify pathogens such as *Listeria monocytogenes, Vibrio cholerae, Vibrio parahaemolyticus, Pseudogulbenkiana* spp., *Salmonella typhimurium, Shigella flexneri, Clostridium perfringens*, and *E. coli* at a limit of detection of 100 CFU/L [56, 61]. Despite its high sensitivity, qPCR has significant drawbacks, such as the inability to provide information on the physiological status of target cells in environmental samples. Humic substances found in environmental samples such as water hinder DNA polymerase activity, and colloidal debris has been reported to have a DNA affinity [62, 63]. There is no universal answer to avert such problems. As a result, the existence of these compounds in environmental samples has the potential to adversely affect the amplification effectiveness of qPCR, which is used to detect small quantities of bacteria [60]. To overcome these issues in qPCR, several compounds such as bovine serum albumin, methoxsalen, dimethyl sulfoxide, and internal amplification controls have been proposed. However, these approaches may have certain drawbacks as well as benefits [64, 65]. Walker et al. [63] established a new qPCR technique for detecting and quantifying *E. coli* that targeted a segment of the ybbW gene, which encodes a potential Allantoin transporter. The *ybbW* gene is part of the *E. coli* "core genome," which means that each gene is found in >95 percent of all sequenced strains. For this work, water samples were taken at monthly intervals from different locations in the southwest of England. The *ybbW*-qPCR was found to be 100% specific towards 87 *E. coli* strains tested. This work also reported that despite the theoretically low detection levels achievable by qPCR, the quantity of *E. coli* DNA has been the key issue in limiting the detection in real samples. This could be addressed in part by filtering greater quantities of water samples, but this is likely to be unfeasible for regular sample analysis and could result in the accumulation of higher inhibitory substance quantities.

In another study, Liu et al. [66] reported designing of the novel oligonucleotide primer set and TaqMan probes targeting the specific virulence genes of twelve common food pathogens such as *E. coli* O157:H7, *Salmonella enterica, L. monocytogenes/ ivanovii*, β-*Streptococcus hemolyticus, Enterococcus faecalis, Yersinia enterocolitica, Shigella sp., P. mirabilis, V. fluvialis, V. parahaemolyticus, S. aureus* and *Campylobacter jejuni*. Liu et al. [66] reported the use of TaqMan in artificially spiked dilution series of each pathogen into meat to detect 12 strains. The TaqMan assays demonstrated expected amplification with no amplification inhibition. In spiked food samples, *V. parahaemolyticus* was found in concentrations ranging from 10^3 to 10^7 CFU/g, while the remaining 11 strains were from 10^4 to 10^7 CFU/g. The qPCR has been touted as a specific and sensitive method with high throughput sample analysis. Smati et al. [66] reported a rapid, sensitive, and reliable qPCR method to quantify *E. coli* phylogroup from 100 healthy human stool specimens and demonstrated the existence of subdominant clones. The new 16S-rRNA-qPCR assay was highly repeatable, with a detection limit of 10^5 CFU/g of feces.

3. Loop mediated isothermal amplification (LAMP) assay

In order to circumvent the use of thermocyclers that entail the time-consuming thermal cycling, an innovative method such as isothermal DNA amplification has been introduced which finds its application in the advanced Research & Development (R & D) unit of the food industry. The LAMP reaction that involves isothermal amplification chemistry has a good range of possible applications, including point-of-care testing with the potential of getting developed into portable diagnostic systems, and quick testing of food products, clinical and environmental samples.

The isothermal characteristics of LAMP enable the simplification of the detection process without involving any costly and complex instrumentation wherein a simple heating block or a precise digital water bath would work. Though conventional PCR and LAMP techniques were reported to be vulnerable to several inhibitors while testing various biological (for example urinary and plant materials) matrices [64], yet LAMP is much less sensitive to amplification inhibitors [64], potentially permitting its application bypassing the general requirement for cultural enrichment or DNA purification.

Despite some disadvantages like its qualitative nature of detection, the LAMP offers several advantages over PCR. LAMP assay emphasizes the requirement of a heating block and obviates the need for a thermal cycler. Unlike PCR that requires DNA extraction from samples for amplification, LAMP assay does not require DNA extraction step. The difficulties in amplifying DNA in PCR from unprocessed urinary samples in the presence of a high concentration of urea were reported by Khan et al. [65]. Therefore the LAMP assay, by rendering the DNA extraction step redundant, has made the process more rapid and facile [67]. The implementation of LAMP does not require any denatured template as due to the use of Bst DNA polymerase from *Geobacillus stearothermophilus* with auto-cycling strand-displacement activity denatured template use has been eliminated. In the LAMP reaction, the nucleic acid amplification takes place at a fixed temperature (isothermal) through repetition of two types of elongation reactions occurring at the loop regions: self-elongation of templates from the stem-loop structure formed at the 3′-terminal and the binding and elongation of new primers to the loop region (**Figure 2**) [68]. LAMP reaction time is merely 60–65 min at 60–65°C involving four

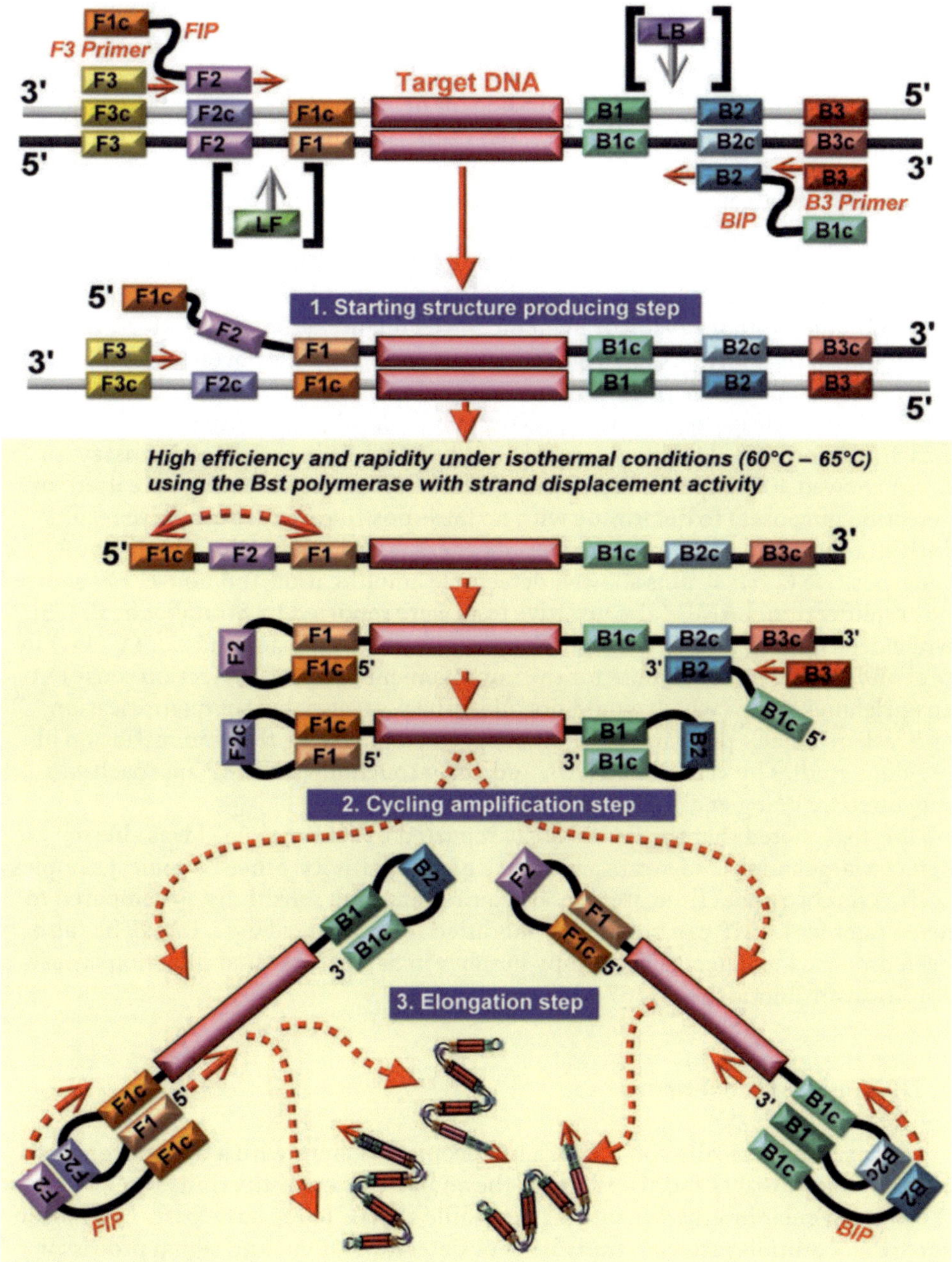

Figure 2.
Schematic representation of the principle steps in a LAMP assay and localization of the eight LAMP primers for specific amplification of target DNA. Adapted from Gallas-Lindemann et al. [68]. Copyright (2017). IntechOpen. Inner primer: FIP (consisting of F1c and F2), BIP (consisting of B1c and B2), typical length ~ 40 bp; outer primers: F3, B3 typical length ~ 20 bp; loop primers: LF and LB, typical length ~ 20 bp.

to six precisely designed primers to amplify DNA targets at the specific amplification temperature [68].

The very purpose of inner primers that consisted of two different sequences was to recognize a sense and antisense sequences of the target viral DNA, and the outer primers were designed to recognize an external sequence of the target viral DNA [69]. Additionally, in the LAMP assay, as an advantage, the identification of a positive reaction does not involve any special processing or electrophoresis. Only the visual observation of color change of the reaction mix in normal light is enabled when the appropriate DNA-binding dye is used. Thus, LAMP positive results could be better detected through visual observation of turbidity changes [70]. This visualization process can be improved by a UV transilluminator. Hill et al. [67] had demonstrated the use of propidium iodide for detecting the LAMP products.

In order to detect generic *E. coli*, *E. coli* O157, or different VTEC virulence genes a number of LAMP assays were developed and discussed in several previous studies [71–74]. In order to develop the LAMP assays for the simultaneous detection of an *E. coli*-specific gene and verotoxin-elaborating genes, and capable of distinguishing between generic *E. coli* and VTEC that would serve the purpose of simultaneous detection both *E. coli* and VTEC simultaneously in beef would allow the simultaneous monitoring of hygienic status/quality of beef. Therefore, the development of multiplex LAMP assay was of paramount importance. In the study, the LAMP assay was designed to detect nonpathogenic *E. coli* targeting the phoA and VTEC targeting the *stx1* and *stx2* without the need for a cumbersome culture enrichment process. The specificity of the *phoA* LAMP-based detection assay for *E. coli* showed 100% specificity (when a total of 58 bacterial strains were used for detection purposes) to determine with no false-positive or false-negative results with strains of any of the other bacterial genera tested. Interestingly only phoA gene-positive *E. coli* strains showed detectable amplification and non-*E. coli* showed no amplification. LAMP-false negative tests were reported by Stratakos et al. [75] while determining non-pathogenic *E. coli* and verocytotoxigenic *E. coli* (VTEC) in beef and bovine feces. Of note for the improvement of LAMP detection sensitivity, an enrichment step (which would not allow the post-enrichment quantification of *E. coli* or VTEC) prior to LAMP was suggested following the demonstration in previous studies by Wang et al. [76], and also a touchdown LAMP approach was suggested by Wang et al. [73].

It is to be noted that the LAMP assay reported by Hill et al. [67] was able to detect a large number of strains with very high sensitivity. Since biological samples such as cerebrospinal fluid and blood require very high sensitivity as compared to urine samples LAMP can be suitably modified for its clinical uses. LAMP has also been proposed to detect a lower copy number in partially treated infections (post-empirical antibiotic doses) [67].

4. DNA-based biosensors

A biosensor typically consists of a bioreceptor element with a transducer. The bioreceptor, interacts specifically with the analyte, whereas the transducer converts the biomolecular interaction into an electronic signal. Three basic parts of a biosensor are recognition material, transducer or detector system, and signal processor [74]. Monitoring the molecular interaction between the DNA-based bioreceptor and the analyte is an essential element of various DNA-based sensing strategies. The measurement methods of DNA–DNA interactions that take place on the various sensor surfaces are gaining much interest to improve sensor performance. The assays are applicable to the determination of low numbers of *E. coli* cells in various matrices. In addition, the molecular detection of *E. coli* using single-stranded nucleic acids or aptamers coupled with the electrochemical impedance spectroscopy for sensing of DNA is a growing field of research and proving to be an alternative method of detection to traditional techniques [74].

Arora et al. [77] reported an electrochemical DNA biosensor for the detection of *E. coli*. In this study, avidin was modified with –COOH and then attached to the polyaniline (PANI)-modified platinum disk by the covalent binding between –COOH and –NH/NH_2 of PANI. Subsequently, the biotin-labeled DNA probe was functionalized on the electrode surface to achieve a LOD of 0.01 ng/uL for *E. coli* genomic DNA. Few studies reported the use of nanomaterials with graphene oxide (GO) to enhance the sensitivity of the DNA biosensor for *E. coli* detection. For example, a DNA biosensor for the detection of *E. coli* O157:H7 eaeA gene based on a novel sensing tag of GOx-Thi-Au@SiO2 nanocomposites is reported [78]. The combined use of GO and Au@SiO2 creates an environment for maintaining the appropriate conformation of DNA. These biosensor modalities led to wide linear response for *E. coli* O157:H7 eaeA gene in the range of 0.02 to 50.0 nM with LoD of 0.01 nM. In addition, Tiwari et al. [79] reported a DNA biosensor for *E. coli* O157:H7 using a DNA probe sequence. The DNA probe was immobilized onto GO modified iron oxide-chitosan hybrid nanocomposite (GIOCh) film. The DNA biosensor resulted in linear response to *E. coli* DNA in the range of 10^{-6} to 10^{-14} M with a LoD of 10^{-14} M.

Since its discovery in the 1980s, the system has demonstrated widespread applications in basic biotechnology research and disease treatment [80, 81]. A pressing need of the hour is the availability of a cost-efficient, rapid and selective molecular diagnostic platform to detect different pathogens and lethal diseases in the early stage of the infection. Quantitative PCR and metagenomic next-generation sequencing (mNGS) are the most commonly explored molecular platforms for the same; however, these methods have their disadvantages and limitations. Clustered Regularly Interspaced Short Palindromic Repeat/associated protein (CRISPR/Cas)-based diagnostic platform for the detection of nucleic acids has progressively demonstrated its potential as an ideal diagnostic approach for pathogens, cancer biomarker, and single-nucleotide polymorphisms (SNPs) detection. CRISPR systems have evolved in prokaryotes as a defensive mechanism against foreign viruses by cleaving their nucleic acids [82–84].

Additionally, the unique cleavage activity of Cas9 is often utilized for the development of ultra-low abundance DNA biosensors. A highly innovative and sensitive CRISPR/Cas9 system was developed by Huang et al. [84] that triggered isothermal exponential amplification reaction (CAS-EXPAR) strategy to detect DNA targets with attomolar (aM) sensitivity and single-base specificity [84]. CAS-EXPAR was primed by the target DNA fragment produced by cleavage of CRISPR/Cas9, and associated with the cyclical amplification reaction to produce numerous DNA replicates capable of getting detected by a real-time SYBR Green fluorescence signal [83].

Recently, Sun et al. [84] reported the detection of *E. coli* O157:H7 based on the CRISPR/Cas9 coupled with metal–organic framework platform (MoF) (**Figure 3**). In this approach, the virulence gene sequences of *E. coli* O157:H7 were identified and spliced by the CRISPR/Cas9 system leading to strand displacement and rolling circle amplification. Subsequently, amplified products were hybridized with the target-specific probes. The virulence genes were detected by the fluorescence quenching caused due to MoF platform. The method showed high sensitivity with LoD of 4.0×10^1 CFU mL^{-1} [84]. Although there is only one reported work available for CRISPR/Cas-based detection of *E. coli*, however, the CRISPR/Cas system can be exploited further for the detection of *E. coli* and other waterborne pathogens using novel strategies.

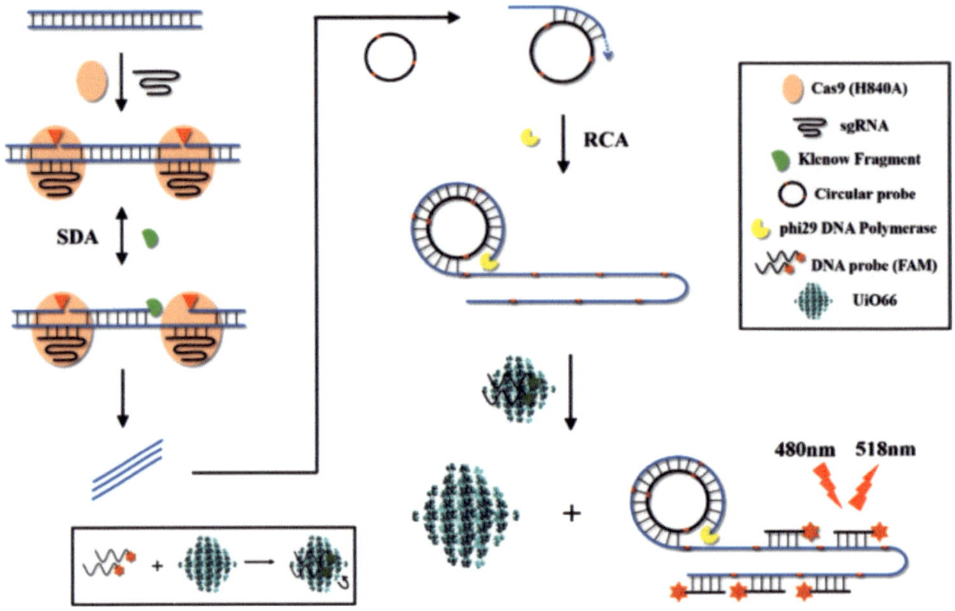

Figure 3.
CRISPR/Cas9 platform coupled with two-step isothermal amplification for detection of E. coli *O157:H7. Reprinted with permission from Sun et al. [84]. Copyright (2020) American Chemical Society.*

5. Conclusion

Molecular diagnostic platforms have become promising alternatives to traditional methods for *E. coli* detection. In particular, LAMP assay and DNA biosensors because of their advantages of lower detection limits, and high reproducibility are preferred for pathogen detection. In this review, we have discussed the recent advances in the development of PCR, LAMP assay, and DNA biosensors platforms applied to *E. coli* detection. In the case of CRISPR/Cas platforms, the major challenge associated with the CRISPR/Cas sensing platform is the time taken to produce the results. Therefore, the future perspective would be to reduce the assay turnaround time for CRISPR/Cas sensing. Nonetheless, CRISPR/Cas sensing platforms possess the potential to overcome the use of conventional molecular diagnostic platforms and become a promising tool for next-generation diagnostic platforms for sensitive and selective detection of DNA in clinical, food, and environmental samples. In the future, more, specific, sensitive, cost-sensitive, and portable biosensors will be required to detect *E. coli*, hence, further leading to controlling and monitoring the waterborne epidemics.

Conflict of interest

The authors declare no conflict of interest.

Author details

Rehan Deshmukh[1] and Utpal Roy[2]*

1 Faculty of Science, School of Biology, MIT World Peace University, Pune, India

2 Department of Biological Sciences, Birla Institute of Technology and Science, Pilani, Goa, India

*Address all correspondence to: utpalroy@gmail.com

References

[1] José Figueras M, Borrego JJ. New perspectives in monitoring drinking water microbial quality. International Journal of Environmental Research and Public Health. 2010;7(12):4179-4202

[2] J. M. Colford et al., "Water quality indicators and the risk of illness at beaches with nonpoint sources of fecal contamination.," Epidemiology, vol. 18, no. 1, pp. 27-35, Jan. 2007.

[3] V. J. Harwood, C. Staley, B. D. Badgley, K. Borges, and A. Korajkic, "Microbial source tracking markers for detection of fecal contamination in environmental waters: Relationships between pathogens and human health outcomes," FEMS Microbiology Reviews 38, 1, pp. 1-40, Jan. 2014.

[4] Gorchev HG, Ozolins G. WHO Guidelines for Drinking-Water Quality. Vol. 38, no. 3. WHO Press; 2011

[5] UNICEF. Somalia - Water, Sanitation and Hygiene - Priority issues. Unicef; 2019. Available from: https://reliefweb.int/report/somalia/somalia-water-sanitation-and-hygiene-assessment-report-december-2019

[6] Graham N. Guidelines for Drinking-Water Quality, 2nd edition, Addendum to Volume 1 – Recommendations, World Health Organisation, Geneva, 1998, 36 pages. Urban Water. 1999;**1**(2):183

[7] Deshmukh RA, Joshi K, Bhand S, Roy U. Recent developments in detection and enumeration of waterborne bacteria: a retrospective minireview. Microbiology. 2016;**5**(6):901-922

[8] Ryoo E. Causes of acute gastroenteritis in Korean children between 2004 and 2019. Clinical and Experimental Pediatrics. 2021;**64**(6):260-268

[9] Stockmann C et al. Detection of 23 gastrointestinal pathogens among children who present with Diarrhea. Journal of the Pediatric Infectious Diseases Society. 2017;**6**(3): 231-238

[10] Y. Kim et al., "Community-acquired Escherichia coli enteritis in Korean children: The clinical application of a stool polymerase chain reaction assay," Infection & Chemotherapy, vol. 49, no. 4, pp. 275-281, Dec. 2017

[11] J. Park et al., "A waterborne outbreak of multiple diarrhoeagenic Escherichia coli infections associated with drinking water at a school camp," International Journal of Infectious Diseases, vol. 66, pp. 45-50, Jan. 2018

[12] A. Rompré, P. Servais, J. Baudart, M. R. De-Roubin, and P. Laurent, "Detection and enumeration of coliforms in drinking water: Current methods and emerging approaches," Journal of Microbiological Methods, vol. 49, no. 1, pp. 31-54, Mar. 2002

[13] Maheux AF et al. Molecular method for detection of total coliforms in drinking water samples. Applied and Environmental Microbiology. 2014;**80**(14):4074-4084

[14] Kaper JB, Nataro JP, Mobley HLT. Pathogenic *Escherichia coli*. Nature Reviews. Microbiology. 2004;**2**(2): 123-140

[15] Lupindu AM. Epidemiology of Shiga toxin-producing Escherichiaï¿¿coli O157:H7 in Africa in review. Southern African Journal of Infectious Diseases. 2018;**33**(1):24-30

[16] I. Son, R. Binet, A. Maounounen-Laasri, A. Lin, T. S. Hammack, and J. A. Kase, "Detection of five Shiga toxin-producing *Escherichia coli* genes with multiplex PCR," Food Microbiology, vol. 40, pp. 31-40, Jun. 2014.

[17] Perna NT et al. Genome sequence of enterohaemorrhagic *Escherichia coli* O157:H7. Nature. 2001;**409**(6819): 529-533

[18] Mead PS et al. Food-related illness and death in the United States. Emerging Infectious Diseases. 1999; **5**(5):607-625

[19] Ferens WA, Hovde CJ. *Escherichia coli* O157:H7: Animal reservoir and sources of human infection. Foodborne Pathogens and Disease. 2011;**8**(4):465-487

[20] Tok S, De Haan K, Tseng D, Usanmaz CF, Ceylan Koydemir H, Ozcan A. Early detection of: *E. coli* and total coliform using an automated, colorimetric and fluorometric fiber optics-based device. Lab on a Chip. 2019;**19**(17):2925-2935

[21] Stender H, Broomer AJ, Oliveira K, Hyldig-nielsen JJ, Sage A, Coull J. Combination of 16S rRNA-targeted oligonucleotide probes with flow cytometry for analyzing mixed microbial populations. Applied and Environmental Microbiology. 2001;**67**(1):142-147

[22] Rompre A, Servais P, Baudart J, De-Roubin M-R, Laurent PP. Detection and enumeration of coliforms in drinking water: current methods and emergin approaches. Journal of Microbiological Methods. 2002;**49**(2002):31-54

[23] Khan IUH et al. Development of a rapid quantitative PCR assay for direct detection and quantification of culturable and non-culturable *Escherichia coli* from agriculture watersheds. Journal of Microbiological Methods. 2007;**69**(3):480-488

[24] Luna GM, Dell 'anno A, Pietrangeli B, Danovaro R. A new molecular approach based on qPCR for the quantification of fecal bacteria in contaminated marine sediments. Journal of Biotechnology. 2011;**157**:446-453

[25] Ludwig W, Schleifer K-H. How quantitative is quantitative PCR with respect to cell counts? Systematic and Applied Microbiology. 2000;**23**(4): 556-562

[26] Foulds IV, Granacki A, Xiao C, Krull UJ, Castle A, Horgen PA. Quantification of microcystin-producing cyanobacteria and *E. coli* in water by 5′-nuclease PCR. Journal of Applied Microbiology. 2002;**93**(5): 825-834

[27] Maheux AF et al. Rapid concentration and molecular enrichment approach for sensitive detection of *Escherichia coli* and Shigella species in potable water samples. Applied and Environmental Microbiology. 2011;**77**(17):6199-6207

[28] Bej AK, DiCesare JL, Haff L, Atlas RM. Detection of *Escherichia coli* and Shigella spp. in water by using the polymerase chain reaction and gene probes for uid. Applied and Environmental Microbiology. 1991; **57**(4):1013-1017

[29] Bae S, Wuertz S. Discrimination of viable and dead fecal Bacteroidales bacteria by quantitative PCR with Propidium monoazide. Applied and Environmental Microbiology. 2009;**75**(9):2940-2944

[30] Y. Liu, A. Gilchrist, J. Zhang, and X.-F. Li, “Detection of viable but nonculturable *Escherichia coli* O157:H7 bacteria in drinking water and river water.,” Applied and Environmental Microbiology, vol. 74, no. 5, pp. 1502-1507, Mar. 2008.

[31] Ercoli L et al. Prevalence and characteristics of verotoxigenic *Escherichia coli* strains isolated from pigs and pork products in Umbria and Marche regions of Italy. International Journal of Food Microbiology. 2016;**232**:7-14

[32] Hamner S et al. Isolation of potentially pathogenic *Escherichia coli*

O157:H7 from the Ganges River. Applied and Environmental Microbiology. 2007;**73**(7):2369-2372

[33] Chassagne L, Pradel N, Robin F, Livrelli V, Bonnet R, Delmas J. Detection of stx1, stx2, and eae genes of enterohemorrhagic *Escherichia coli* using SYBR Green in a real-time polymerase chain reaction. Diagnostic Microbiology and Infectious Disease. 2009;**64**(1): 98-101

[34] Alperi A, Figueras MJ. Human isolates of Aeromonas possess Shiga toxin genes (stx1 and stx2) highly similar to the most virulent gene variants of *Escherichia coli*. Clinical Microbiology and Infection. 2010;**16**(10):1563-1567

[35] K. E. Hyma et al., "Evolutionary genetics of a new pathogenic Escherichia species: Escherichia albertii and related Shigella boydii strains," Journal of Bacteriology, vol. 187, no. 2, pp. 619-628, Jan. 2005.

[36] Deer DM, Lampel KA. Development of a multiplex real-time PCR assay with internal amplification control for the detection of Shigella species and enteroinvasive *Escherichia coli*. Journal of Food Protection. 2010;**73**(9):1618-1625

[37] Feng P, Lum R, Chang GW. Identification of uidA gene sequences in beta-D-glucuronidase-negative *Escherichia coli*. Applied and Environmental Microbiology. 1991;**57**(1):320-323

[38] Frampton EW, Restaino L. Methods for *Escherichia coli* identification in food, water and clinical samples based on beta-glucuronidase detection. The Journal of Applied Bacteriology. 1993;**74**(3):223-233

[39] Feng P, Lum R, Chang GW. Identification of uidA gene sequences in beta-D-glucuronidase-negative *Escherichia coli*. Applied and Environmental Microbiology. 1991;**57**(1):320-323

[40] Molina F, López-Acedo E, Tabla R, Roa I, Gómez A, Rebollo JE. Improved detection of *Escherichia coli* and coliform bacteria by multiplex PCR. BMC Biotechnology. 2015;**15**:48

[41] Deshmukh RA, Bhand S, Roy U. A novel molecular quantitative method for rapid and sensitive detection of *Escherichia coli* from roof-harvested rainwater. Analytical Methods. 2019;**11**(25):3155-3167

[42] Fricker EJ, Fricker CR. Application of the polymerase chain reaction to the identification of *Escherichia coli* and coliforms in water. Letters in Applied Microbiology. 1994;**19**(1):44-46

[43] Li B, Liu H, Wang W. Multiplex real-time PCR assay for detection of *Escherichia coli* O157:H7 and screening for non-O157 Shiga toxin-producing *E. coli*. BMC Microbiology. 2017;**17**(1):215

[44] Ibekwe AM, Leddy M, Murinda SE. Potential human pathogenic bacteria in a mixed urban watershed as revealed by pyrosequencing. PLoS One. 2013;**8**(11): e79490

[45] Nguyen TT, Van Giau V, Vo TK. Multiplex PCR for simultaneous identification of *E. coli* O157:H7, Salmonella spp. and L. monocytogenes in food. 3 Biotech. 2016;**6**(2):205

[46] Ochoa TJ, Contreras CA. Enteropathogenic *Escherichia coli* infection in children. Current Opinion in Infectious Diseases. 2011;**24**(5): 478-483

[47] Botkin DJ, Galli L, Sankarapani V, Soler M, Rivas M, Torres AG. Development of a multiplex PCR assay for detection of Shiga toxin-producing *Escherichia coli*, enterohemorrhagic *E. coli*, and enteropathogenic *E. coli* strains.

Frontiers in Cellular and Infection Microbiology. 2012;**2**:8

[48] Wang Z et al. Development of a multiplex PCR assay for the simultaneous and rapid detection of six pathogenic bacteria in poultry. AMB Express. 2019;**9**(1):1-11

[49] Toma C et al. Multiplex PCR assay for identification of human diarrheagenic *Escherichia coli*. Journal of Clinical Microbiology. 2003;**41**(6): 2669-2671

[50] Chen Q et al. Rapid genetic typing of diarrheagenic *Escherichia coli* using a two-tube modified molecular beacon based multiplex real-time PCR assay and its clinical application. Annals of Clinical Microbiology and Antimicrobials. 2014;**13**(1):30

[51] Ramírez-Castillo FY et al. Waterborne pathogens: Detection methods and challenges. Pathogens. 2015;**4**(2):307-334

[52] Picard FJ et al. Internal control for nucleic acid testing based on the use of purified bacillus atrophaeus subsp. globigii Spores. Journal of Clinical Microbiology. 2009;**47**(3):751-757

[53] Girones R et al. Molecular detection of pathogens in water - The pros and cons of molecular techniques. Water Research. 2010;**44**(15):4325-4339

[54] Law JW-F, Ab Mutalib N-S, Chan K-G, Lee L-H. Rapid methods for the detection of foodborne bacterial pathogens: Principles, applications, advantages and limitations. Frontiers in Microbiology. 2014;**5**(January):770

[55] Bellin T, Pulz M, Matussek A, Hempen HG, Gunzer F. Rapid detection of enterohemorrhagic *Escherichia coli* by real-time PCR with fluorescent hybridization probes. Journal of Clinical Microbiology. 2001;**39**(1):370-374

[56] Ishii S, Nakamura T, Ozawa S, Kobayashi A, Sano D, Okabe S. Water quality monitoring and risk assessment by simultaneous multipathogen quantification. Environmental Science & Technology. 2014;**48**(9):4744-4749

[57] J. S. Way, K. L. Josephson, S. D. Pillai, M. Abbaszadegan, C. P. Gerba, and I. L. Pepper, "Specific detection of Salmonella spp. by multiplex polymerase chain reaction," Applied and Environmental Microbiology, vol. 59, no. 5, pp. 1473-1479, May 1993.

[58] Sharma VK, Carlson SA. Simultaneous detection of Salmonella strains and *Escherichia coli* O157:H7 with fluorogenic PCR and single-enrichment-broth culture. Applied and Environmental Microbiology. 2000;**66**(12):5472-5476

[59] Bustin SA et al. The MIQE guidelines: Minimum information for publication of quantitative real-time PCR experiments. Clinical Chemistry. 2009;**55**(4):611-622

[60] Straub TM, Pepper IL, Gerba CP. Removal of PCR inhibiting substances in sewage sludge amended soil. Water Science and Technology. 1995;**31**(5-6): 311-315

[61] Klerks MM, Zijlstra C, Van Bruggen AHC. Comparison of real-time PCR methods for detection of Salmonella enterica and *Escherichia coli* O157:H7, and introduction of a general internal amplification control. Journal of Microbiological Methods. 2004;**59**(3):337-349

[62] Shanks OC et al. Quantitative PCR for detection and enumeration of genetic markers of bovine fecal pollution. Applied and Environmental Microbiology. 2008;**74**(3):745-752

[63] Walker DI et al. A highly specific *Escherichia coli* qPCR and its comparison with existing methods for

environmental waters. Water Research. 2017;**126**:101-110

[64] Schrader C, Schielke A, Ellerbroek L, Johne R. PCR inhibitors - occurrence, properties and removal. Journal of Applied Microbiology. 2012;**113**(5):1014-1026

[65] Khan G, Kangro HO, Coates PJ, Heath RB. Inhibitory effects of urine on the polymerase chain reaction for cytomegalovirus DNA. Journal of Clinical Pathology. 1991;**44**(5):360-365

[66] Liu Y, Cao Y, Wang T, Dong Q, Li J, Niu C. Detection of 12 common food-borne bacterial pathogens by taq man real-time PCR using a single set of reaction conditions. Frontiers in Microbiology. 2019;**10**(FEB):222

[67] Hill J et al. Loop-mediated isothermal amplification assay for rapid detection of common strains of *Escherichia coli*. Journal of Clinical Microbiology. 2008;**46**(8):2800-2804

[68] Gallas-Lindemann C, Sureshkumar P, Noack MJ, Sotiriadou I. Loop-mediated isothermal amplification: An advanced method for the detection of giardia. Current Topics Giardiasis. 2017

[69] Notomi T, Mori Y, Tomita N, Kanda H. Loop-mediated isothermal amplification (LAMP): Principle, features, and future prospects. Journal of Microbiology. 2015;**53**(1):Jan

[70] Mori Y, Nagamine K, Tomita N, Notomi T. Detection of loop-mediated isothermal amplification reaction by turbidity derived from magnesium pyrophosphate formation. Biochemical and Biophysical Research Communications. 2001;**289**(1):150-154

[71] Kouguchi Y, Fujiwara T, Teramoto M, Kuramoto M. Homogenous, real-time duplex loop-mediated isothermal amplification using a single fluorophore-labeled primer and an intercalator dye: Its application to the simultaneous detection of Shiga toxin genes 1 and 2 in Shiga toxigenic *Escherichia coli* isolates. Molecular and Cellular Probes. 2010;**24**(4):190-195

[72] Ravan H, Amandadi M, Sanadgol N. A highly specific and sensitive loop-mediated isothermal amplification method for the detection of *Escherichia coli* O157: H7. Microbial Pathogenesis. 2016;**91**:161-165

[73] Wang Y et al. Rapid and sensitive detection of Shigella spp. and Salmonella spp. by multiple endonuclease restriction real-time loop-mediated isothermal amplification technique. Frontiers in Microbiology. 2015;**6**:1400

[74] Wu Q, Zhang Y, Yang Q, Yuan N, Zhang W. Review of electrochemical DNA biosensors for detecting food borne pathogens. Sensors (Switzerland). 2019;**19**(22):4916

[75] Stratakos AC, Linton M, Millington S, Grant IR. A loop-mediated isothermal amplification method for rapid direct detection and differentiation of nonpathogenic and verocytotoxigenic *Escherichia coli* in beef and bovine faeces. Journal of Applied Microbiology. 2017;**122**(3):817-828

[76] F. Wang, L. Jiang, and B. Ge, “Loop-mediated isothermal amplification assays for detecting Shiga toxin-producing *Escherichia coli* in ground beef and human stools,” Journal of Clinical Microbiology, vol. 50, no. 1, pp. 91-97, Jan. 2012.

[77] Arora K, Prabhakar N, Chand S, Malhotra BD. *Escherichia coli* genosensor based on polyaniline. Analytical Chemistry. 2007;**79**(16):6152-6158

[78] Li Y et al. A novel electrochemical DNA biosensor based on HRP-mimicking hemin/G-quadruplex

wrapped GOx nanocomposites as tag for detection of *Escherichia coli* O157: H7. Biosensors & Bioelectronics. 2015;**63**:1-6

[79] Tiwari I, Singh M, Pandey CM, Sumana G. Electrochemical genosensor based on graphene oxide modified iron oxide-chitosan hybrid nanocomposite for pathogen detection. Sensors and Actuators B: Chemical. 2015;**206**:276-283

[80] Makarova KS et al. Evolutionary classification of CRISPR–Cas systems: A burst of class 2 and derived variants. Nature Reviews Microbiology. 2020;**18**(2). Nature Research:67-83

[81] Li SY et al. CRISPR-Cas12a-assisted nucleic acid detection. Cell Discovery. 2018;**4**(1):20

[82] Hsu PD, Lander ES, Zhang F. Development and applications of CRISPR-Cas9 for genome engineering. Cell. 2014;**157**(6):1262-1278

[83] Huang M, Zhou X, Wang H, Xing D. Clustered regularly interspaced short palindromic repeats/Cas9 triggered isothermal amplification for site-specific nucleic acid detection. Analytical Chemistry. 2018;**90**(3):2193-2200

[84] Sun X et al. CRISPR-Cas9 triggered two-step isothermal amplification method for *E. coli* O157:H7 detection based on a metal-organic framework platform. Analytical Chemistry. 2020;**92**(4):3032-3041

Section 2

Escherichia coli and Pathogenicity

Chapter 5

The Biology and the Evolutionary Dynamics of Diarrheagenic *Escherichia coli* Pathotypes

Opeyemi U. Lawal, Valeria R. Parreira and Lawrence Goodridge

Abstract

Escherichia coli is a commensal of the gastrointestinal tract of humans and animals, and a leading cause of gastroenteritis, bloodstream, and urinary tract infection, among others. Pathogenic *E. coli* causing diarrhea is delineated into six different types (pathotypes) based on the type of infection they cause. While these pathotypes have similar mechanisms to colonize the intestinal epithelial layers and cause diseases, they differ in their capacity to acquire virulence, resistance determinants, and other accessory genes essential for niche adaptation. The advent of whole-genome sequencing technologies has greatly enhanced our understanding of the physiology, emergence, and global spread of multidrug-resistant and pathogenic clones of *E. coli*. In this chapter, we provided a snapshot of the resistome and virulome, as well as their contributions to the ecological adaptation, evolution, and dissemination of *E. coli* pathotypes.

Keywords: *E. coli*, pathotypes, antimicrobial resistance, genetic lineages, pathogenicity, genomics

1. Introduction

Escherichia coli inhabits and adapts to different hosts, a quest that resulted in the acquisition and loss of genes, which further drive diversity in this bacterium and contribute to the evolution of harmless strains to pathogenic lifestyles [1]. While *E. coli* is an integral part of the microbiota of different hosts, it can also cause severe infections in humans and animals [2, 3]. A subgroup of *E. coli* that are pathogenic can cause a broad range of human diseases due to evolution that resulted in the development of patho-features enabling it to adapt and survive in different environments. These environments range from the gastrointestinal tract to extraintestinal sites such as the urinary tract, or meninges, [4] in addition to fecal contamination of food that could cause enteric infection resulting from food poisoning or contamination [5]. Based on the type of infection they cause, pathogenic *E. coli* are divided into intestinal or diarrheagenic *E. coli* (DEC) that cause diarrheal illness and extraintestinal *E. coli* (ExPEC) that are implicated in infections such as urinary tract infections [3]. Diarrheal illness constitutes a public health burden and is a leading cause of mortality worldwide, causing >300 million illnesses and about 200,000 deaths annually, particularly in children in developing countries, including sub-Saharan and Southeast Asian countries (**Figure 1**) [6, 7].

The treatment of *E. coli* associated illness is toppled by its growing resistance to antibiotics, culminated by either the acquisition of resistance determinants or mutations that encodes for low uptake and tolerance to a higher concentration of the antimicrobials. Hence, *E. coli* could serve as a major reservoir of resistance genes not only for other *E. coli* strains but also for Enterobacteriaceae [8]. In addition, virulence determinants and genes that are associated with stringent response in nutrient low environments could also be acquired, thereby contributing to the survival and persistence of this bacterium in its environment [9]. Transmission of these antibiotic-resistant or pathogenic *E. coli* strains between different hosts, particularly in animals and humans could be through several routes such as direct contact with fecal-contaminated samples or other secretions from animals, or via the consumption of contaminated food [5].

Assessing the antimicrobial resistance, virulence, and transmission dynamics of *E. coli* requires characterization of this bacterium. A widely accepted classic method for characterizing *E. coli* is the serotyping technique that is based on the Kauffman classification scheme, where the O (somatic) polysaccharides and H (flagellar) surface antigens are determined [10, 11]. Other methods of typing and assessing the genetic relatedness and detecting outbreaks of *E. coli* strains are pulsed-field gel electrophoresis (PFGE) [12], multilocus enzyme electrophoresis (MLEE) [13], multilocus variable-number tandem repeat analysis (MVLA) [14], or multilocus sequence typing (MLST) [15]. These methods have proven to be effective in the epidemiological investigation of pathogenic *E. coli* [16] and the assessment of the emergence and dissemination of multidrug-resistant clones. However, none of these methods can accurately define the evolutionary relationships between *E. coli* strains, hence the need for a tool with a higher resolution. The advent of whole-genome sequencing (WGS) technologies has greatly enhanced not only the epidemiological investigation of outbreaks and the global spread of multidrug resistant and pathogenic clones of *E. coli* [17], but also our understanding of the physiology and evolutionary history of how some pathogenic strains evolve from commensal *E.coli* strains.

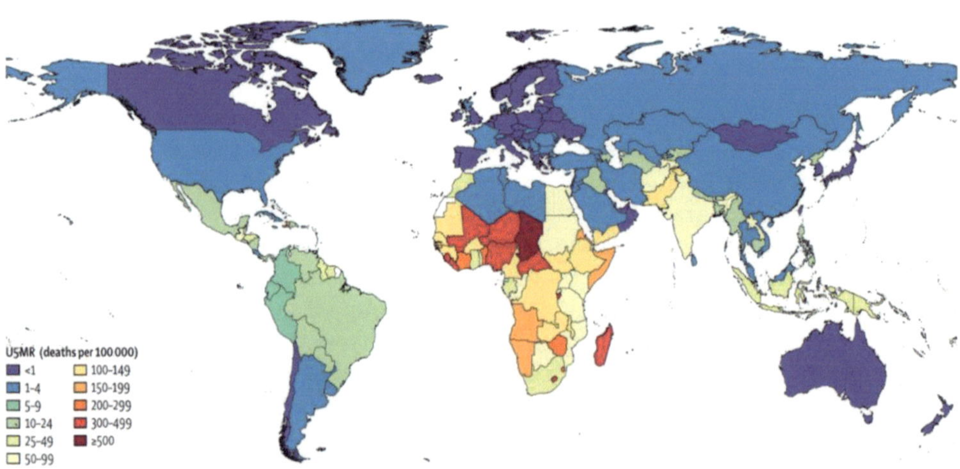

Figure 1.
Global mortality rate from diarrhea in children under 5 years in 2016. Data represent the analysis of diarrhea burden in 195 countries in 1990–2016, showing the regions most affected by the illness. Reprinted from Troeger et al. [6] which was published under Creative Commons License.

Based on clinical manifestation, presence of specific virulence determinants and phylogenetic profiles, diarrheagenic *E. coli* are categorized into six main pathotypes namely, enterotoxigenic *E. coli* (ETEC), enterohaemorrhagic *E. coli* (EHEC) or Shiga toxin-producing *E. coli* (STEC), enteropathogenic *E. coli* (EPEC), enteroaggregative *E. coli* (EAEC), entero-invasive *E. coli* (EIEC), and diffusely adherent *E. coli* (DAEC) [3, 7]. This chapter provides a snapshot of the biology of *E. coli* by focusing on the resistome, virulome, as well as the population structure of diarrheagenic *E. coli* pathotypes.

2. Enterotoxigenic *E. coli* (ETEC)

2.1 An overview of ETEC

Enterotoxigenic *E.* coli (ETEC) is a major cause of travelers' diarrhea, with a high prevalence in developing countries and responsible for about 200 million cases of diarrheal illnesses and 100,000 deaths each year [18, 19]. The incidence of ETEC could be recovered from symptomatic and asymptomatic carriers and is most common in younger children with a high mortality rate in this group. In regions such as Africa, South America, and Southeast Asia the incidence of ETEC-related infection is estimated to be at least one to two episodes per year [3] further reinforcing the significance of this pathotype in that region. While ETEC is not limited to humans, it is also a common cause of edema and post-weaning diarrheal diseases in food production animals such as cattle, pigs, and sheep [20]. ETEC are transmitted through the fecal-oral route by contaminated food as well as surface and groundwater in developing countries with limited access to clean water (**Figure 2**) [3, 6].

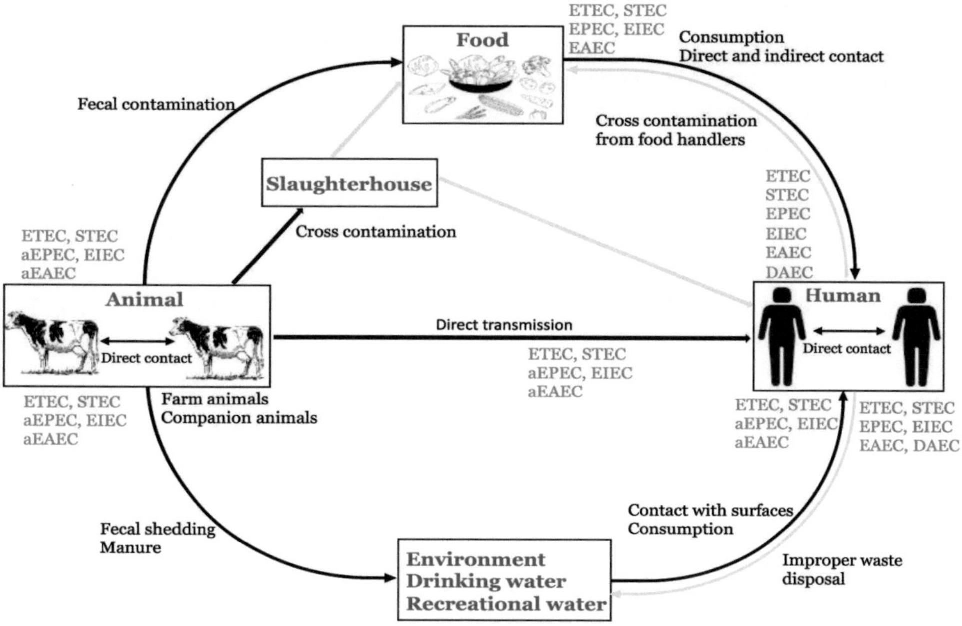

Figure 2.
Dissemination and transmission routes of pathogenic Escherichia coli *pathotypes. Solid black arrows represent a direct transmission while gray arrows depict an indirect transmission of pathogenic* E. coli *pathotypes. ETEC: Enterotoxigenic* E. coli; *STEC: Shiga toxin-producing* E. coli; *EPEC: Enteropathogenic* E. coli; *EAEC: Enteroaggregative* E. coli; *EIEC: Entero-invasive* E. coli; *DAEC: Diffusely adherent* E. coli.

There is a great genetic diversity in ETEC as more than 100 different O antigens have been reported to be associated with clinical isolates. Among these, the O6 serogroup is most common, and geographically diverse among all ETEC serogroups, and has been implicated in multiple outbreaks in different countries [21]. Additionally, at least 34 H antigens are also associated with this pathotype. Among the serotypes as determined by the combination of O and H antigens, O6:H16 (heat-stable (ST) or heat-labile (LT) toxin), O148:H28 (ST), O167:H5, O153:H45 (ST), O169: H41 (ST only) are frequently isolated from humans, animals, environmental matrices, and from outbreaks in developing countries [22–24]. ETEC produces one or more colonization factors that facilitate its attachment to specific receptors on the mucosal layer of the small intestine of humans and animals, and secretes enterotoxins that cause electrolyte imbalance in the intestinal lumen resulting in dehydration, metabolic acidosis, and diarrheal [3, 18]. The ST toxin is a nonimmunogenic small protein molecule, but the LT toxin is structurally homologous and exhibits a similar mechanism of action to cholera toxin produced by *Vibrio cholerae* [23, 25].

2.2 Virulome of ETEC

ETEC employs an array of genetic factors that are either chromosomal or plasmid-borne that mediate colonization and adherence to the intestinal epithelium, proliferation within the host, and evasion of host defense mechanisms (**Table 1**) [26].

2.2.1 Colonization and adhesion

Colonization and adhesion are the primary and essential steps in the pathogenesis of pathogens. ETEC is not an exception as the colonization of the host intestinal epithelium by this pathotype is mediated by plasmid-borne genes that encode adhesins and one or more colonization factors (CFs) namely, pilus or pilus-related adhesins [23]. Pili are hair-like appendages on the cell surface of bacteria where they mediate the attachment of bacteria to surfaces.

They are composed of protein subunits (pilins) that are structurally polymeric and are almost exclusively plasmid-borne (**Figure 3**) [27]. ETEC CFs are designated as CS (coli surface antigens) followed by a number, except for CFA/I and PCFO71 [27]. Presently, at least 30 CFs have been reported in ETEC of human origin. It is estimated that about 50% of strains in this pathotype carry one or more CFs that are not detectable, suggesting that there could be more CFs that are yet to be discovered and characterized [18, 23]. The co-expression of one or more CFs with toxin-encoding genes has been described. For example, CFA/I + LT, CS7 with LT, CS5 + CS6 with LT + ST, CS2 + CS3 with LT + STh, among others [28, 29]. In the prototypical ETEC strain H10407, the production of CFA/I is mediated by *cfaABCE* operon that is tightly regulated by CfaD, a transcriptional regulator that triggers its expression.

Other plasmid-encoded genetic factors that have been reported to play a significant role in the pathogenesis of ETEC include a class I SPATE (serine protease autotransporters of the Enterobacteriaceae) EatA that digests EtpA secreted by ETEC, thereby promoting the adhesion of flagella to the host receptor [30, 31]. ETEC can invade the host cell with two chromosomally encoded genes *tia* and *tibA*. The former (*tia* gene) is borne on a 46-kb pathogenicity island (PAI). The expression of these genes was reported to be associated with adhesion and invasion of ETEC in host cells [27]. Likewise, a *leoA* gene encoding GTPase is reported to be associated with virulence in ETEC (**Table 1**) [32].

E. coli pathotypes	Main reservoir	Clinical presentations	Virulence factors
ETEC	Humans and animals	Watery noninflammatory diarrhea, adherence to small intestinal epithelium, nutrient malabsorption	Colonization factors (CFs), EatA, Tia, TibA, LeoA, ST, LT, EAST1
STEC	Animals	Watery bloody diarrhea, hemorrhagic colitis, Hemorrhagic Uremic syndrome, attaching-effacing lesions in the large intestine	LEE, Saa, Paa, EhaA, LpfA, OI-7, ST, EhxA, ToxB, EspP, KatP
EPEC	Humans and animals	Shigella-like toxin, watery and/or bloody diarrhea, noninflammatory diarrhea, attaching-effacing lesions in the colon, nutrient malabsorption	LEE, pEAF, Bfp, OI-22, EAST1
EAEC	Humans and animals	Clump intestinal cells; mucoid watery diarrhea with persistent inflammation, nutrient malabsorption, postinfectious irritable bowel syndrome	AA, Afp, CapU, Air, Shf, AatA, Pic, EAST1, ShET1, Pet, SigA, SepA, pAA, HlyE
EIEC	Humans and animals	Bacillary dysentery, watery diarrhea with or without blood and leukocytes, inflammation of the large intestinal epithelium	pINV, SepA, SigA, Sat, ShET2
DAEC	Humans	Watery diarrhea, persistent diarrhea, chronic inflammatory colon disease	Afa/Dr. Adhesins, Sat, Pet, SenB, HlyE

Legend: ETEC: Enterotoxigenic E. coli; *STEC: Shiga toxin-producing* E. coli; *EPEC: Enteropathogenic* E. coli; *EAEC: Enteroaggregative* E. coli; *EIEC: Entero-invasive* E. coli; *DAEC: Diffusely adherent* E. coli.

Table 1.
Summary of the clinical characteristics and virulence factors of Escherichia coli *pathotypes.*

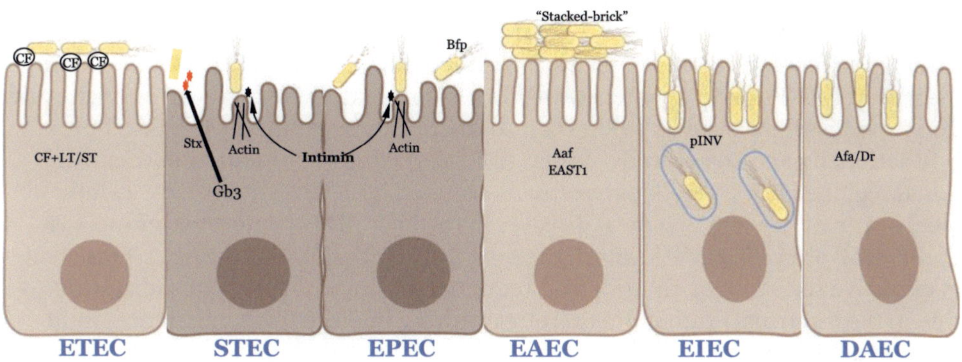

Figure 3.
Colonization and adherence patterns of diarrheagenic Escherichia coli *to the host epithelium. Enterotoxigenic* E. coli *(ETEC) uses colonization factors (CFs) to attach to host intestinal mucosa. Shiga toxin-producing* E. coli *(STEC) and Enteropathogenic* E. coli *(EPEC) attach to the intestinal epithelial cells and efface microvilli, forming characteristic A/E lesions. EPEC also forms microcolonies using bundle-forming pili (Bfp) resulting in a localized adherence pattern. Enteroaggregative* E. coli *(EAEC) forms a biofilm matrix on the intestinal mucosa that promotes the formation of a "stacked brick" adherence pattern. Enteroinvasive* E. coli *(EIEC)/*Shigella *are intracellular pathogens that penetrate the intestinal epithelium through M cells to gain access to the submucosa. Diffusely adherent* E. coli *(DAEC) is scattered over the surfaces of intestinal cells, resulting in a diffuse adherence pattern.*

2.2.2 Enterotoxin secretion

One of the salient features that define ETEC is its ability to produce two types of enterotoxins, ST or LT [23]. STs are non-antigenic small enterotoxins that are frequent in human diseases, found in about 80% of ETEC either singly or in combination with LT [18, 33]. STs are classified into two different classes (STa and STb) based on their structure and function. STa is soluble in methanol and protease-resistant. It is frequent in human diseases and encoded by *estA* genes, whereas STb is insoluble in methanol and sensitive to protease, and causes disease only in animals and is encoded by *estB* gene [34]. Based on host specificity, STa is further designated into two genetic variants namely STp and STh. The former (STp) is 18 amino acids in length and produced by ETEC strains of porcine, bovine, and human origin, while STh is 19 amino acids long and exclusive in ETEC strains of human origin [29]. Recently, six genetic variants of STa encoding gene (*estA*) have been reported, where *estA1*, *estA5*, and *estA6* are common in ETEC strains of porcine origin and *estA2*, *estA3*/4 and *estA7* are frequent in isolates of human origin (STh), while *estA5* gene is described to be frequent in ETEC strains causing disease both in animals and humans, especially traveler's diarrhea in adults [29, 34]. Secretion of STh and STp in the intestinal epithelium of the host requires the efflux protein TolC [35].

Unlike, STs, LTs are hexameric and strongly immunogenic that are encoded by the *eltAB* operon [29]. LTs have two subtypes: LT-I and LT-II, both of which have been reported in ETEC strains causing diarrhea in humans and in different species of post-weaned animals. LT-Is are plasmid-borne and highly similar to cholera toxin produced by *V. cholera* [3, 36]. Conversely, LT-II is chromosomal and has been hypothesized to be prophage encoded [3, 36]. LT-II is classified into LT-IIa, LT-II, and LT-IIc, with LT-IIc being the more frequent in LT-II ETEC strains [36].

LTs promote the adherence of ETEC to host intestinal epithelial cells and evade the host defense mechanisms by inhibiting the expression of antimicrobial peptides produced by the hosts, in addition to the activation of host signaling pathways [3]. Another virulence factor encoding enterotoxin in ETEC strains is enteroaggregative heat-stable toxin (EAST1). EAST1 toxin is heat-stable and 38 amino acids long encoded by *astA* gene that is commonly plasmid-borne [37]. ETEC strains producing EAST1 toxin have been recovered from humans and animals. This toxin was reported to have originated from EAEC but it is prevalent in ETEC [38, 39]. While the role of EAST1 toxin in enteric infection is not clear, there has been evidence and direct associational studies linking this toxin to diarrheal illness [38]. EAST1 toxin is functionally and structurally similar to STa, sharing 50% identity in their functional regions [38]. Overall, enterotoxins secreted by ETEC strains have a similar mechanism of causing diarrheal diseases in the host. ETEC enterotoxins increase cyclic AMP or cyclic GMP levels in the intestinal epithelium of the host. This results in excessive secretion of chloride and reduction in the adsorption of sodium chloride in the intestinal epithelium thereby resulting in electrolyte imbalance, fluid loss and dehydration [29, 40].

2.3 Antibiotic resistance in ETEC

Since the first isolation of ETEC in Kolkata about five decades ago [41], the emergence and increase in multidrug-resistant strains have been reported. A homogenous and high antibiotic susceptibility pattern was observed for ETEC strains at a time but the treatment of travelers' diarrhea with different classes of antimicrobials such as macrolides (erythromycin and azithromycin), fluoroquinolones (norfloxacin, ofloxacin, ciprofloxacin), tetracycline (doxycycline), rifamycin and sulfamethoxazole-trimethoprim that are used to treat other types of infections [22] may have also contributed to the emergence of antimicrobial resistance in this pathotype [22]. Another contributor could be the indiscriminate use of antibiotics for the treatment of diarrheas caused by viral agents that are sometimes misdiagnosed because they present similar symptoms [3].

There are several studies from different countries assessing the antibiotic resistance profile and distribution of resistance determinants in ETEC. In a study, the antimicrobial resistance profile among patients with recent travel history to ETEC endemic regions between 2001 and 2004 reported that up to 60% of the ETEC isolates were resistant to sulfamethoxazole-trimethoprim, tetracycline, and/or ampicillin [42]. Ciprofloxacin resistance was reported to markedly increase from 1% to 8% within 10 years (1994–2004) in patients [42] which clearly suggests a rapid emergence of resistance with time in this pathotype. In a recent study on the WGS analyses of eight strains representing the major ETEC lineages that are causing diarrheal diseases in humans around the globe, all the strains showed resistance and carried resistance determinants to at least two of the 14 antibiotics tested, with resistance to penicillin, norfloxacin and chloramphenicol being the most common. In this study, two plasmids designated (pAvM_E1441_17 and pAvM_E2980_15) carried resistance determinants to mercury (*mer* operon) and multiple antibiotics including streptomycin (*aadA1-like, strA,* and *strB*) and ampicillin (*bla*TEM-1b, *ampC*) [43].

ETEC in animals, however, may be slightly different. In a study of 112 ETEC isolates recovered from pigs in Canada over a two-decade period (1978–2000), *tetB* gene that encodes resistance to tetracycline was the most common and found in 80% of the collection [44]. Another interesting observation from this study was the increase in the determinants encoding resistance to gentamicin *(aac(3)-IV),* kanamycin *(aph(3′)-Ia)* and trimethoprim (*dhfrV*), while others appear to be either consistent or decrease over time [44].

2.4 Population structure of ETEC

ETEC strains are epidemiologically and phenotypically diverse and exhibit high genetic diversity. In addition to being polyphyletic, the distribution of ETEC lineages is not restricted by geography [45]. Several reports on the phylogenetic analyses of strains from the human origin using MLEE and MLST, and well as CF-toxin-based phylogeny showed that this pathotype might have evolved multiple times through clonal expansion and probably due to lack of common clonal lineage [46, 47]. In spite of the genomic diversity among strains in this pathotype, Turner and colleagues [48] reported ETEC to be associated with sequence type 10 (ST10). In a broader evolutionary study of a large collection of 1019 ETEC isolates from humans in 13 countries using MLST, 42 clonal groups were observed with evidence for horizontal gene exchange of plasmid-encoded CF genes between the lineages [46]. Since the advent of next-generation sequencing technologies, the study of the population structure of ETEC has improved the understanding of the genetic diversity and evolution of the pathotype [24, 49, 50].

A global collection of ETEC isolates from humans collected over a period of three decades (1980–2011) in 20 countries and representing four continents was assessed for genetic relatedness using WGS-based single nucleotide polymorphism (SNP) [49]. Indeed, ETEC strains are genetically diverse as they were reported to be distributed across different *E. coli* phylogenetic groups (A, B1, B2, D, and E) (**Figure 4**), an observation that is also in accordance with the structure defined by MLST [46, 48, 51]. An interesting finding from the study that could be attributed to the higher resolution of WGS was the identification of ETEC-specific clusters (L1-L14) that clustered geographically diverse strains that were phylogenetic related and associated with specific plasmid-encoded virulence determinants. The L1 and L2 clustered the commonly found ETEC strains expressing O6 antigen and carried similar profiles for CF and LT and ST enterotoxins, suggesting that these plasmid-encoded virulence determinants could be important to understand the evolutionary histories of these clusters [24, 49].

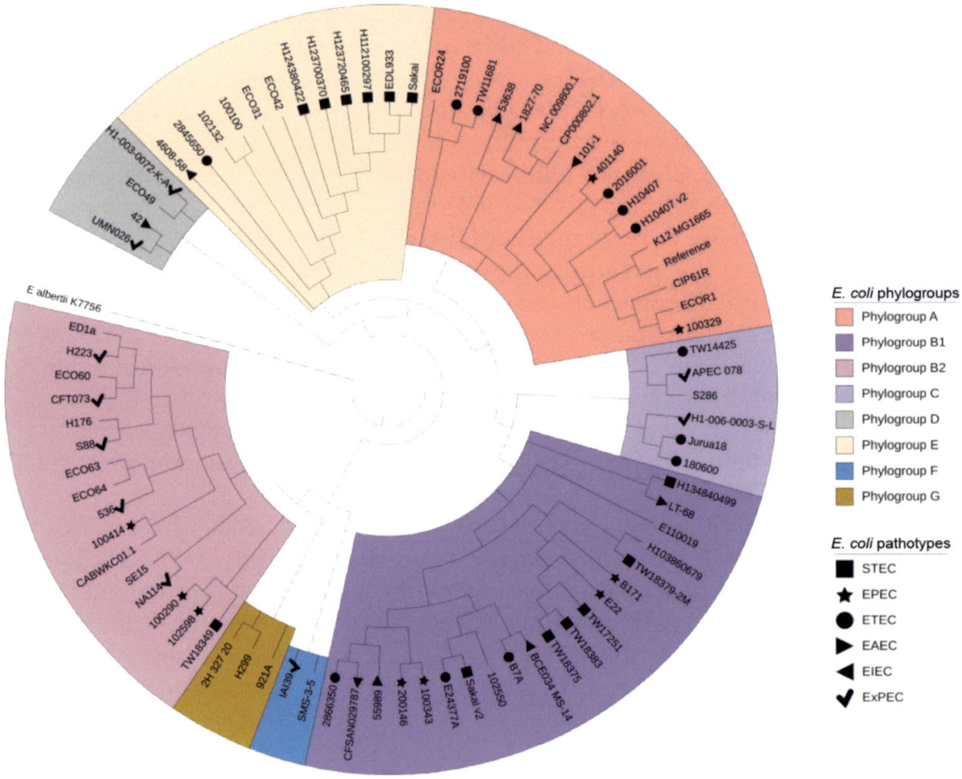

Figure 4.
Single nucleotide polymorphism based Phylogenetic tree of Escherichia coli. *Draft genomes of* E. coli *pathotypes were downloaded from NCBI and core genomes were defined and aligned against* E. coli *K12. SNPs from the core genome alignment were called using Snippy (https://github.com/tseemann/snippy). Concatenated SNPs alignment cured of recombination (https://github.com/sanger-pathogens/gubbins) was used to construct a phylogenetic tree and visualized using iTOL (http://itol.embl.de/). Different colors and shapes depict* E. coli *phylogroups and pathotypes, respectively.*

Similarly, in a local study on the phylogenomic diversity of 94 ETEC isolates from Bangladesh [24], a polyphyletic scenario and a direct correlation between lineages and virulence profiles and CFs were noted. Using comparative genomic tools, the authors identified six novel CF variants. However, the experimental validation of these CFs would be important to decipher their association with other virulence determinants as well as their interaction with the host cells.

3. Shiga toxin-producing *E. coli* (STEC)

3.1 An overview of STEC

Shiga toxin-producing *E. coli* (STEC), also known as verocytotoxin-producing *E. coli* (VTEC) was first identified as a pathogen in 1982 and is defined by the presence of Shiga toxin genes (*stx1* and *stx2*) that are known to be encoded on lambdoid prophages [52]. STEC had since emerged as a major enteric foodborne and zoonotic pathogen causing gastroenteritis and enterocolitis. STEC-related infection could progress to hemolytic-uremic syndrome (HUS) that in some cases could be fatal or result in renal failure in children (**Table 1**) [3]. Enterohemorrhagic *E. coli* (EHEC) is associated with hemorrhagic colitis which makes it a subset of STEC pathotype [3, 53]. STEC causes more than one million cases annually with about 30% of such cases found in the United States [54]. According to Blanco et al. [55], 435 STEC serotypes strains have been identified but only a few are frequently associated with human infection among which are O26, O45, O103, O111, O121, O145, and O157 [56, 57]. STEC O157:H7 has been widely studied most likely because it is frequently implicated in STEC foodborne outbreaks. Although a series of outbreaks that are mostly caused by STEC O157:H7 have been reported globally in developed and developing countries, some are believed to go undetected or underreported [54, 58].

The main reservoir for STEC strains causing infection in humans are known to be ruminant food production animals including cattle, sheep, and goats. These animals are asymptomatic carriers and shedders as the vascular receptor that facilitate the transportation of the Shiga toxins to organs are absent. This *E. coli* pathotype is also common in the gastrointestinal tracts of poultry, pigs as well as some companion animals such as dogs and cats [59]. Other STEC asymptomatic carriers have been reported in other ruminant and monogastric animals, as well as in insects, suggesting their roles in the food contamination, dissemination, and transmission of different strains of this pathotype to humans (**Figure 2**) [59, 60]. Since STEC is part of the intestinal flora of food-production animals and is readily shed through feces, direct contact with contaminated environmental matrices and/or the consumption of contaminated products including undercooked meat, unpasteurized dairy products, vegetables, and/or water are a potential route of transmission of different strains of this pathotype to humans [61, 62]. STEC transmission due to direct contact with infected person or animals or their environments or products has been documented [63].

3.2 Virulome of STEC

STEC carries genes encoding adhesins and enterotoxins that are either chromosomal, on PAIs, or plasmid-borne. These determinants mediate colonization, attachments, and invasion of host cells (**Table 1**) [64].

3.2.1 Colonization and adherence

Ingestion of contaminated food or direct contact with contaminated environmental matrices or infected persons or animals precedes STEC-related diseases. Colonization and attachment/adherence of STEC to intestinal epithelium is mediated by several genetic factors some of which are carried by the locus of enterocyte effacement (LEE) PAI [65]. LEE locus which is also often present in EPEC strains [66] encodes a type III secretion system (T3SS) that plays a role in the secretion and translocation of virulence-associated genetic factors into host cells [67]. One of these genetic factors is *eae* gene that encodes intimin, an adhesin that is essential for the attachment of STEC to the host intestinal mucous membrane and facilitates the production of attaching-and-effacing (A/E) lesion (**Figure 3**) [65]. The injection of its translocated intimin receptor (Tir) into the host cells and the interaction between this protein (Tir), proteins that form the needle component of the T3SS (EspADB), and intimin facilitates the induction of lesions [65, 68]. Intimin has at least 30 reported subtypes [69] and some of these variants are associated with specific serotypes. Oftentimes, intimin encoding gene (*eae*) subtype $\gamma 1$ is associated with STEC O157:H7 and O145:H28 [69, 70], while subtypes $\beta 1$, ε, and θ are frequently found in O26:H11, O103:H2 and O111:H8 STEC strains, respectively [69, 70]. Intimin is reported to be common in clinical strains of STEC with a prevalence that could range from 70 to 90% [71–73].

LEE is unarguably important for the pathogenesis of STEC strains, but STEC LEE-negative strains/serotypes (e.g. O103:H21) have been implicated in infections [74].

This implies that there are several other virulence-associated factors that are carried on PAIs or mobile genetic elements mediating colonization and adherence to host cells in these strains [74]. For example, a gene (*saa*) encoding autoagglutinating adhesin was isolated from a large plasmid of a LEE-negative STEC O113:H21 implicated in an outbreak. The expression of this gene is described to enhance adherence of STEC strains to HEp-2 cells [75]. Other protein-encoding genes reported to promote the colonization and adherence of STEC strains to mucosal membrane include *paa* that encodes attachment of bacterial cells to enterocytes in pigs [76], *ehaA* a STEC autotransporter [77], *lpfA,* long polar fimbriae [78] that are both involved in the attachment of STEC to surfaces. Additionally, "O" islands (OI) that encode macrophage toxin and ClpB-like chaperone (OI-7), urease clusters (OI-43 and OI-48), two toxins and PagC-like virulence factor (OI-122) among others have been reported to be linked with virulence in STEC [57, 79].

3.2.2 Cytotoxic toxin production

STEC-related infections in humans are reported to be associated with the presence and expression of several virulence determinants, with the phage-encoded Shiga toxin genes Stx1 (Stx1a) and Stx2 (Stx2a) being the main virulence factors [64, 80]. Stx1 consists of 293 amino acids while the Stx2a is longer by only four amino acids. At least, 16 subtypes of these two toxins have been described based on amino acid differences and the level of cytotoxicity. Stx1 contains four variants encoded by *stx1a*, *stx1c*, *stx1d* and *stx1e*, whereas Stx2 comprised 12 subtypes encoded by *stx2a*, *stx2b*, *stx2c*, *stx2d*, *stx2e*, *stx2f*, *stx2g*, *stx2h*, *stx2i*, *stx2j*, *stx2k* and *stx2l* [80]. While Stx1a has been implicated in human infection, Stx2a, Stx2c, and Stx2d are the major subtypes that are frequently associated with hemorrhagic colitis and HUS [64]. However, Stx2a and Stx2d subtypes are described to exhibit higher cytotoxicity relative to Stx2b and Stx2c in a mouse model [64, 81]. The interaction of Stxs with the host cell receptor is very complex and is based on characteristics of the environment of the receptor in the plasma membrane [64]. Stxs bind to the globotriaosylceramide Gb3, an insoluble molecule that has multiple binding sites and comprised a lipid component. The interaction between these two molecules (Stx and Gb3) is described to be important in the uptake of the toxin (**Figure 3**). Stxs are ribotoxins that disrupt protein synthesis within the host cell and provoke apoptosis [64, 81].

STEC strains can carry *stx1* or *stx2* genes or both [2]. In a study of 351 STEC strains from bovine feces, the great majority of the strains (82%) carried *stx2* while 18% carried *stx*1. Both genes were found only in ~3% of the collection [2]. In another study involving 220 STEC strains from humans and animals, *stx1* and *stx2* were found in 15% and 53%, respectively, while both genes were found in 32% of the isolates [82]. Stxs subtypes are heterogeneously distributed in the population, but specific variants have been reported to be host-specific. For example, Stx2e is less cytotoxic and sporadic in human diseases, and is commonly associated with edema diseases in weaned pigs [83]. Likewise, Stx1c is reported to be associated with STEC of ovine origin [84]. Indeed, the severity of STEC infection has been noted to be directly proportional to a number of Stx types or subtypes carried by the infecting strains [85]. While the production of only Stxs has been described to cause HUS, the infection is however exacerbated when associated with other virulence determinants including the LEE [21]. In addition to Stx, other toxins or hemolysins have been reported to be associated with STEC virulence. These include the hemolysin, encoded by the *ehxA* or *hlyA* gene, that are usually found on megaplasmid pO113 and/or pO157 and linked to cytotoxic effects on endothelial cells that may also promote the development of HUS [86, 87]. Other virulence determinants carried on these plasmids include *toxB* that is essential for adherence of STEC to host cells [88], *espP* that encodes an extracellular protease and *katP* that is associated with catalase-peroxidase production important for oxidative stress response [89].

3.3 Antibiotic resistance in STEC

Treatment of STEC infections with antibiotics is not encouraged as this might exacerbate the disease by activating the lytic cycle of the phage carrying Shiga toxin that could aggravate tissue damage in infected individuals. Antibiotics such as rifaximin, fosfomycin, azithromycin, and meropenem that do not encourage the release of Shiga toxin have been used for the treatment of early onset of STEC infection to prevent the progression of the diseases to HUS [90].

Several studies on the prevalence of antibiotic resistance in STEC from different countries and host or environments have reported that resistance to beta-lactams, sulfonamide, tetracycline, and trimethoprim are common STEC, while multidrug-resistance is more frequent in non-O157 than O157:H7 serotypes [91]. For STEC O157, resistance to ampicillin and cephalothin is common in strains of human origin, whereas tetracycline and sulphamethoxazole resistances are frequent in strains of animal origin [92]. In a study involving 54 STEC strains recovered from cattle and pigs, genetic determinants that encode resistance to trimethoprim (*dfrA1*), tetracycline (*tetA* and *tetB*), beta-lactam (bla_{TEM-1}), and aminoglycoside (*aac(6)-Ib*) were found in the great majority (≥81%) of the isolates, while chloramphenicol resistance gene (*cat1*) was also carried in more than 50% of the collection [93]. Likewise, in a 15-year surveillance study of STEC in Sweden [94], 70 antibiotic resistance determinants that were associated with 10 different classes of antibiotics were found in 184 STEC isolates, where 50% of these genes were present in all isolates. Six resistance determinants to fluoroquinolone (*crp*, *hns*, *acrB*, *marA*, *mdtM*, and *emrA*) were found to be frequent. Equally, *emrE* that encodes resistance to multiple antibiotics was associated with STEC O157:H7, whereas *fosA7, sat-1,* and $bla_{TEM-150}$ and *dfrA5* were associated with non-STEC O157 serotype.

3.4 Population structure of STEC

Genetic relatedness of STEC isolates from different hosts and countries have been studied using different molecular tools that ranged from serotyping, PFGE, conventional MLST, and WGS. Evidence for transmission and dissemination of different STEC serogroups and clones have also been documented. Unlike ETEC that evolved multiple times through clonal expansion, STEC appears to have evolved by parallel evolution. Indeed, phylogenetic analyses of STEC strains have shown that isolates form multiple distinct clonal lineages, where strains with the same serotype and virulence content were nested together in the cluster [95]. STEC strains are spread across *E. coli* phylogroups and the great majority belonged to phylogroup B1 (**Figure 4**) [96]. STEC O157:H7 are further delineated into three lineages, I, I/II, and II [97, 98] that are disseminated globally. Lineage I is predominant among clinical isolates of human origin while lineage II is more prevalent in animals [99]. Intra-lineage diversity is apparent as lineages varied in the adherence and virulence determinant expression, Stx-encoding bacteriophage (Stxϕ) insertion sites, *stx2* expression, and stress resistance [97]. This intra-clonal diversity is hypothesized to have been a consequence of the global spread of a single clone and geographic expansion [97]. Interestingly, a time-dependent clonal replacement and geographical-dependent clonal expansion of lineages and sub-lineages of STEC O157:H7 have been reported [97, 100]. The STEC O157:H7 lineage I/II that was predominant in human infection in the 1980s in the UK declined and was replaced by sub-lineage Ic in the 1990s. Also, in the past few years, this region has reported the replacement of the dominant sub-lineage (Ic) by sub-lineage IIb, a phenomenon they reported to have been a consequence of the acquisition of prophage encoding *stx2a* [97, 100].

In a recent study using WGS to understand the population dynamics of 757 STEC O157:H7 isolates from humans and animals from four continents, seven clades were reported and designated as A-G [101]. The most recent common ancestor of the isolates in this study was reported to have originated in the Netherlands in the late 19th century (1890) and then spread to other parts of the world. Although isolates were clustered on a geographical basis, there was an admixture of strains from different hosts suggesting transmission events between them [101]. The pangenome analyses of these isolates also showed that STEC O157:H7 from humans and animals differed in phage-related protein content. The molecular epidemiology of non-STEC O157:H7 is equally important especially considering their roles in outbreaks. From 1995 to 2017, a total of 674 outbreaks by non-O157 STEC strains were reported worldwide, where O26:H11 was predominant during this period [102]. Other serogroups implicated in these outbreaks include O26:H11, O45, O103:H25, O104:H4, O111:H8, O121, and O145:NM [102]. MLST-based phylogenetic analysis of 894 non-STEC isolates from patients over a period of 18 years (2001–2018) in Michigan revealed that the great majority of the isolates (95%) belonged to one clade [103]. Although the information on the evolutionary dynamics of STEC is inexhaustible, studies focusing on identifying new genetic factors associated with ecological adaptation of different lineages are elusive. Further studies should focus on this area.

4. Enteropathogenic *E. coli* (EPEC)

4.1 An overview of EPEC

Enteropathogenic *E. coli* (EPEC) is a pathotype that causes infrequent diarrheal diseases in adults and has also been implicated in gastroenteritis outbreaks in children in health care settings [104, 105]. EPEC was the first pathotype described in 1955 to refer to *E. coli* causing infantile diarrheal and implicated in a few outbreaks between the 1940s and 1950s [106]. Infection from this pathotype is frequent in children under two years living in low- and middle-income countries, and is the second leading cause of death among this age group, amounting to about 1.5 million deaths annually [105]. Like other DEC, the onset of EPEC-related diarrheal is characterized by acute watery stool which if it persists could result in loss of electrolytes and malabsorption of nutrients in children [3, 107]. Infection caused by EPEC strains is not limited to humans as they have also been implicated as a causative agent of diarrheal illness in young calves (**Figure 2**) [108].

EPEC strains are previously classified solely based on the combination of the three immunogenic structures O, H, and K antigens but the diversity observed for these antigens rendered serotyping unreliable rapid diagnostic tool for this pathotype [17, 107]. However, as recommended in 1987 by World Health Organization, 12 serogroups; O26, O55, O86, O111, O114, O119, O125, O126, O127, O128, O142, and O158 belonged to EPEC pathotype. In addition to six others; O39, O88, O103, O145, O157, and O158 have been classified and belonged to this pathotype although some of these serogroups consist of *E. coli* strains from different serotypes [108, 109]. EPEC strains are classified as either motile (H^+) or non-motile (H^-). Among EPEC strains with flagellar associated antigens, H2 and H6 are the most frequent, whereas others that are less common include H7, H8, H9, H12, H21, H27, H25, and H34 [3, 107].

EPEC pathotype is defined based on the carriage of LEE locus that mediates the induction of A/E localized lesions [110], a feature that is shared with some STEC strains. However, the inability to produce Shiga toxins or other enterotoxins differentiate EPEC pathotype from EHEC/STEC strains [3, 10]. Additionally, based on the presence or absence of *E. coli* adherence factor plasmid (pEAF), EPEC pathotype is sub-grouped into two subtypes; typical (tEPEC) and atypical (aEPEC) [111]. Relative to tEPEC that is regarded to be more virulent, aEPEC group is reported to be highly diverse and more prevalent in diarrheal illness in children [111, 112]. Several O and/or H antigens of aEPEC strains are nontypeable. Of the typeable serogroups belonging to this subtype, O51 is the most frequent followed by five others (O145, O26, O55, O111, and O119) [3]. aEPEC O55:H7 is closely related with STEC O157:H7 and from the evolutionary perspective, the latter is believed to have evolved and diverged 400 years ago from the ancestor of the former [113]. EPEC like other diarrheagenic *E. coli* pathotypes is transmitted through the fecal-oral route as well as contact with contaminated surfaces or secretions. While humans are believed to be the major reservoir for tEPEC strains, aEPEC strains are present both in healthy individuals and in animals (**Figure 2**) [111, 112, 114].

4.2 Virulome of EPEC

Since EPEC does not produce Shiga toxin or other enterotoxins, the major feature the pathogenic strains in this pathotype employ is their ability to attach tightly to the host mucosal membrane, destroy microvilli, and induce the formation of lesions (**Table 1**) [115]. In addition, EPEC carries other genes encoding proteins that have been linked to colonization and adherence to host cells [116, 117].

4.2.1 Colonization and adherence

The defining characteristic of EPEC is the carriage of LEE locus that is essential for inducing A/E lesions, causing localized lesions by attaching closely to the surface of the intestinal epithelial cells. Like some STEC strains, all the EPEC strains carry *eae* and *tir* genes as well as T3SS that is able to inject a large number of effector proteins into the host cell [3, 111]. Studies have shown that the presence of LEE locus in EPEC strains is enough to cause infection in the host even in aEPEC-related infection scenario [118].

The ~80 kb pEAF plasmid that defines tEPEC carries *per* and *bfp* operons (**Table 1**) [119, 120]. The *per* operon (*perABC*) is plasmid-borne and contains *perA* that encodes a regulator that activates the transcription of *bfp* operon that encodes the type IV pili called bundle-forming pilus (BFP) (**Figure 3**) [119, 120]. The *bfpA* gene which encodes the bundling of the major structure of BFP and 13 other genes are carried on the pEAF plasmid [116]. The carriage of this plasmid has been described to be essential in the localized adherence of EPEC to intestinal epithelium in the host [118]. tEPEC strains carry *lifA* gene that encodes lymphocyte inhibitory factor, a large surface protein that is described to promote the intestinal colonization of mice by *Citrobacter rodentium* [3]. Although pEAF plasmid is absent in aEPEC strains, they often carry virulence determinants typical of STEC strains most likely because they share a common ancestor [113]. Afset et al. [117] identified 12 genes that were statistically associated with aEPEC-related diarrhea in children. Of note are *efa1/lifA* genes that are located on OI-122, as well as *lpfA* gene previously reported in STEC [121]. Likewise, *astA* gene that encodes EAST1, an ST-like toxin that is present in ETEC is also carried by EPEC, being more prevalent in aEPEC than in tEPEC strains implicated in diarrhea [122, 123].

4.3 Antibiotics resistance in EPEC

Although EPEC-related infection could resolve itself or simply by oral rehydration therapy that replenishes the lost fluid, the persistence of this infection may necessitate the use of antibiotics. In this case, especially in adults, the recommended antimicrobial is trimethoprim/sulfamethoxazole, norfloxacin, or ciprofloxacin [124]. However, studies on antibiotic resistance of EPEC strains from different sources and countries have shown high resistance of this pathotype to ampicillin, cefpodoxime, nalidixic acid, trimethoprim, and tetracycline [125, 126]. While resistance to the great majority of these antibiotics is reported to be frequent in tEPEC, trimethoprim resistance is more common in aEPEC strains [127].

In a global study of 185 aEPEC isolates collected from healthy and diarrheal children living in seven sites in sub-Saharan Africa and South Asia, at least 55% of the isolates showed phenotypic resistance to ampicillin, trimethoprim, trimethoprim/sulphamethoxazole, and tetracycline, while streptomycin resistance was reported in 43% of the isolates. Shockingly, more than 50% of the isolates were resistant to three or more of the tested antibiotics [128]. The study also reported point mutations in genes that are associated with resistance to quinolone (*gyrA*, *parC*) and nitrofurantoin (*nfsA*) in addition to over forty different antibiotics resistance genes reported. Equally, more than 50% of the isolates carried at least four resistance determinants that include bla_{TEM} (ampicillin), *strA* and *strB* (streptomycin), *sul2* (sulphonamides), and *dfr genes* (trimethoprim/sulfamethoxazole). These resistance determinants were found singly or co-localized on plasmids (pCERC1, pCERC2) or in transposons (Tn*6029*).

4.4 Population structure of EPEC

The acquisition of LEE and pEAF has been the defining evolutionary phenomenon for EPEC pathotypes [3, 129]. While tEPEC that carries pEAF plasmid is believed to be less diverse, aEPEC is greatly heterogeneous. The loss of pEAF plasmid in aEPEC and its close relatedness with LEE-positive STEC in serotypes, genetic characteristics, virulence properties, and reservoirs make serotype-based lineage definition unreliable [111, 130]. Based on the conventional MLEE and MLST, EPEC strains belong to six clonal lineages (EPEC1–EPEC6) that were represented among the EPEC strains worldwide [129, 131]. The whole genome-based phylogeny reported nine more EPEC lineages designated as EPEC7-EPEC15 [104, 132]. These phylogenomic EPEC lineages belonged to four *E. coli* phylogroups (A, E, B1, and B2) (**Figure 4**), where the great majority were found in B1 and B2 [104, 133], suggesting a clear genetic heterogeneity within this pathotype.

The close relatedness of aEPEC to other pathotypes could play a significant role in the diversity within this pathotype. This EPEC subtype can also include tEPEC that have lost the pEAF plasmid and LEE-positive STEC strains that have lost the Stx encoding bacteriophage during transmission events between hosts, within-host evolution, interaction with the host microbiota, or selective pressure in the environment [104, 130]. This could be a possible explanation why some aEPEC strains would cluster with other pathotypes. Indeed, phylogenomic analyses of 106 Brazilian and 221 global aEPEC genomes showed that isolates were clustered into the previously reported phylogroups for this pathotype and phylogroup D. Additionally, 42.5% of the isolates belonged to the four previously defined EPEC lineages [129, 131], while the remaining isolates were found in EPEC11-EPEC14 phylogenomic lineages, suggesting a gradual and continuous clonal expansion of this pathotype [132]. Of note, dissemination of the phylogenomic lineages of EPEC pathotype is not restricted by geography. Conversely, in a multicentre study involving seven sites in developing countries, EPEC isolates from sub-Saharan countries (The Gambia and Kenya) were clustered into two EPEC lineages (EPEC5 and EPEC10) in phylogroup A [104]. Overall, EPEC represents a pathotype that is still undergoing clonal expansion due to the occurrence of novel phylogenomic lineages with distinct accessory gene content and their pathogenic potential.

5. Enteroaggregative *E. coli* (EAEC)

5.1 An overview of EAEC

Enteroaggregative *E. coli* (EAEC) is implicated in epidemic diarrheal illnesses, being a causative agent of traveler's diarrhea, persistent diarrhea in children in EAEC endemic areas, and in immunocompromised patients, particularly in human immunodeficiency virus (HIV) patients [3, 134]. EAEC was first described in 1987 by comparing adherence patterns of *E. coli* isolates to HEp-2 cells, where it showed a stacked-brick aggregative phenotype. EAEC strains are able to infect the colon and/or small bowel of their host where they disrupt the intestinal epithelium and result in loss of electrolytes, watery diarrhea with or without blood and mucus, vomiting, among other symptoms [134]. Persistent EAEC-related diarrhea could result in chronic intestinal inflammation that induces the production of fecal lactoferrin and interleukin (IL-8), and malabsorption of nutrients [135]. Studies on the development of postinfectious irritable bowel disease syndrome in acute EAEC-related diarrhea have been documented [135] but the role of EAEC is not fully understood.

EAEC strains are identified using a molecular probe AA that hybridizes with a region of pAA plasmid encoding an ATP binding cassette transporter apparatus which translocates dispersion across the bacterial cell membrane [136]. Isolates that carry the *aggR* gene that encodes autoagglutination that are associated with persistent diarrhea in patients are able to hybridize with the AA probe. In addition, EAEC has a different adherence pattern, and not all HEp-2 adherent EAEC strains isolated from humans with diarrhea carried *aggR*. Hence, EAEC was classified into two subtypes: typical (*aggR* positive) and atypical (*aggR* negative) subtype [134, 137]. Humans are the reservoir for typical EAEC (tEAEC), whereas atypical EAEC (aEAEC) strains have been isolated from young calves, piglets, and horses as well as companion animals, suggesting the role of animals as a reservoir for this subtype (**Figure 2**) [3].

Although some serotypes including O126:H27, O111:H21, O125, O44:H18 are frequently isolated from EAEC strains, the autoagglutinating phenotype by some EAEC strains complicates the serotyping of this pathotype [3, 138]. In several studies, EAEC strains are often described as nontypeable or as "O?" or O-rough. In a study of EAEC strains from children in Germany, 14 out of 16 isolates that were typeable belonged to different serotypes [139]. Likewise, in a study in the UK, 97 out of 143 EAEC strains that were typeable belonged to more than 40 different O-types [140]. While serotyping is no longer a dependable diagnostic tool for EAEC strains causing diarrheal illness [3, 138], a specific Shiga toxin producing EAEC serotype O104:H4 is associated with a series of outbreaks worldwide [141, 142].

5.2 Virulome of EAEC

EAEC strains that are implicated in diarrheal illness employ several virulence factors that initiate colonization, promote persistence through adherence to mucosal layers of the intestine, and enterotoxin and cytotoxin secretion (**Table 1**) [134].

5.2.1 Colonization and adherence

EAEC colonizes the intestinal epithelium of the host using aggregative adhesion fimbriae (AAFs) that also activate the host inflammatory responses and afimbrial adhesins [143]. So far, five AAF variants have been described and are encoded by *aggA* (AAF/I), *aafA* (AAF/II), *agg3A* (AAF/III), *agg4A* (AAF/IV), and *agg5A* (AAF/V) that are regulated by the transcriptional activator *aggR*, borne on EAEC plasmid pAA [143]. AggR also regulates the expression of a type VI Secretion System (T6SS) and a chromosomal PAI encoded by *aaiA-aaiP* operon [144] as well all other virulence genes involved in the aggregation and toxin production in pathogenic strains of EAEC (**Figure 3**) [143, 144]. However, in EAEC strains where AAF is absent, an aggregate-forming pili (AFP), a type VI pilus that is encoded by *afp* operon was reported to be responsible for the establishment of a similar aggregative adhesion pattern (**Figure 3**) [144]. Other virulence determinants associated with colonization and adherence of EAEC include *air* gene that codifies for an enteroaggregative immunoglobulin repeat protein and *capU* that encodes a hexosyltransferase homolog, as well as *shf* and *aatA* that have been linked to biofilm formation [3, 144].

5.2.2 Enterotoxin and cytotoxin secretion

EAEC produces enterotoxins and cytotoxins including EAST1 and colonization factors encoded by *astA* and *pic* genes, respectively [37, 145]. The latter (*pic*) is often associated with *set1*A and *set1B* encoding two subunits of *Shigella* enterotoxin 1 (ShET1) that are linked to the induction intestinal secretion during infection [145]. This pathotype also carries genes encoding class I cytotoxic SPATE protein family that includes autotransporter proteases encoded by *sigA* and *sepA* and plasmid-borne toxin encoded by *pet* [144, 146]. Also, EAEC strains produce dispersin, an anti-aggregation protein that is encoded by *aatPABCD* located on plasmid pAA, and promotes the dispersion of bacteria in the mucosal layer of the intestine [147]. A gene *hlyE* encoding a hemolytic pore-forming toxin that has a cytotoxic effect on cultured cells has also been reported to be present in some EAEC strains, although the role of this gene in the pathogenicity of EAEC is still unclear [3, 148]. Some EAEC strains carry Stx2a phage-encoding Shiga toxin that is associated with HUS in STEC-related infection [141, 146].

The prevalence of the virulence determinants varies with studies and EAEC subtypes [144, 146]. For example, in a study, *pic* gene was reported to be the most prevalent, present in only 47%, while *sepA* and *sigA* were present in less than 15% of the studied isolates [149]. In the study, the authors also noted that *pet* and *pic* genes were associated with tEAEC, whereas *sepA* was associated with aEAEC.

5.3 Antibiotic resistance in EAEC

EAEC-related diseases such as travelers' diarrhea where antimicrobial therapy is proposed, fluoroquinolones, azithromycin, and rifaximin are often recommended. In immunocompromised patients that require chemoprophylaxis, fluoroquinolones are also considered [150]. For Shiga toxin producing EAEC O104: H4 related infections, azithromycin which has been shown to inhibit *stx* expression in *in-vitro* assay is seldomly used [3, 150].

Although a highly successful treatment rate is achieved with these antibiotics, EAEC strains that are resistant to multiple antibiotics have emerged in different regions [150]. Studies on the resistance of EAEC strains from Southeast Asia,

India, Africa, and Latin America with travelers' diarrhea showed that more than 50% of the isolates were resistant to ampicillin, sulphamethoxazole, and tetracycline [150, 151]. In a similar study in Iran, 78% and 60% of the extended-spectrum beta-lactamase (ESBL) producing EAEC strains carried the transposable bla_{TEM} and $bla_{CTX\text{-}M}$ genes, respectively [152]. Also, plasmid-mediated quinolone resistance (PMQR) genes (*qnr*) that encode resistance to quinolone have been identified in EAEC in different studies [153, 154]. In England, among the 155 EAEC strains from diarrhea patients in 2015–2016 [155] showing antibiotic-resistant phenotypes, 43 genetic determinants that encode resistance to seven different classes of antibiotics were identified, with $bla_{TEM\text{-}1}$ being the most common (40%) followed by *sul2* (37%) and *strA-strB* (32%). Undoubtedly, the rise in antibiotic resistance in this pathotype should be a concern for public health.

5.4 Population structure of EAEC

EAEC subtypes are defined based on the presence of virulence plasmid pAA that carries *aggR* gene that regulates the expression of other virulence determinants located on the plasmid. This and the high serotype diversity, as well as other accessory genes contribute to the high heterogeneity noted for this pathotype. An earlier study on the phylogenetic analysis of EAEC revealed that isolates belonged to multiple lineages [156]. A similar observation was noted with MLST where 150 Nigerian EAEC strains were clustered into 96 STs [157]. Indeed, EAEC strains are known to belong and spread across four *E. coli* phylogroups (A, B1, B2, and D) (**Figure 4**) with diverse serotypes [144, 156].

In a recent study [146], of the 97 EAEC strains analyzed using MLST, 42% were reported to belong to phylogroup B1, while the majority of the few strains that belonged to phylogroup A lack the AAF-associated genes. Although serotype diversity is high in this pathotype, this study also noted that EAEC strains that belonged to phylogroup D were clustered into three serotype-specific lineages (lineage 1–3). All strains in lineage 1 were O166:H15 and belonged to ST349, lineage 2 consisted of serogroups O44, O73, and O17/O77 in combination with either H18 or H34 and ST130, while lineage 3 carried O153:H30 serotype and ST38 [146]. Contrarily, in India, EAEC strains implicated in diarrhea were more prevalent in phylogroup D [158] suggesting that the diversity in the pathotype is not limited by geography. While EAEC subtypes are believed to differ in their virulence determinants content, this hypothesis and comparative phylogenetic analysis of aEAEC and tEAEC are underexplored. Also, large-scale phylogenomic and phylogeographic analyses of this pathotype are scarce. Further studies should focus on this.

6. Entero-invasive *E. coli* (EIEC)

6.1 An overview of EIEC

Enteroinvasive *E. coli* (EIEC) pathotype causes bacillary dysentery in humans worldwide characterized by abdominal cramps, bloody and mucous diarrhea [159]. The incidence of EIEC-related diseases varies by geographic region but is highly frequent in developing countries. In developed countries, EIEC-related infections are mainly travelers' diarrheal cases in people with recent travel history to endemic regions [160]. The first EIEC strain belonging to serotype O124 was reported in 1947 [161]. In the later years, some of the bacterial species implicated in dysentery that were previously classified as *Shigella* were renamed as EIEC [162]. EIEC and *Shigella* spp. share several serogroups, phenotypic and other genotypic characteristics, which often makes it challenging to discriminate between the two genera in clinical samples. Like some *Shigella spp.*, most EIEC strains are non-motile and lack the ability to decarboxylate lysine or ferment lactose [3]. EIEC invades the human intestinal epithelial layer where it induces dysentery syndrome that is characterized by watery stool containing blood, mucus, and leukocytes, symptoms that are similar to those presented by *Shigella* spp. associated infection (**Figures 2** and **3**) [159].

At least, 20 serotypes have been assigned to this pathotype [159] among which some of the EIEC-associated O antigens including O28, O112ac, O121, O124, O143, O144, O152, and O167, are identical to O antigens present in *Shigella* spp. [3, 159]. Humans are the major reservoir of EIEC strains and transmission occurs through the fecal-oral route from the ingestion of contaminated foods or water and person-to-person contact [159, 160]. The incidence rate and morbidity for EIEC are less or underreported but it appears to follow a similar trend as *Shigella* [3, 159]. Although EIEC strains cause sporadic cases of infection, they are also implicated in outbreaks. Of note are outbreaks of EIEC linked to O96:H19 strain that was traced to cooked vegetables and salads that were contaminated by asymptomatic food handlers in 2012 and 2014 in Italy [163] and the United Kingdom [164], respectively. Recently, the first case of EIEC outbreak in the US in about half a century was caused by EIEC serotype O8:H19 [160].

6.2 Virulome of EIEC

EIEC strains cause infection in humans by their ability to invade the colon mucosa layer with the expression of essential virulence determinants that mediate colonization, adherence, and invasion of the intestinal epithelial cells of the host (**Table 1**). These genes that are also shared with *Shigella* are located on the chromosome or virulence plasmid.

6.2.1 Colonization and adherence

The colonization, adherence, and invasion of intestinal epithelial cells by EIEC are mediated by genetic factors encoded by genes on a plasmid, pINV. pINV is a virulence plasmid found in EIEC that encodes the type III secretion system necessary for attachment, invasion of the host cell, and intercellular spread. This plasmid is structurally and functionally similar to those in *Shigella* strains [159], and with the replication (*rep*) and conjugation (*tra*) regions in IncFIIA plasmids. pINV had large deletions in the *tra* region which makes it incapable of self-transfer by conjugation but can be mobilized by other conjugative plasmids. Among the numerous functional insertion sequences present in this plasmid is the IS*1111* family, but only defective copies of the IS family are found in *Shigella* pINV plasmids [159]. pINV carries a PAI-structure that is composed of gene clusters encoding a T3SS apparatus (Mxi and Spa), its effector proteins (IpaB, IpaC, and IpaD) with their chaperons (IpgA, IpgC, IpgE, and Spa15), and two global transcriptional regulators (VirB and MxiE) that activate and regulate the expression of most of the virulence genes [3, 159]. All T3SS effectors are carried on the pINV except for a few effector proteins of the IpaH family that are chromosomal (**Figure 3**).

Also carried on pINV are genes that encode IcsA, a protein that facilitates the bacterial movement inside the cytoplasm, VirA, a GTPase-activating protein, and RnaG, a small RNA that negatively control the expression of *icsA* gene [159, 165], as well as the gene encoding OspG and OspF proteins which facilitate the evasion of the host innate immune response. All EIEC isolates are reported to carry this plasmid as it is essential for the pathogenesis and pathoadaptation of this pathotype. However, loss of this genetic element has been reported in some EIEC strains [159, 165].

6.2.2 Cytotoxins production

EIEC strains also carry a plasmid-borne gene, *sen* that mediates a novel 63-kDa enterotoxin (ShET2) [166]. A mutation in this gene was reported to cause a substantial loss in the enterotoxic ability of EIEC strain. Although the role of this gene in the pathogenesis of EIEC is not fully understood, toxins are known to be important in the induction of watery diarrhea during *E. coli* infection. Additionally, plasmid encoding enterotoxigenic and cytotoxigenic factors namely *pic*, *sepA*, *sigA*, and *sat* that belong to SPATEs family and that are reported to contribute to intestinal fluid accumulation in an animal model are carried by EIEC strains. Nonetheless, these genes are not carried by all EIEC strains. Two different studies on the prevalence of the virulence genes among EIEC reported that *sen*, *sigA,* and *pic* were found in at least 70%, 64%, and 27% of the isolates, respectively [167, 168], whereas *sat* gene was found only in 15% of the collection [167].

6.3 Antibiotic resistance in EIEC

EIEC-related infection is self-limiting that could be managed with rehydration to replenish the loss electrolyte. Zinc supplementation and nutritional therapy with iron-rich green plantain have also been shown to reduce the severity and the duration of diarrheal illness. However, in rare cases of severe symptoms antimicrobial treatment therapy has been reported to be effective [3, 10]. Since *Shigella* and EIEC present similar symptoms and are often misdiagnosed, similar antimicrobials include azithromycin (macrolide), ceftriaxone, (cephalosporin), and ciprofloxacin (fluoroquinolone) are recommended [3, 159].

Like other *E. coli* pathotypes and *Shigella*, there is an emergence of multidrug-resistant EIEC strains. In a study of EIEC isolates from adults with enteric infection in Cameroon, high resistance to ampicillin and sulfamethoxazole-trimethoprim was noted in 57.14% and 71.43%, respectively. Resistance determinants to ampicillin (bla_{TEM} and bla_{Oxa}) were found in 28.57% of the isolates. Additionally, *cat1* and *cat2* genes were noted in chloramphenicol resistant strains while *tetA*, *tetB* encoding resistance to tetracycline, *dfr12, dfr7, dfr1a* to sulphamethoxazole-trimethoprim, and *sul1* gene to sulfonamide were present in more than 85% of the EIEC isolates [169]. In a large-scale study of *Shigella* and EIEC isolates from eight countries in four continents between 1971 and 1999, 48% of EIEC isolates were resistant to tetracycline [170]. In another study, an EIEC O164 strain isolated from a traveler with diarrhea in Japan was found to be resistant to streptomycin, spectinomycin, co-trimoxazole, and ampicillin, with reduced susceptibility to ciprofloxacin [171]. In this strain, resistance determinant for trimethoprim (*dfrXII*), streptomycin and spectinomycin (*aadA2*), and an ORF of unknown function was carried on a class 1 integron located on a transferable plasmid. While ampicillin resistance gene bla_{TEM} was detected, the reduced susceptibility to ciprofloxacin was reported to be due to a single mutation P158-to-S in *parC*.

6.4 Population structure of EIEC

EIEC pathotype is diverse and highly specialized due to the carriage of large virulence plasmid, a genetic element that is shared with *Shigella*. The pINV plasmid that is the hallmark of EIEC lacks the ability for autonomous horizontal transfer. Although there is no consensus on the evolution of pINV in Shigella/

EIEC, it has been hypothesized that this genetic element was probably acquired in an ancestral *E. coli* prior to the diversification of the two bacterial species and the emergence of different Shigella/EIEC lineages. Conversely, Shigella/EIEC strains could have evolved from different *E. coli* strains that had acquired the pINV independently from other pINV-carrying Shigella/EIEC or from an unknown donor [172].

A phylogenetic analysis of 32 EIEC strains based on four housekeeping genes (*trpA, trpB, pabB,* and *putP*) revealed four clusters (clusters 4–7) where most of the O antigens were found in a single cluster [173]. EIEC cluster 4 comprised strains with serotype O28, O29, O124, O136, and O164. Serotypes O124, O135, O152, and O164 belonged to cluster 5, and O143 and O167 to cluster 6 while cluster 7 had only O144 [173]. This is similar to a SNP-based phylogeny described in another study [174]. WGS alignment-based phylogeny of 20 EIEC isolates revealed that the great majority of the strains belonged to three distinct EIEC lineages (lineage 1–3) that belonged to three different *E. coli* phylogroups (A, B1, and E) (**Figure 4**). All the EIEC strains in lineage 1 (phylogroup E) were all serotype O143:H26, whereas, in other lineages the serotypes were diverse. EIEC lineage 3 (phylogroup B1) was reported to be globally disseminated as the strains from six different countries were clustered together in this lineage [175].

Insertion sequences were recently reported to contribute to the population structure of EIEC. A recent study [176] on the evolutionary dynamics of *Shigella* and EIEC lineages identified the genetic factors driving strain-to-strain variation within each population and contributing to functional gene loss within and between species. In the study, the author found that all *Shigella* and EIEC lineages had higher IS copy numbers relative to other *E. coli* pathotypes indicative of IS expansion in these lineages. The authors also found that *Shigella* and EIEC lineages carried the same five ISs (IS*1*, IS*2*, IS*4*, IS*911,* and IS*600*) indicative of a parallel expansion of these IS types, although at a high degree in *Shigella.* The data also suggests that *Shigella* and EIEC lineages underwent an expansion of their native IS*1* alleles and that pINV is a potential source for the introduction of other ISs (IS*2*, IS*4*, IS*600,* and IS*911*) that are rare in *E. coli* into *Shigella* and EIEC lineages [176].

In a comparative pangenome analysis of EIEC with *Shigella* and other *E. coli* pathotypes, seven gene clusters were identified to be enriched in EIEC strains but absent in all other *E. coli* pathotypes and *Shigella* strains. These included genes that encode a putative pyruvate kinase, a periplasmic protein, and some uncharacterized proteins. However, when *Shigella* isolates were excluded, the authors identified 96 gene clusters that were present in more than half of the EIEC strains. A total of 87 gene clusters were reported when EIEC and *Shigella* genomes combined were compared to other *E. coli* pathotypes. Among these were plasmid-associated genes encoding a hypothetical toxin-antitoxin system and putative proteins hypothesized to be involved in conjugal transfer [175].

The EIEC lineages were reported to have distinct phenotypic and genotypic features. Lan et al. [173] reported that EIEC strains belonging to cluster 4 lack mucate fermentation ability, whereas strains in cluster 6 were able to utilize acetate and ferment mucate [173]. Likewise, Hazen et al. [175] identified up to 155 gene clusters that were exclusive in EIEC strains belonging to one phylogroup. Additionally, 12–155 gene clusters were also reported to be lineage-specific in the EIEC pathotype [175]. Protein-encoding genes that are linked to transcriptional regulation, metabolism, and transport, and a colicin were exclusive in EIEC lineage 1, whereas genes that encode membrane protein, the aerobactin siderophore receptor, and hypothetical proteins were exclusive for EIEC lineage 2. Genes encoding several transcriptional regulators and hypothetical proteins were limited to EIEC lineage 3 [175].

7. Diffusely adherent *E. coli* (DAEC)

7.1 An overview of DAEC

Diffusely adherent *E. coli* (DAEC) pathotype is a group of *E. coli* causing diarrhea that can attach to host cells but not in a localized or A/E adherence pattern [3]. *E. coli* strains belonging to this pathotype binds to the entire surface of the epithelial HEp-2 cells in a scattered pattern termed diffuse adherence. Although the adherence pattern is unique, this pathotype is difficult to classify or identify, a possible reason for the scarce epidemiological studies on this group [177].

DAEC is widespread and associated with diarrhea in both developing and industrialized countries around the world [3]. DAEC strains are associated with watery diarrhea in children under 5 years and can persist resulting in an increase in severity of disease in this age group [177]. DAEC has also been implicated in extraintestinal infections such as UTI, pregnancy complications [3, 177]. It has been speculated that the asymptomatic carriage of DAEC by this age group and adults can lead to chronic inflammatory colon disease such as Crohn's disease [3, 177]. Meanwhile, there is no universal detection method for this pathotype but based on DNA hybridization of fimbrial encoding *daaC* gene probe and adherence pattern to HEp-2 cells of 221 diarrheagenic *E. coli* from different age-groups in Brazil [178], DAEC was identified and shown to be associated with diarrhea in children under 12 months of age in this region. The authors also noted that the presence of DAEC in younger children was not associated with diarrhea, suggesting that the association would probably be based on geographic regions [178]. The asymptomatic carriage by different age groups and the lack of epidemiological data from different regions undermines the development of a universal identification method for this pathotype.

While it is unclear how DAEC is transmitted or its reservoir, the fact that there are asymptomatic carriers could suggest humans as the main reservoir and the fecal-oral route as the primary means of transmission (**Figure 2**). Information regarding serotypes associated with DAEC is scarce. A study of 112 DAEC isolates from diarrheal and asymptomatic individuals in Brazil reported 45 different serotypes, of which 19 were exclusive in patients with diarrhea [179], whereas in another study [180] the serotypes were nontypeable.

7.2 Virulome of DAEC

DAEC strains infect the intestinal epithelium of the host by expressing surface-exposed adhesins that mediate colonization and attachment which allow them to resist host clearance mechanisms [177]. There are several virulence factors that mediate this process, and they include fimbria or afimbrial structures, adhesins, and secretion of cytotoxic toxins that promote the invasion of the host cells (**Table 1**) [177].

7.2.1 Colonization and adherence

DAEC pathotype carries genes that encode Afa/Dr adhesins [3, 177]. Afa/Dr family includes Afa, Dr, and F1845 adhesins that are both afimbrial (such as AfaE-I and AfaE-III), and fimbrial (such as F1845 and Dr) adhesive structures on the bacterial surface encoded by the *afa*, *dra*, and *daa* operons, respectively [177, 181]. These adhesins have been found not only in DAEC but also in UPEC (Uropathogenic *E. coli*), indicating that strains that produce Afa/Dr adhesins may cause both intestinal and extraintestinal infections. Afa/Dr adhesins bind to human decay-accelerating factor (hDAF) and carcinoembryonic antigen-related cell adhesion molecules that induce receptor clustering resulting in a partial internalization of bacterial cells (**Figure 3**). These adhesins can induce the production of cytokine IL-8 and result in intestinal inflammation, loss of microvilli structure, and watery diarrhea [177]. There are two different subclasses of atypical DAEC; a subclass that contains all the adhesins typical of the Afa/Dr family of adhesins in another *E. coli* pathotype such as diffusely adherent EPEC, while the other subclass does not bind hDAF and expresses a different array of adhesins on its surface, including AfaE-VII, AfaE-VIII, AAF-I, AAF-II, and AAF-III and still able to induce proinflammation [177].

The prevalence of the genes constituting the operons that encode the Afa/Dr adhesins have been reported to vary in DAEC pathotype. In a study, the prevalence of the Afa/Dr adhesin family encoding genes *afaE-1, afaE-2, afaE-3, afaE-5* and *daaE* were reported in 64.3%, 14.2%, 28.6%, 21.5% and 21.5% of DAEC isolates, respectively [182], while in another study [183], the prevalence of these genes were 44%, 10%, 2%, 2% and 6%, respectively.

7.2.2 Secretion of cytotoxins

DAEC secretes a class I SPATE toxin that is called secreted autotransporter toxin (Sat), encoded by *sat* gene. Sat is reported to have enterotoxic activity in an animal model, and mediate the induction of fluid accumulation, loss of microvilli, inflammation, and polymorphonuclear lymphocytes (PMNL) infiltration, like the LT effect of ETEC (**Figure 3**) [184]. Although *sat* gene is reported to be equally expressed in DAEC strains from diarrheal and asymptomatic adults, this gene is significantly associated with DAEC-related diarrhea in children [183]. For example, in two studies, *sat* gene was identified in 44–63% of DAEC strains collected from children with diarrhea while it was found in 0–20% of DAEC strains from asymptomatic children [183–185]. Noteworthy, *sat* gene is not exclusive for DAEC pathotype. In fact, it is prevalent in other *E. coli* pathotypes including EAEC and UPEC [186]. Other virulence factors that have been reported in DAEC include *pet* [187], *astA* [187], and *senB* [188] genes that encode enterotoxins and *hlyE* gene that encodes alpha-hemolysin [186]. However, the role of these genes in the pathogenesis of DAEC still remains unclear.

7.3 Antibiotic resistance in DAEC

Oral rehydration solution therapy is the only recommended treatment for DAEC-related watery diarrhea. However, there are reports of antibiotic resistance in this pathotype. In a study of 112 DAEC strains isolated from children with watery diarrhea in Brazil [179], all DAEC isolates were susceptible to five antibiotics including gentamicin, ofloxacin, and nalidixic acid while 70% were resistant to three or more antibiotics and 50% showed resistance to either ampicillin, co-trimoxazole, streptomycin, sulfonamide, or tetracycline. Additionally, 20% of the strains were resistant to chloramphenicol [179]. A similar observation was noted in a study from Iran where 75–100% of DAEC strains from pediatric diarrhea were resistant to ampicillin, cefotaxime, and trimethoprim-sulfamethoxazole [182].

7.4 Population structure of DAEC

DAEC is a heterogeneous group that has also been implicated in extraintestinal infections such as UTI. Despite its implication in diarrhea in children, studies on its population structure are limited. The few studies available on the phylogenetic analysis of DAEC strains using MLEE reported that they are distributed among all of the phylogroups [47, 156, 189]. Conversely, a study of 31 DAEC strains from diarrhea and asymptomatic carriers in Peru reported that 87% of the isolates belonged to phylogroup D [190]. A large-scale genomic analysis of DAEC strains would be important to understand the population structure, determine dissemination and transmission dynamics of genetic lineages of this pathotype, as well as identify novel virulence determinants and other genetic factors that contribute to its patho-adaptation in the intestinal epithelium.

8. Conclusion

Much progress has been made on the biology and the genomic epidemiology of diarrheagenic *E. coli* pathotypes since the development of WGS technologies. In particular, there is a notable advancement in the timely detection of outbreaks and the understanding of the population structure of some of the DEC pathotypes. However, the distinct phenotypes underlying the genomic signatures that drive the evolution of pathogenic *E. coli* are not fully understood. Large-scale genomic analyses of different *E. coli* pathotypes are scarce, hence, the genetic factors that define each pathovar and specific lineages are still underexplored. These should form the direction for future studies to better understand the evolutionary dynamics of *E. coli* pathotypes.

Author details

Opeyemi U. Lawal, Valeria R. Parreira and Lawrence Goodridge*
Canadian Research Institute for Food Safety, Department of Food Science, University of Guelph, Ontario, Canada

*Address all correspondence to: goodridl@uoguelph.ca

References

[1] Frazão N, Sousa A, Lässig M, Gordo I. Horizontal gene transfer overrides mutation in *Escherichia coli* colonizing the mammalian gut. Proceedings of the National Academy of Sciences. 2019;**116**:17906-17915

[2] Capps KM, Ludwig JB, Shridhar PB, Shi X, Roberts E, et al. Identification, Shiga toxin subtypes and prevalence of minor serogroups of Shiga toxin-producing *Escherichia coli* in feedlot cattle feces. Scientific Reports. 2021;**11**. Epub ahead of print 21 April 2021. DOI: 10.1038/s41598-021-87544-w

[3] Croxen MA, Law RJ, Scholz R, Keeney KM, Wlodarska M, et al. Recent advances in understanding enteric pathogenic *Escherichia coli*. Clinical Microbiology Reviews. 2013;**26**:822-880

[4] Kim KS. Human meningitis-associated *Escherichia coli*. EcoSal Plus. 2016;7. Epub ahead of print May 2016. DOI: 10.1128/ecosalplus.ESP-0015-2015

[5] Devasia RA, Jones TF, Ward J, Stafford L, Hardin H, et al. Endemically acquired foodborne outbreak of enterotoxin-producing *Escherichia coli* serotype O169:H41. The American Journal of Medicine. 2006;**119**:168.e7-168.e10

[6] Troeger C, Blacker BF, Khalil IA, Rao PC, Cao S, et al. Estimates of the global, regional, and national morbidity, mortality, and aetiologies of diarrhoea in 195 countries: A systematic analysis for the Global Burden of Disease Study 2016. The Lancet Infectious Diseases. 2018;**18**:1211-1228

[7] Winstead A, Hunter JC, Griffin PM. *Escherichia coli*, Diarrheagenic. In: Chapter 4—2020 Yellow Book | Travelers' Health | CDC. Available from: https://wwwnc.cdc.gov/travel/yellowbook/2020/travel-related-infectious-diseases/escherichia-coli-diarrheagenic. [Accessed: 16 August 2021]

[8] Poirel L, Madec J-Y, Lupo A, Schink A-K, Kieffer N, et al. Antimicrobial Resistance in *Escherichia coli*. Microbiology Spectrum. 2018;**6**. Epub ahead of print 27 July 2018. DOI: 10.1128/microbiolspec.ARBA-0026-2017

[9] Irving SE, Choudhury NR, Corrigan RM. The stringent response and physiological roles of (pp)pGpp in bacteria. Nature Reviews. Microbiology. 2021;**19**:256-271

[10] Kaper JB, Nataro JP, Mobley HLT. Pathogenic *Escherichia coli*. Nature Reviews. Microbiology. 2004;**2**:123-140

[11] Beutin L, Delannoy S, Fach P. Genetic diversity of the fliC genes encoding the flagellar antigen H19 of *Escherichia coli* and application to the specific identification of enterohemorrhagic E. coli O121:H19. Applied and Environmental Microbiology. 2015;**81**:4224-4230

[12] Ejrnaes K, Sandvang D, Lundgren B, Ferry S, Holm S, et al. Pulsed-field gel electrophoresis typing of *Escherichia coli* strains from samples collected before and after pivmecillinam or placebo treatment of uncomplicated community-acquired urinary tract infection in women. Journal of Clinical Microbiology. 2006;**44**:1776

[13] Pupo GM, Lan R, Reeves PR, Baverstock PR. Population genetics of *Escherichia coli* in a natural population of native Australian rats. Environmental Microbiology. 2000;**2**:594-610

[14] Naseer U, Olsson-Liljequist BE, Woodford N, Dhanji H, Cantón R, et al. Multi-locus variable number of tandem repeat analysis for rapid and accurate typing of virulent multidrug resistant

Escherichia coli clones. PLoS One. 2012;**7**:e41232

[15] Adiri RS, Gophna U, Ron EZ. Multilocus sequence typing (MLST) of *Escherichia coli* O78 strains. FEMS Microbiology Letters. 2003;**222**:199-203

[16] Schürch AC, Arredondo-Alonso S, Willems RJL, Goering RV. Whole genome sequencing options for bacterial strain typing and epidemiologic analysis based on single nucleotide polymorphism versus gene-by-gene–based approaches. Clinical Microbiology and Infection. 2018;**24**:350-354

[17] Joensen KG, Scheutz F, Lund O, Hasman H, Kaas RS, et al. Real-time whole-genome sequencing for routine typing, surveillance, and outbreak detection of verotoxigenic *Escherichia coli*. Journal of Clinical Microbiology. 2014;**52**:1501-1510

[18] Crofts AA, Giovanetti SM, Rubin EJ, Poly FM, Gutiérrez RL, et al. Enterotoxigenic *E. coli* virulence gene regulation in human infections. Proceedings of the National Academy of Sciences. 2018;**115**:E8968-E8976

[19] Gupta SK, Keck J, Ram PK, Crump JA, Miller MA, et al. Part III. Analysis of data gaps pertaining to enterotoxigenic *Escherichia coli* infections in low and medium human development index countries, 1984-2005. Epidemiology and Infection. 2008;**136**:721-738

[20] Hartadi EB, Effendi MH, Plumeriastuti H, Sofiana ED, Wibisono FM, et al. A review of enterotoxigenic *Escherichia coli* infection in piglets: Public health importance. Systematic Reviews in Pharmacy. 2020;**11**:12

[21] Pattabiraman V, Katz LS, Chen JC, McCullough AE, Trees E. Genome wide characterization of enterotoxigenic *Escherichia coli* serogroup O6 isolates from multiple outbreaks and sporadic infections from 1975-2016. PLoS One. 2018;**13**:e0208735

[22] Begum YA, Talukder KA, Azmi IJ, Shahnaij M, Sheikh A, et al. Resistance pattern and molecular characterization of enterotoxigenic *Escherichia coli* (ETEC) strains isolated in Bangladesh. PLoS One. 2016;**11**:e0157415

[23] Qadri F, Svennerholm A-M, Faruque ASG, Sack RB. Enterotoxigenic *Escherichia coli* in developing countries: Epidemiology, microbiology, clinical features, treatment, and prevention. Clinical Microbiology Reviews. 2005;**18**:465-483

[24] Sahl JW, Sistrunk JR, Baby NI, Begum Y, Luo Q, et al. Insights into enterotoxigenic *Escherichia coli* diversity in Bangladesh utilizing genomic epidemiology. Scientific Reports. 2017;**7**:3402

[25] Sánchez J, Holmgren J. Virulence factors, pathogenesis and vaccine protection in cholera and ETEC diarrhea. Current Opinion in Immunology. 2005;**17**:388-398

[26] Duan Q, Yao F, Zhu G. Major virulence factors of enterotoxigenic *Escherichia coli* in pigs. Annales de Microbiologie. 2012;**62**:7-14

[27] Vipin Madhavan TP, Sakellaris H. Colonization factors of enterotoxigenic *Escherichia coli*. In: Advances in Applied Microbiology. Amsterdam, The Netherlands: Elsevier. pp. 155-197

[28] Chung S, Kwon T, Bak Y-S, Park JJ, Kim C-H, et al. Comparative genomic analysis of enterotoxigenic *Escherichia coli* O159 strains isolated from diarrheal patients in Korea. Gut Pathogens. 2019;**11**:9

[29] Joffré E, von Mentzer A, Svennerholm A-M, Sjöling Å. Identification of new heat-stable (STa)

enterotoxin allele variants produced by human enterotoxigenic *Escherichia coli* (ETEC). International Journal of Medical Microbiology. 2016;**306**: 586-594

[30] Bhakat D, Mondal I, Chatterjee NS. EatA, a non-classical virulence factor, of Enterotoxigenic *Escherichia coli* (ETEC) is modulated by the host factors during pathogenesis. International Journal of Infectious Diseases. 2020;**101**:3-4

[31] Kumar P, Luo Q, Vickers TJ, Sheikh A, Lewis WG, et al. EatA, an immunogenic protective antigen of enterotoxigenic *Escherichia coli*, Degrades Intestinal Mucin. Infection and Immunity. 2014;**82**:500-508

[32] Liu F, Yang X, Wang Z, Nicklasson M, Qadri F, et al. Draft genomes of four enterotoxigenic *Escherichia coli* (ETEC) clinical isolates from China and Bangladesh. Gut Pathogens. 2015;**7**:10

[33] Ajami NJ, Youmans BP, Highlander SK, DuPont HL, Petrosino JF, et al. Development and accuracy of quantitative real-time polymerase chain reaction assays for detection and quantification of enterotoxigenic *Escherichia coli* (ETEC) heat labile and heat stable toxin genes in travelers' diarrhea samples. The American Journal of Tropical Medicine and Hygiene. 2014;**90**:124-132

[34] Wang H, Zhong Z, Luo Y, Cox E, Devriendt B. Heat-stable enterotoxins of enterotoxigenic *Escherichia coli* and Their impact on host immunity. Toxins. 2019;**11**:24

[35] Zhu Y, Luo Q, Davis SM, Westra C, Vickers TJ, et al. Molecular determinants of enterotoxigenic *Escherichia coli* heat-stable toxin secretion and delivery. Infection and Immunity. 2018;**86**:15

[36] Jobling MG, Holmes RK. Type II heat-labile enterotoxins from 50 diverse *Escherichia coli* isolates belong almost exclusively to the LT-IIc family and may be prophage encoded. PLoS One. 2012;**7**:e29898

[37] Ménard L-P, Dubreuil JD. Enteroaggregative *Escherichia coli* heat-stable enterotoxin 1 (EAST1): A new toxin with an old twist. Critical Reviews in Microbiology. 2002;**28**:43-60

[38] Turner SM, Scott-Tucker A, Cooper LM, Henderson IR. Weapons of mass destruction: Virulence factors of the global killer Enterotoxigenic *Escherichia coli*. FEMS Microbiology Letters. 2006;**263**:10-20

[39] Dubreuil JD. EAST1 toxin: An enigmatic molecule associated with sporadic episodes of diarrhea in humans and animals. Journal of Microbiology. 2019;**57**:541-549

[40] Zhang W, Berberov EM, Freeling J, He D, Moxley RA, et al. Significance of heat-stable and heat-labile enterotoxins in porcine colibacillosis in an additive model for pathogenicity studies. Infection and Immunity. 2006;**74**:3107-3114

[41] Gross RJ, Rowe B. *Escherichia coli* diarrhoea. The Journal of Hygiene. 1985;**95**:531-550

[42] Mendez Arancibia E, Pitart C, Ruiz J, Marco F, Gascon J, et al. Evolution of antimicrobial resistance in enteroaggregative *Escherichia coli* and enterotoxigenic *Escherichia coli* causing traveller's diarrhoea. The Journal of Antimicrobial Chemotherapy. 2009;**64**:343-347

[43] von Mentzer A, Blackwell GA, Pickard D, Boinett CJ, Joffré E, et al. Long-read-sequenced reference genomes of the seven major lineages of enterotoxigenic *Escherichia coli* (ETEC) circulating in modern time. Scientific Reports. 2021;**11**:9256

[44] Maynard C, Fairbrother JM, Bekal S, Sanschagrin F, Levesque RC, et al.

Antimicrobial resistance genes in enterotoxigenic *Escherichia coli* O149:K91 isolates obtained over a 23-year period from pigs. Antimicrobial Agents and Chemotherapy. 2003;**47**:3214-3221

[45] Steinsland H, Valentiner-Branth P, Aaby P, Mølbak K, Sommerfelt H. Clonal relatedness of enterotoxigenic *Escherichia coli* strains isolated from a cohort of young children in Guinea Bissau. Journal of Clinical Microbiology. 2004;**42**:3100-3107

[46] Steinsland H, Lacher DW, Sommerfelt H, Whittam TS. Ancestral lineages of human enterotoxigenic *Escherichia coli*. Journal of Clinical Microbiology. 2010;**48**:2916-2924

[47] Escobar-Páramo P, Clermont O, Blanc-Potard A-B, Bui H, Le Bouguénec C, et al. A specific genetic background is required for acquisition and expression of virulence factors in *Escherichia coli*. Molecular Biology and Evolution. 2004;**21**:1085-1094

[48] Turner SM, Chaudhuri RR, Jiang Z-D, DuPont H, Gyles C, et al. Phylogenetic comparisons reveal multiple acquisitions of the toxin genes by enterotoxigenic *Escherichia coli* strains of different evolutionary lineages. Journal of Clinical Microbiology. 2006;**44**:4528-4536

[49] von Mentzer A, Connor TR, Wieler LH, Semmler T, Iguchi A, et al. Identification of enterotoxigenic *Escherichia coli* (ETEC) clades with long-term global distribution. Nature Genetics. 2014;**46**:1321-1326

[50] Sahl JW, Sistrunk JR, Fraser CM, Hine E, Baby N, et al. Examination of the enterotoxigenic *Escherichia coli* population structure during human infection. mBio. 2015;**6**. Epub ahead of print July 2015. DOI: 10.1128/mBio.00501-15

[51] Sahl JW, Steinsland H, Redman JC, Angiuoli SV, Nataro JP, et al. A comparative genomic analysis of diverse clonal types of enterotoxigenic *Escherichia coli* reveals pathovar-specific conservation. Infection and Immunity. 2011;**79**:950-960

[52] Scotland SM, Smith HR, Willshaw GA, Rowe B. Vero cytotoxin production in strain of *Escherichia coli* is determined by genes carried on bacteriophage. The Lancet. 1983;**322**:216

[53] Delannoy S, Chaves BD, Ison SA, Webb HE, Beutin L, et al. Revisiting the STEC testing approach: Using espK and espV to make enterohemorrhagic *Escherichia coli* (EHEC) detection more reliable in beef. Frontiers in Microbiology. 2016;7. Epub ahead of print 22 January 2016. DOI: 10.3389/fmicb.2016.00001

[54] FAO, WHO. Attributing illness caused by Shiga toxin-producing *Escherichia coli* (STEC) to specific foods. In: Microbiological Risk Assessment. Rome, Italy: FAO; 2019. p. 74

[55] Blanco M, Blanco JE, Mora A, Rey J, Alonso JM, et al. Serotypes, virulence genes, and intimin types of shiga toxin (verotoxin)-producing *Escherichia coli* isolates from healthy sheep in Spain. Journal of Clinical Microbiology. 2003;**41**:1351-1356

[56] Hughes JM, Wilson ME, Johnson KE, Thorpe CM, Sears CL. The emerging clinical importance of non-O157 shiga toxin--producing *Escherichia coli*. Clinical Infectious Diseases. 2006;**43**:1587-1595

[57] Karama M, Mainga AO, Cenci-Goga BT, Malahlela M, El-Ashram S, et al. Molecular profiling and antimicrobial resistance of Shiga toxin-producing *Escherichia coli* O26, O45, O103, O121, O145 and O157 isolates from cattle on cow-calf operations in South Africa. Scientific Reports. 2019;**9**:11930

[58] Centers for Disease Control and Prevention (CDC). National Shiga

Toxin-producing *Escherichia coli* (STEC) Surveillance Overview. Atlanta, Georgia, US: US Department of Health and Human Services; 2012

[59] Kim J-S, Lee M-S, Kim JH. Recent updates on outbreaks of shiga toxin-producing *Escherichia coli* and its potential reservoirs. Frontiers in Cellular and Infection Microbiology. 2020;**10**. Epub ahead of print 2020. DOI: 10.3389/fcimb.2020.00273

[60] Persad AK, LeJeune JT. Animal reservoirs of shiga toxin-producing *Escherichia coli*. Microbiology Spectrum. 2014;**2**. Epub ahead of print 15 August 2014. DOI: 10.1128/microbiolspec.EHEC-0027-2014

[61] Chekabab SM, Paquin-Veillette J, Dozois CM, Harel J. The ecological habitat and transmission of *Escherichia coli* O157:H7. FEMS Microbiology Letters. 2013;**341**:1-12

[62] Wang LYR, Jokinen CC, Laing CR, Johnson RP, Ziebell K, et al. Multi-year persistence of verotoxigenic *Escherichia coli* (VTEC) in a closed Canadian beef herd: A cohort study. Frontiers in Microbiology. 2018;**9**. Epub ahead of print 31 August 2018. DOI: 10.3389/fmicb.2018.02040

[63] Zschock M, Hamann HP, Kloppert B, Wolter W. Shiga-toxin-producing *Escherichia coli* in faeces of healthy dairy cows, sheep and goats: Prevalence and virulence properties. Letters in Applied Microbiology. 2000;**31**:203-208

[64] Melton-Celsa AR. Shiga toxin (Stx) classification, structure, and function. Microbiology Spectrum. 2014;**2**. Epub ahead of print 15 August 2014. DOI: 10.1128/microbiolspec.EHEC-0024-2013

[65] Farfan MJ, Torres AG. Molecular mechanisms that mediate colonization of shiga toxin-producing *Escherichia coli* strains. Infection and Immunity. 2012;**80**:903-913

[66] Cho S-H, Oh K-H, Kim S-H, Oh H-B, Park M-S. Distribution of virulence genes and their association of serotypes in pathogenic *Escherichia coli* isolates from diarrheal patients in Korea. Osong Public Health and Research Perspectives. 2010;**1**:29-35

[67] Makino S-I, Tobe T, Asakura H, Watarai M, Ikeda T, et al. Distribution of the secondary type III secretion system locus found in enterohemorrhagic *Escherichia coli* O157:H7 isolates among shiga toxin-producing *E. coli* strains. Journal of Clinical Microbiology. 2003;**41**:2341-2347

[68] Campellone K. Tails of two Tirs: Actin pedestal formation by enteropathogenic E. coli and enterohemorrhagic *E. coli* O157:H7. Current Opinion in Microbiology. 2003;**6**:82-90

[69] Yang X, Sun H, Fan R, Fu S, Zhang J, et al. Genetic diversity of the intimin gene (eae) in non-O157 Shiga toxin-producing *Escherichia coli* strains in China. Scientific Reports. 2020;**10**:3275

[70] Bibbal D, Loukiadis E, Kérourédan M, Peytavin de Garam C, Ferré F, et al. Intimin gene (*eae*) subtype-based real-time PCR strategy for specific detection of shiga toxin-producing *Escherichia coli* serotypes O157:H7, O26:H11, O103:H2, O111:H8, and O145:H28 in cattle feces. Applied and Environmental Microbiology. 2014;**80**:1177-1184

[71] Fierz L, Cernela N, Hauser E, Nüesch-Inderbinen M, Stephan R. Characteristics of shigatoxin-producing *Escherichia coli* strains isolated during 2010-2014 from human infections in Switzerland. Frontiers in Microbiology. 2017;**8**:1471

[72] Carroll KJ, Harvey-Vince L, Jenkins C, Mohan K, Balasegaram S. The epidemiology of Shiga toxin-producing *Escherichia coli* infections in the South East of England: November 2013–March 2017 and significance for clinical and public health. Journal of Medical Microbiology. 2019;**68**:930-939

[73] Hua Y, Bai X, Zhang J, Jernberg C, Chromek M, et al. Molecular characteristics of *eae*-positive clinical Shiga toxin-producing *Escherichia coli* in Sweden. Emerging Microbes & Infections. 2020;**9**:2562-2570

[74] Balière C, Rincé A, Delannoy S, Fach P, Gourmelon M. Molecular profiling of shiga toxin-producing *Escherichia coli* and enteropathogenic *E. coli* strains isolated from french coastal environments. Applied and Environmental Microbiology. 2016;**82**:3913-3927

[75] Paton AW, Srimanote P, Woodrow MC, Paton JC. Characterization of saa, a novel autoagglutinating adhesin produced by locus of enterocyte effacement-negative shiga-toxigenic *Escherichia coli* strains that are virulent for humans. Infection and Immunity. 2001;**69**:6999-7009

[76] Batisson I, Guimond M-P, Girard F, An H, Zhu C, et al. Characterization of the novel factor Paa involved in the early steps of the adhesion mechanism of attaching and effacing *Escherichia coli*. Infection and Immunity. 2003;**71**:4516-4525

[77] Wells TJ, Sherlock O, Rivas L, Mahajan A, Beatson SA, et al. EhaA is a novel autotransporter protein of enterohemorrhagic *Escherichia coli* O157:H7 that contributes to adhesion and biofilm formation. Environmental Microbiology. 2008;**10**:589-604

[78] Doughty S, Sloan J, Bennett-Wood V, Robertson M, Robins-Browne RM, et al. Identification of a novel fimbrial gene cluster related to long polar fimbriae in locus of enterocyte effacement-negative strains of enterohemorrhagic *Escherichia coli*. Infection and Immunity. 2002;**70**:6761-6769

[79] Ju W, Shen J, Toro M, Zhao S, Meng J. Distribution of pathogenicity islands OI-122, OI-43/48, and OI-57 and a high-pathogenicity island in shiga toxin-producing *Escherichia coli*. Applied and Environmental Microbiology. 2013;**79**:3406-3412

[80] EFSA BIOHAZ Panel, Koutsoumanis K, Allende A, Alvarez-Ordóñez A, Bover-Cid S, et al. Pathogenicity assessment of Shiga toxin-producing *Escherichia coli* (STEC) and the public health risk posed by contamination of food with STEC. EFSA Journal. 2020;**18**. Epub ahead of print January 2020. DOI: 10.2903/j.efsa.2020.5967

[81] Fuller CA, Pellino CA, Flagler MJ, Strasser JE, Weiss AA. Shiga toxin subtypes display dramatic differences in potency. Infection and Immunity. 2011;**79**:1329-1337

[82] Pradel N, Livrelli V, De Champs C, Palcoux J, Reynaud A, Cheutz F, et al. Prevalence and characterization of shiga toxin-producing *Escherichia coli* isolated from cattle, food, and children during a one-year prospective study in France. Journal of Clinical Microbiology. 2000;**38**(3):1023-1031

[83] Meng Q, Bai X, Zhao A, Lan R, Du H, et al. Characterization of Shiga toxin-producing *Escherichia coli* isolated from healthy pigs in China. BMC Microbiology. 2014;**14**:5

[84] Brett KN, Ramachandran V, Hornitzky MA, Bettelheim KA, Walker MJ, et al. stx_{1c} is the most common shiga toxin 1 subtype among shiga toxin-producing *Escherichia coli* isolates from sheep but not among

isolates from Cattle. Journal of Clinical Microbiology. 2003;**41**:926-936

[85] Byrne L, Adams N, Jenkins C. Association between Shiga toxin–producing *Escherichia coli* O157:H7 *stx* gene subtype and disease severity, England, 2009-2019. Emerging Infectious Diseases. 2020;**26**:2394-2400

[86] Jiang C, An T, Wang S, Wang G, Si W, et al. Role of the ehxA gene from *Escherichia coli* serotype O82 in hemolysis, biofilm formation, and in vivo virulence. Canadian Journal of Microbiology. 2015;**61**:335-341

[87] Schmidt H, Beutin L, Karch H. Molecular analysis of the plasmid-encoded hemolysin of *Escherichia coli* O157:H7 strain EDL 933. Infection and Immunity. 1995;**63**:1055-1061

[88] Tatsuno I, Horie M, Abe H, Miki T, Makino K, et al. toxB gene on pO157 of enterohemorrhagic *Escherichia coli* O157:H7 is required for full epithelial cell adherence phenotype. Infection and Immunity. 2001;**69**:6660-6669

[89] Brunder W, Schmidt H, Karch H. KatP, a novel catalase-peroxidase encoded by the large plasmid of enterohaemorrhagic *Escherichia coli* O157:H7. Microbiology. 1996;**1996**(142):3305-3315

[90] Mir RA, Kudva IT. Antibiotic-resistant Shiga toxin-producing *Escherichia coli*: An overview of prevalence and intervention strategies. Zoonoses and Public Health. 2019;**66**:1-13

[91] Pan Y, Hu B, Bai X, Yang X, Cao L, et al. Antimicrobial resistance of non-O157 shiga toxin-producing *Escherichia coli* isolated from humans and domestic animals. Antibiotics. 2021;**10**:74

[92] Schroeder CM, Zhao C, DebRoy C, Torcolini J, Zhao S, et al. Antimicrobial resistance of *Escherichia coli* O157 isolated from humans, cattle, swine, and food. Applied and Environmental Microbiology. 2002;**68**:576-581

[93] Galarce N, Sánchez F, Fuenzalida V, Ramos R, Escobar B, et al. Phenotypic and genotypic antimicrobial resistance in non-O157 Shiga toxin-producing *Escherichia coli* isolated from cattle and swine in Chile. Frontiers in Veterinary Science. 2020;**7**:367

[94] Bai X, Zhang J, Hua Y, Jernberg C, Xiong Y, et al. Genomic insights into clinical Shiga toxin-producing *Escherichia coli* strains: A 15-year period survey in Jönköping, Sweden. Frontiers in Microbiology. 2021;**12**:627861

[95] Ogura Y, Ooka T, Iguchi A, Toh H, Asadulghani M, et al. Comparative genomics reveal the mechanism of the parallel evolution of O157 and non-O157 enterohemorrhagic *Escherichia coli*. Proceedings of the National Academy of Sciences of the United States of America. 2009. Epub ahead of print 5 October 2009. DOI: 10.1073/pnas.0903585106

[96] Coura FM, de Araújo DS, Mussi JMS, Silva MX, Lage AP, et al. Characterization of virulence factors and phylogenetic group determination of *Escherichia coli* isolated from diarrheic and non-diarrheic calves from Brazil. Folia Microbiologia (Praha). 2017;**62**: 139-144

[97] Dallman TJ, Ashton PM, Byrne L, Perry NT, Petrovska L, et al. Applying phylogenomics to understand the emergence of Shiga-toxin-producing *Escherichia coli* O157:H7 strains causing severe human disease in the UK. Microbial Genomics. 2015;**1**:e000029

[98] Ingle DJ, Gonçalves da Silva A, Valcanis M, Ballard SA, Seemann T, et al. Emergence and divergence of major lineages of Shiga-toxin-producing *Escherichia coli* in Australia. Microbial Genomics. 2019;**5**:e000268

[99] Yang Z, Kovar J, Kim J, Nietfeldt J, Smith DR, et al. Identification of common subpopulations of non-sorbitol-fermenting, β-glucuronidase-negative *Escherichia coli* O157:H7 from bovine production environments and human clinical samples. Applied and Environmental Microbiology. 2004;**70**:6846-6854

[100] Cowley LA, Dallman TJ, Jenkins C, Sheppard SK. Phage predation shapes the population structure of shiga-toxigenic *Escherichia coli* O157:H7 in the UK: An evolutionary perspective. Frontiers in Genetics. 2019;**0**. Epub ahead of print 2019. DOI: 10.3389/fgene.2019.00763

[101] Franz E, Rotariu O, Lopes BS, MacRae M, Bono JL, et al. Phylogeographic analysis reveals multiple international transmission events have driven the global emergence of *Escherichia coli* O157:H7. Clinical Infectious Diseases. 2019;**69**:428-437

[102] Valilis E, Ramsey A, Sidiq S, DuPont HL. Non-O157 Shiga toxin-producing *Escherichia coli*—A poorly appreciated enteric pathogen: Systematic review. International Journal of Infectious Diseases. 2018;**76**:82-87

[103] Blankenship HM, Mosci RE, Dietrich S, Burgess E, Wholehan J, et al. Population structure and genetic diversity of non-O157 Shiga toxin-producing *Escherichia coli* (STEC) clinical isolates from Michigan. Scientific Reports. 2021;**11**:4461

[104] Hazen TH, Donnenberg MS, Panchalingam S, Antonio M, Hossain A, et al. Genomic diversity of EPEC associated with clinical presentations of differing severity. Nature Microbiology. 2016;**1**:1-9

[105] Wardlaw T, Salama P, Brocklehurst C, Chopra M, Mason E. Diarrhoea: Why children are still dying and what can be done. The Lancet. 2010;**375**:870-872

[106] Goosney DL, Gruenheid S, Finlay BB. Gut feelings: Enteropathogenic *E. coli* (EPEC) interactions with the host. Annual Review of Cell and Developmental Biology. 2000;**16**:173-189

[107] Mare AD, Ciurea CN, Man A, Tudor B, Moldovan V, et al. Enteropathogenic *Escherichia coli*—A summary of the literature. Gastroenterology Insights. 2021;**12**:28-40

[108] Monaghan Á, Byrne B, Fanning S, Sweeney T, McDowell D, et al. Serotypes and virulence profiles of atypical enteropathogenic *Escherichia coli* (EPEC) isolated from bovine farms and abattoirs. Journal of Applied Microbiology. 2013;**114**:595-603

[109] Scaletsky ICA. Enteropathogenic *Escherichia coli*. London, The United Kingdom: IntechOpen. Epub ahead of print 18 September 2019. DOI: 10.5772/intechopen.82861

[110] Elliott SJ, Sperandio V, Girón JA, Shin S, Mellies JL, et al. The locus of enterocyte effacement (LEE)-encoded regulator controls expression of both LEE- and non-LEE-encoded virulence factors in enteropathogenic and enterohemorrhagic *Escherichia coli*. Infection and Immunity. 2000;**68**: 6115-6126

[111] Trabulsi LR, Keller R, Gomes TAT. Typical and atypical enteropathogenic *Escherichia coli*. Emerging Infectious Diseases. 2002;**8**:508-513. DOI: 10.321/eid0805

[112] Hu J, Torres AG. Enteropathogenic *Escherichia coli*: Foe or innocent bystander? | Elsevier Enhanced Reader. Clinical Microbiology and Infection. 2015;**21**:729-734

[113] Zhou Z, Li X, Liu B, Beutin L, Xu J, et al. Derivation of *Escherichia coli* O157:H7 from Its O55:H7 precursor. PLoS One. 2010;**5**:e8700

[114] Moura RA, Sircili MP, Leomil L, Matté MH, Trabulsi LR, et al. Clonal relationship among atypical enteropathogenic *Escherichia coli* strains isolated from different animal species and humans. Applied and Environmental Microbiology. 2009;**75**:7399-7408

[115] Donnenberg MS. Interactions between enteropathogenic *Escherichia coli* and epithelial cells. Clinical Infectious Diseases. 1999;**28**:451-455

[116] Khursigara C, Abul-Milh M, Lau B, Girón JA, Lingwood CA, et al. Enteropathogenic *Escherichia coli* virulence factor bundle-forming pilus has a binding specificity for phosphatidylethanolamine. Infection and Immunity. 2001;**69**:6573-6579

[117] Afset JE, Bruant G, Brousseau R, Harel J, Anderssen E, et al. Identification of virulence genes linked with diarrhea due to atypical enteropathogenic *Escherichia coli* by DNA microarray analysis and PCR. Journal of Clinical Microbiology. 2006;**44**:3703-3711

[118] Clarke SC, Haigh RD, Freestone PPE, Williams PH. Virulence of enteropathogenic *Escherichia coli*, a global pathogen. Clinical Microbiology Reviews. 2003;**16**:365-378

[119] Lara-Ochoa C, González-Lara F, Romero-González LE, Jaramillo-Rodríguez JB, Vázquez-Arellano SI, et al. The transcriptional activator of the bfp operon in EPEC (PerA) interacts with the RNA polymerase alpha subunit. Scientific Reports. 2021;**11**:8541

[120] Tobe T, Sasakawa C. Role of bundle-forming pilus of enteropathogenic *Escherichia coli* in host cell adherence and in microcolony development. Cellular Microbiology. 2001;**3**:579-585

[121] Morabito S, Tozzoli R, Oswald E, Caprioli A. A mosaic pathogenicity island made up of the locus of enterocyte effacement and a pathogenicity island of *Escherichia coli* O157:H7 is frequently present in attaching and effacing *E. coli*. Infection and Immunity. 2003;**71**:3343-3348

[122] Dulguer MV, Fabbricotti SH, Bando SY, Moreira-Filho CA, Fagundes-Neto U, et al. Atypical enteropathogenic *Escherichia coli* strains: Phenotypic and genetic profiling reveals a strong association between enteroaggregative *E. coli* heat-stable enterotoxin and diarrhea. The Journal of Infectious Diseases. 2003;**188**:1685-1694

[123] Silva LE, Souza TB, Silva NP, Scaletsky IC. Detection and genetic analysis of the enteroaggregative *Escherichia coli* heat-stable enterotoxin (EAST1) gene in clinical isolates of enteropathogenic *Escherichia coli* (EPEC) strains. BMC Microbiology. 2014;**14**:135

[124] DuPont HL. Persistent diarrhea: A clinical review. JAMA. 2016;**315**:2712

[125] Karami P, Bazmamoun H, Sedighi I, Mozaffari Nejad AS, Aslani MM, et al. Antibacterial resistance patterns of extended spectrum β-lactamase—producing enteropathogenic *Escherichia coli* strains isolated from children. Arab Journal of Gastroenterology. 2017;**18**:206-209

[126] Sirous M, Hashemzadeh M, Keshtvarz M, Amin M, Shams N, et al. Molecular characterization and antimicrobial resistance of enteropathogenic *Escherichia coli* in children from Ahvaz, Iran. Jundishapur Journal of Microbiology. 2020;**13**. Epub ahead of print 28 September 2020. DOI: 10.5812/jjm.100877

[127] Scaletsky IC, Souza TB, Aranda KR, Okeke IN. Genetic elements associated with antimicrobial resistance in enteropathogenic *Escherichia coli* (EPEC) from Brazil. BMC Microbiology. 2010;**10**:25

[128] Ingle DJ, Levine MM, Kotloff KL, Holt KE, Robins-Browne RM. Dynamics of antimicrobial resistance in intestinal *Escherichia coli* from children in community settings in South Asia and sub-Saharan Africa. Nature Microbiology. 2018;**3**:1063-1073

[129] Lacher DW, Steinsland H, Blank ET, Donnenberg MS, Whittman TS. Molecular evolution of typical enteropathogenic *Escherichia coli*: Clonal analysis by multilocus sequence typing and virulence gene allelic profiling. Journal of Bacteriology. 2007;**189**:342-350

[130] Hazen TH, Sahl JW, Fraser CM, Donnenberg MS, Scheutz F, et al. Refining the pathovar paradigm via phylogenomics of the attaching and effacing *Escherichia coli*. Proceedings of the National Academy of Sciences. 2013;**110**:12810-12815

[131] Beutin L, Orskov I, Orskov F, Zimmermann S, Prada J, et al. Clonal diversity and virulence factors in strains of *Escherichia coli* of the classic enteropathogenic serogroup 0114. The Journal of Infectious Diseases. 1990;**162**:1329-1334

[132] Hernandes RT, Hazen TH, dos Santos LF, Richter TKS, Michalski JM, et al. Comparative genomic analysis provides insight into the phylogeny and virulence of atypical enteropathogenic *Escherichia coli* strains from Brazil. PLoS Neglected Tropical Diseases. 2020;**14**:e0008373

[133] Ingle DJ, Tauschek M, Edwards DJ, Hocking DM, Pickard DJ, et al. Evolution of atypical enteropathogenic E. coli by repeated acquisition of LEE pathogenicity island variants. Nature Microbiology. 2016;**1**:1-9

[134] Okeke IN, Nataro JP. Enteroaggregative *Escherichia coli*. The Lancet. Infectious Diseases. 2001;**1**:10

[135] Sobieszczanska BM, Osek J, Wasko-Czopnik D, Dworniczek E, Jermakow K. Association of enteroaggregative *Escherichia coli* with irritable bowel syndrome | Elsevier Enhanced Reader. Clinical Microbiology and Infection. 2007;**13**:404-407

[136] Huang DB, Mohanty A, DuPont HL, Okhuysen PC, Chiang T. A review of an emerging enteric pathogen: Enteroaggregative *Escherichia coli*. Journal of Medical Microbiology. 2006;**2006**(55):1303-1311

[137] Estrada-Garcia T, Perez-Martinez I, Bernal-Reynaga R, Zaidi MB. Enteroaggregative *Escherichia coli*: A pathogen bridging the north and south. Current Tropical Medicine Reports. 2014;**1**(2):88-96. Epub ahead of print 25 March 2014. DOI: 10.1007/s40475-014-0018-7

[138] Kaur P, Chakraborti A, Asea A. Enteroaggregative *Escherichia coli*: An emerging enteric food borne pathogen. Interdisciplinary Perspectives on Infectious Diseases. 2010;**2010**:e254159

[139] Huppertz H-I, Rutkowski S, Aleksic S, Karch H. Acute and chronic diarrhoea and abdominal colic associated with enteroaggregative *Escherichia coli* in young children living in western Europe. The Lancet. 1997;**349**:1660-1662

[140] Jenkins C, Chart H, Willshaw GA, Cheasty T, Smith HR. Genotyping of enteroaggregative *Escherichia coli* and identification of target genes for the detection of both typical and atypical strains. Diagnostic Microbiology and Infectious Disease. 2006;**55**:13-19

[141] Kimata K, Lee K, Watahiki M, Isobe J, Ohnishi M, et al. Global distribution of epidemic-related Shiga toxin 2 encoding phages among enteroaggregative *Escherichia coli*. Scientific Reports. 2020;**10**:11738

[142] Boisen N, Melton-Celsa AR, Scheutz F, O'Brien AD, Nataro JP. Shiga

toxin 2a and Enteroaggregative *Escherichia coli*—A deadly combination. Gut Microbes. 2015;**6**:272-278

[143] Jonsson R, Struve C, Boisen N, Mateiu RV, Santiago AE, et al. Novel aggregative adherence fimbria variant of enteroaggregative *Escherichia coli*. Infection and Immunity. 2015;**83**:1396-1405

[144] Dias RCB, Tanabe RHS, Vieira MA, Cergole-Novella MC, dos Santos LF, et al. Analysis of the virulence profile and phenotypic features of typical and atypical enteroaggregative *Escherichia coli* (EAEC) isolated from diarrheal patients in Brazil. Frontiers in Cellular and Infection Microbiology. 2020;**0**. Epub ahead of print 2020. DOI: 10.3389/fcimb.2020.00144

[145] Henderson IR, Czeczulin J, Eslava C, Noriega F, Nataro JP. Characterization of pic, a secreted protease of Shigella flexneri and Enteroaggregative *Escherichia coli*. Infection and Immunity. 1999;**67**:5587-5596

[146] Boisen N, Østerlund MT, Joensen KG, Santiago AE, Mandomando I, et al. Redefining enteroaggregative *Escherichia coli* (EAEC): Genomic characterization of epidemiological EAEC strains. PLoS Neglected Tropical Diseases. 2020;**14**:e0008613

[147] Johnson T, Nolan LK. Pathogenomics of the virulence plasmids of *Escherichia coli*. Microbiology and Molecular Biology Reviews. 2009;**73**:750-774

[148] Mueller M, Grauschopf U, Maier T, Glockshuber R, Ban N. The structure of a cytolytic α-helical toxin pore reveals its assembly mechanism. Nature. 2009;**459**:726-730

[149] Andrade FB, Abreu AG, Nunes KO, Gomes TAT, Piazza RMF, et al. Distribution of serine protease autotransporters of Enterobacteriaceae in typical and atypical enteroaggregative *Escherichia coli*. Infection, Genetics and Evolution. 2017;**50**:83-86

[150] Guiral E, Gonçalves Quiles M, Muñoz L, Moreno-Morales J, Alejo-Cancho I, et al. Emergence of resistance to quinolones and β-lactam antibiotics in enteroaggregative and enterotoxigenic *Escherichia coli* causing traveler's diarrhea. Antimicrobial Agents and Chemotherapy. 2019;**63**. Epub ahead of print February 2019. DOI: 10.1128/AAC.01745-18

[151] Ochoa TJ, Ruiz J, Molina M, del Valle LJ, O MV, et al. High frequency of antimicrobial resistance of diarrheagenic *E. coli* in Peruvian infants. The American Journal of Tropical Medicine and Hygiene. 2009;**81**:296-301

[152] Khoshvaght H, Haghi F, Zeighami H. Extended spectrum betalactamase producing Enteroaggregative *Escherichia coli* from young children in Iran. Gastroenterology and Hepatology from Bed to Bench. 2014;**7**:131-136

[153] Riveros M, Riccobono E, Durand D, Mosquito S, Ruiz J, et al. Plasmid-mediated quinolone resistance genes in enteroaggregative *Escherichia coli* from infants in Lima, Peru. International Journal of Antimicrobial Agents. 2012;**39**:540-542

[154] Herrera-León S, Llorente MT, Sánchez S. Plasmid-mediated quinolone resistance in different diarrheagenic *Escherichia coli* pathotypes responsible for complicated, noncomplicated, and traveler's diarrhea cases. Antimicrobial Agents and Chemotherapy. 2016;**60**:1950-1951

[155] Do Nascimento V, Day MR, Doumith M, Hopkins KL, Woodford N, et al. Comparison of phenotypic and WGS-derived antimicrobial resistance profiles of enteroaggregative *Escherichia*

coli isolated from cases of diarrhoeal disease in England, 2015-16. The Journal of Antimicrobial Chemotherapy. 2017;**72**:3288-3297

[156] Czeczulin JR, Whittam TS, Henderson IR, Navarro-Garcia F, Nataro JP. Phylogenetic analysis of enteroaggregative and diffusely adherent *Escherichia coli*. Infection and Immunity. 1999;**67**:2692-2699

[157] Okeke IN, Wallace-Gadsden F, Simons HR, Matthews N, Labar AS, et al. Multi-locus sequence typing of enteroaggregative *Escherichia coli* isolates from Nigerian children uncovers multiple lineages. PLoS One. 2010;**5**:e14093

[158] Modgil V, Mahindroo J, Narayan C, Kalia M, Yousuf M, et al. Comparative analysis of virulence determinants, phylogroups, and antibiotic susceptibility patterns of typical versus atypical Enteroaggregative *E. coli* in India. PLOS Neglected Tropical Diseases. 2020;**14**:e0008769

[159] Pasqua M, Michelacci V, Di Martino ML, Tozzoli R, Grossi M, et al. The intriguing evolutionary journey of enteroinvasive *E. coli* (EIEC) toward pathogenicity. Frontiers in Microbiology. 2017;**8**. Epub ahead of print 5 December 2017. DOI: 10.3389/fmicb.2017.02390

[160] Herzig CTA. Notes from the field: Enteroinvasive *Escherichia coli* outbreak associated with a Potluck Party — North Carolina, June–July 2018. Morbidity and Mortality Weekly Report is a Weekly. 2019;**68**. Epub ahead of print 2019. DOI: 10.15585/mmwr.mm6807a5

[161] Ewing WH, Gravatti JL. Shigella types encountered in the mediterranean area 1. Journal of Bacteriology. 1947;**53**:191-195

[162] Rowe B, Gross RJ, Woodroof DP. Proposal to recognise seroovar 145/46 (synonyms: 147, *Shigella* 13, *Shigella sofia*, and *Shigella manolovii*) as a new *Escherichia coli* O group, O164. International Journal of Systematic Bacteriology. 1977;**27**(1):15-18

[163] Escher M, Scavia G, Morabito S, Tozzoli R, Maugliani A, et al. A severe foodborne outbreak of diarrhoea linked to a canteen in Italy caused by enteroinvasive *Escherichia coli*, an uncommon agent. Epidemiology and Infection. 2014;**142**:2559-2566

[164] Newitt S, MacGregor V, Robbins V, Bayliss L, Chattaway MA, et al. Two linked enteroinvasive *Escherichia coli* outbreaks, Nottingham, UK, June 2014. Emerging Infectious Diseases. 2016;**22**:1178-1184

[165] Fung CC, Octavia S, Mooney A-M, Lan R. Virulence variations in Shigella and enteroinvasive *Escherichia coli* using the *Caenorhabditis elegans* model. FEMS Microbiology Letters. 2015;**362**:1-5

[166] Nataro JP, Seriwatana J, Fasano A, Maneval DR, Guers LD, et al. Identification and cloning of a novel plasmid-encoded enterotoxin of enteroinvasive *Escherichia coli* and Shigella strains. Infection and Immunity. 1995;**63**:4721-4728

[167] Farajzadeh-Sheikh A, Savari M, Ahmadi K, Nave HH, Shahin M, et al. Distribution of genes encoding virulence factors and the genetic diversity of enteroinvasive *Escherichia coli* (EIEC) isolates from patients with diarrhea in Ahvaz, Iran. Infection and Drug Resistance. 2020;**13**:119-127

[168] Hosseini Nave H, Mansouri S, Taati Moghadam M, Moradi M. Virulence gene profile and multilocus variable-number tandem-repeat analysis (MLVA) of enteroinvasive *Escherichia coli* (EIEC) isolates from patients with diarrhea in Kerman, Iran. Jundishapur Journal of Microbiology. 2016;**9**:e33529

[169] Marbou WJT, Jain P, Samajpati S, Halder G, Mukhopadhyay AK, et al.

Profiling virulence and antimicrobial resistance markers of enterovirulent *Escherichia coli* from fecal isolates of adult patients with enteric infections in West Cameroon. Osong Public Health and Research Perspectives. 2020;**11**:216-230

[170] Hartman AB, Essiet II, Isenbarger DW, Lindler LE. Epidemiology of tetracycline resistance determinants in Shigella spp. and Enteroinvasive *Escherichia coli*: Characterization and dissemination of tet(A)-1. Journal of Clinical Microbiology. 2003;**41**:1023-1032

[171] Ahmed AM, Miyoshi S, Shinoda S, Shimamoto T. Molecular characterization of a multidrug-resistant strain of enteroinvasive *Escherichia coli* O164 isolated in Japan. Journal of Medical Microbiology. 2005;**2005**(54):273-278

[172] Escobar-Páramo P, Giudicelli C, Parsot C, Denamur E. The evolutionary history of Shigella and Enteroinvasive *Escherichia coli* revised. Journal of Molecular Evolution. 2003;**57**:140-148

[173] Lan R, Alles MC, Donohoe K, Martinez MB, Reeves PR. Molecular evolutionary relationships of enteroinvasive *Escherichia coli* and *Shigella* spp. Infection and Immunity. 2004;**72**:5080-5088

[174] Pettengill EA, Pettengill JB, Binet R. Phylogenetic analyses of Shigella and Enteroinvasive *Escherichia coli* for the identification of molecular epidemiological markers: Whole-genome comparative analysis does not support distinct genera designation. Frontiers in Microbiology. 2016;**0**. Epub ahead of print 2016. DOI: 10.3389/fmicb.2015.01573

[175] Hazen TH, Leonard SR, Lampel KA, Lacher DW, Maurelli AT, et al. Investigating the relatedness of enteroinvasive *Escherichia coli* to other E. coli and Shigella isolates by using comparative genomics. Infection and Immunity. 2016;**84**:2362-2371

[176] Hawkey J, Monk JM, Billman-Jacobe H, Palsson B, Holt KE. Impact of insertion sequences on convergent evolution of Shigella species. PLoS Genetics. 2020;**16**:e1008931

[177] Servin AL. Pathogenesis of human diffusely adhering *Escherichia coli* expressing Afa/Dr adhesins (Afa/Dr DAEC): Current insights and future challenges. Clinical Microbiology Reviews. 2014;**27**:823-869

[178] Scaletsky ICA, Fabbricotti SH, Carvalho RLB, Nunes CR, Maranhão HS, et al. Diffusely adherent *Escherichia coli* as a cause of acute diarrhea in young children in Northeast Brazil: A case-control study. Journal of Clinical Microbiology. 2002;**40**:645-648

[179] Lopes LM, Fabbricotti SH, Ferreira AJP, Kato MAMF, Michalski J, et al. Heterogeneity among strains of diffusely adherent *Escherichia coli* isolated in Brazil. Journal of Clinical Microbiology. 2005;**43**:1968-1972

[180] Ochoa TJ, Rivera FP, Bernal M, Meza R, Ecker L, et al. Detection of the CS20 colonization factor antigen in diffuse-adhering *Escherichia coli* strains. FEMS Immunology and Medical Microbiology. 2010;**60**:186-189

[181] Le Bouguénec C, Servin AL. Diffusely adherent *Escherichia coli* strains expressing Afa/Dr adhesins (Afa/Dr DAEC): Hitherto unrecognized pathogens. FEMS Microbiology Letters. 2006;**256**:185-194

[182] Javadi K, Mohebi S, Motamedifar M, Hadi N. Characterization and antibiotic resistance pattern of diffusely adherent *Escherichia coli* (DAEC), isolated from paediatric diarrhoea in Shiraz, southern Iran. New Microbes and New Infections. 2020;**38**:100780

[183] Mansan-Almeida R, Pereira AL, Giugliano LG. Diffusely adherent *Escherichia coli* strains isolated from children and adults constitute two different populations. BMC Microbiology. 2013;**13**:1-14

[184] Taddei CR, Moreno ACR, Fernandes Filho A, Montemor LPG, Martinez MB. Prevalence of secreted autotransporter toxin gene among diffusely adhering *Escherichia coli* isolated from stools of children. FEMS Microbiology Letters. 2003;**227**: 249-253

[185] Meza-Segura M, Estrada-Garcia T. Diffusely Adherent *Escherichia coli*. In: Torres AG (editor). *Escherichia coli in the Americas*. Cham: Springer International Publishing. pp. 125-147

[186] Li D, Shen M, Xu Y, Liu C, Wang W, et al. Virulence gene profiles and molecular genetic characteristics of diarrheagenic *Escherichia coli* from a hospital in western China. Gut Pathogen. 2018;**10**:35

[187] Spano LC, da Cunha KF, Monfardini MV, de Cássia Bergamaschi Fonseca R, Scaletsky ICA. High prevalence of diarrheagenic *Escherichia coli* carrying toxin-encoding genes isolated from children and adults in southeastern Brazil. BMC Infectious Diseases. 2017;**17**:773

[188] Meza-Segura M, Zaidi MB, Vera-Ponce de León A, Moran-Garcia N, Martinez-Romero E, et al. New insights into DAEC and EAEC pathogenesis and phylogeny. Frontiers in Cellular and Infection Microbiology. 2020;**10**. Epub ahead of print 2020. DOI: 10.3389/fcimb.2020.572951

[189] Denamur E, Clermont O, Bonacorsi S, Gordon D. The population genetics of pathogenic *Escherichia coli*. Nature Reviews. Microbiology. 2021;**19**:37-54

[190] Mosquito S, Pons MJ, Riveros M, Ruiz J, Ochoa TJ. Diarrheagenic *Escherichia coli* phylogroups are associated with antibiotic resistance and duration of diarrheal episode. Scientific World Journal. 2015;**2015**:e610403

Chapter 6

Escherichia coli O157:H7 and Its Effect on Human Health

Mojtaba Mohseni, Benyamin Djawadi and Noushin Khazaei

Abstract

Escherichia coli (*E. coli*) has many serotypes. The O157:H7 *E. coli* serotype is the most prominent serotype of enterohemorrhagic *E. coli*. It produces the Shiga toxin, which is one of the most important virulent factors discovered till today and has different subtypes with different antigenic and molecular traits. Consumption of contaminated water, milk or even eating an uncooked raw meat can cause bloody diarrhea that can end up in a life-threatening disease, such as hemolytic uremic syndrome (HUS). This is a condition that affects endothelial cells in the blood vessels and leads to thrombocytopenic purpura (TTP) that can cause blood clots formation in small blood vessels. The *E. coli* O157:H7 can be isolated from patient's stool and be identified by serological tests such as enzyme-linked immunosorbent assay (ELISA) and immunoblotting methods. This special strain of *E. coli* can be used as a biological weapon, as it can be so dangerous and has the ability to spread easily form person to person.

Keywords: *E. coli* O157:H7, Stx, STEC, HUS, foodborne illness

1. Introduction

For the first time in the 1970s early work showed that special *E. coli* strains could produce a certain type of a toxin called verotoxin and was named after that because of its's effect on Vero cells they can be classified according to virulence factors such as toxins into different groups [1].

These toxins were also called Shiga toxins (Stx) because of their relation to the toxin produced by *Shigella dysenteriae* type 1. The group of *E. coli* strains producing these toxins is referred to as Shiga-toxin producing *E. coli* (STEC), or verocytotoxin producing *E. coli* (VTEC). The genes encoding Stx can be often carried by bacteriophages and plasmids [2].

E. coli O157:H7 is the most common strain of STEC, but there are many other strains of STEC as well. Anyone can get STEC infection and the whole processes begins when a person eat or drink any contaminated product, particularly raw or undercooked meat. The term enterohemorrhagic *E. coli* (EHEC) is used to designate a subset of STEC that cause severe diseases in humans, including hemorrhagic colitis (HC) (bloody diarrhea) and the hemolytic uremic syndrome (HUS).

2. History

Since the beginning of August 1982 specimens obtained from four patients located in the United States of America, who were suffering from an unusual bloody diarrheal illness started suddenly with abdominal pain within 24 hours followed by watery diarrhea, led to identification of *Escherichia coli* serotype O157:H7. All patients recovered within 7 days.

By further examination of stool samples from different cases of this type of diarrheal illness which nowadays designated as "hemorrhagic colitis," for the first time CDC associated the 1993 large outbreak with undercooked hamburgers served at fast-food chain restaurants in Oregon and Michigan. Hemorrhagic colitis is characterized by severe abdominal pain, grossly bloody diarrhea, and even fever [3].

E. coli O157:H7 is the most commonly identified member of STEC and is becoming as a best-known emerging pathogen in the United States causing foodborne diseases [4, 5].

However, experiences have established a diversity of sources for *E. coli* O157:H7, including apple juice and cider, vegetables such as lettuce, raw milk, and processed foods such as salami [6].

3. Virulence factor

As already we know, STEC is a zoonotic pathogen that is responsible for severe outbreaks worldwide. The main virulence factor of STEC is the production of Shiga toxins 1 and 2. There are additional factors like plasmid-encoded enterohemolysin (EhxA), an autoagglutinating adhesin (Saa), a catalase-peroxidase (KatP), an extracellular serine protease (EspP) that can damage the intestinal tissue or even some factors related to the adhesion to bovine colon like intimin which can induce a characteristic histopathological lesion defined as "attaching and effacing" (A/E) [7–9].

Shiga toxins are encoded by *stx1* and *stx2* genes which are carried by lysogenic phages. They belong to the family of AB5 protein, contains active A subunit and 5 B subunits responsible for binding to cellular receptor available in organs as kidney, brain, liver, and pancreas [10].

These toxins that are produced in colon besides causing local damage can travel via bloodstream to its's target organs such as kidney and play an important role in causing HC and HUS (**Figure 1**) [11].

The damaged caused by toxins is because of inhibition of protein synthesis which leads to apoptosis of endothelial cells [12].

Stx-phages are highly mobile genetic elements which can be transferred through horizontal gene transfer to other *Enterobacteriaceae* [13].

The expression of these genes (especially *stx2*) is affected by environmental conditions such as stress and temperature [14, 15].

The whole cluster of other virulence factors is encoded by chromosomal region called the locus of enterocyte effacement (LEE) presents in many STEC strains, which are responsible for the attaching and causing lesions. In a large proportion of STEC, a plasmid encoding several putative virulence factors like hemolysin can also be found [16].

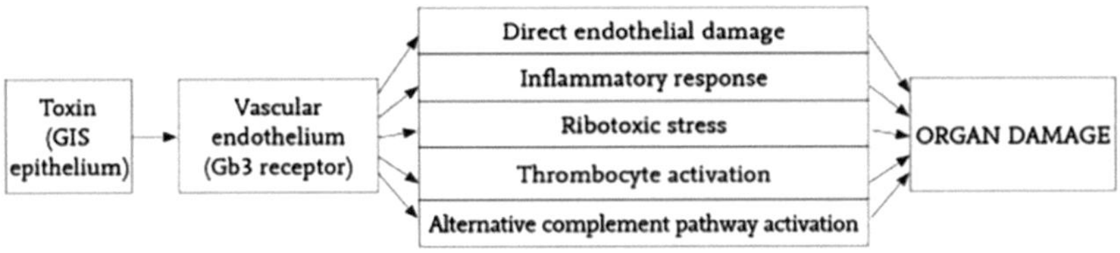

Figure 1.
The effects of Stx in STEC-HUS caused by enterohemorrhagic E. coli. *GIS, gastrointestinal system; Gb3, globotriaosylceramide; Stx, Shiga-toxin.*

4. Infection source and colonization management

E. coli O157:H7 can cycle through the environment and food chain via water, soil, and insect. But cattle and *Salmonella enterica* serovar Typhimurium are considered to be the main reservoir for STEC [17–19]. In the United States, between 1998 and 2005, a majority of the STEC outbreaks were related to contaminated food and occurred in the period from May to October [20].

Considerable effort has been done to inhibit or facilitate infection of animals with STEC O157:H7, because of the readily transmission of pathogen strains such as EHEC in the farm environment and animals can even represent as vectors [21, 22].

However, illnesses caused by contaminated meat product still occur. But great effort has recently been placed on developing new strategies to control the widespread of distribution of EHEC serotypes, O157 and even non-O157 in cattle population to maintain their healthy condition and finally to decrease such illnesses in human [23].

Another practice for controlling is, by the use of beneficial bacteria often referred to as probiotics. Probiotics can interfere with pathogenic strains by producing metabolites that are inhibitory to STEC O157:H [24].

Some strains of *E. coli* strains can produce colicins that are inhibitory to STEC O157:H7 [25].

5. Detection of *E. coli* O157:H7

5.1 Cultured-based detection

O157 STEC can usually be differentiated from most commensal *E. coli* by their ability to ferment sorbitol when plated on a sorbitol-containing agar.

For isolation this strain, samples are plated onto a selective and differential media, such as sorbitol-MacConkey agar (SMAC), cefixime tellurite-sorbitol MacConkey agar (CT-SMAC), CHROMagar O157, or Rainbow agar. After incubation for 16–24 hours at 37°C, the plate is being examined for possible O157 colonies, which are colorless on SMAC or CT-SMAC and are mauve or pink on CHROMagar O157 [26].

Non-motile flagella-less (H-) sorbitol-fermenting STEC O157 might not grow on CT-SMAC agar because of their susceptibility to tellurite [27].

In the laboratory, culture and biochemical analysis is considered as the "gold standard" for the identification of STEC. Selective media, such as SMAC and CT-SMACK can be used to identified O157:H7 STEC because of the ability to ferment sorbitol within 24 hours [28].

Guidance for public health laboratories on the isolation and characterization of Shiga toxin-producing *Escherichia coli* (STEC) from clinical specimens are given in **Table 1** [29].

5.2 Nucleic-acid-based detection

Recently, PCR methods (like real-time PCR and conventional PCR) have been developed to test the samples for the presence of Stx genes [30].

This method is inexpensive and easy to perform. During the protocol multiple primer sets in a single PCR reaction in order to detect different types and subtypes of Stx genes in a certain sample. But we have to keep in mind, the detection of Stx mRNA is not possible because they have not been expressed yet and there's a possibility of having a false negative test. Also analysts must be aware of the presence of cryptic bacteriophages which are prophages that have become trapped within a bacterial genome [31].

Medium	Characteristics	Properties	Morphology
Cefixime-tellurite sorbitol MacConkey agar (CT-SMAC)	Selective and differential distinguishes O157 from other fecal *E. coli*	Inhibits *Proteus mirabilis*, non-O157 STEC, and other sorbitol non-fermenting strains	O157 STEC appear clear, non-O157 STEC appear pink, and other normal enteric bacteria appear pink
CHROMagar™ O157	Selective and differential distinguishes O157 from other fecal *E. coli*	The chromogen mixture consists of artificial substrates	O157 STEC appear mauve, non-O157 STEC appear steel blue or blue green, and other enteric bacteria appear colorless
Rainbow® agar	Selective and differential distinguishes O157 from other fecal *E. coli*	Tellurite and novobiocin reduce the number of bacteria other than *E. coli* O157:H7 that will grow	O157 STEC appear black/gray, non-O157 STEC appear purple or violet, and other enteric bacteria appear pink
Sorbitol MacConkey agar (SMAC)	Modified MacConkey agar distinguishes O157 from other fecal *E. coli*	Primary carbon source sorbitol supports growth of non-O157 STEC	O157 STEC appear clear, non-O157 STEC appear pink, and other enteric bacteria appear pink

Table 1.
Guidance for public health laboratories on the isolation and characterization of Shiga toxin-producing Escherichia coli *(STEC) from clinical specimens [29].*

5.3 Detection by monoclonal antibodies

Detection of Shiga-toxin in clinical samples has been approved by the FDA [27].

These kits can detect Shiga-toxin in the enrichment samples, although none of them can distinguish the seven subtypes of Stx2 or the three subtypes of Stx.

In 2015, researchers designed sandwich ELISAs capable of detecting and distinguishing between stx2 subtypes a, c, and d [32].

These antibodies provide a significant way to test the samples as fast as possible, even including samples from beef and pure culture [33].

5.4 O- and H-antigen determination

The most common method used in clinical laboratories when samples suspected to O157:H7 are being tested, is the O-antigen determination which is run by latex agglutination. These latex particles are coated with antibodies against the O157 antigen and when they are mixed with bacterial growth, O157 STEC bacteria will bind to the latex particle to produce visible agglutination which means positive reaction [34].

H7-specific antisera for latex agglutination are available for O157 but unlike the previous method, detection of flagellar antigens may be difficult and usually it is being done for non-O157 outbreaks [34].

6. Epidemiology and outbreaks

STEC infection causes a wide spectrum of illnesses, such as non-bloody diarrhea, hemolytic uremic syndrome (HUS), and hemorrhagic colitis (HC) [35].

Many non-O157:H7 STEC strains may also cause HUS but the majority of diarrhea-associated HUS cases in the US are caused by infection with O157:H7 STEC [36].

STEC are found in the intestines of healthy animals and are easily transmitted to humans by consumption of contaminated food or water, or even through direct contact with infected animals or persons [37].

Undercooked beef especially ground beef plays an important role in many O157:H7 STEC outbreaks, although other foods including unpasteurized juice, raw milk, and raw produces (e.g., lettuce) have been implicated in outbreaks too [38–40].

For the years 1998 and 1999 data about in implicated vehicles in outbreaks of *E. coli* O157:H7 exist and are given in **Table 2**.

CDC officials in several states, and the U.S. Food and Drug Administration (FDA) have collected data to investigate a multistate outbreak of *E. coli* infections linked to cake mix. As of July 27, 2021, 16 people infected with the outbreak strain of *E. coli* O121 have been reported from 12 states. Illnesses started on dates ranging from February 26, 2021 to June 21, 2021 and this outbreak is over right now [41].

In December 22, 2020 the FDA and CDC investigated a multistate outbreak of *E. coli* O157:H7 infection linked to leafy greens, a total of 40 people infected with the outbreak strain of *E. coli* O157:H7 were reported from 19 states. Illness started on dates ranging from August 10, 2020 to October 31, 2020. Ill people ranged in age from 1 to 85 years [42].

Outbreaks of *E. coli* infection linked to leafy greens (which often eaten raw with no cooking) including various types of lettuce such as romaine or iceberg lettuce, spinach, and mesclun mix in Canada and United States, are known as critical issues since 2008 [43].

Of the 57 *E. coli* infection outbreaks identified, 48 were attributed to *E. coli* O157 and the most of the causative agents (45 of the 48 outbreaks) were identified as *E. coli* O157:H7 and the other nine outbreaks were attributed to non-O157 *E. coli* [44].

Vehicle	1998	1999	Total
Ground beef/hamburger	10	9	19
Roast beef	0	2	2
Lettuce	1	3	4
Coleslaw	2	1	3
Salad	1	1	2
Milk	2	0	2
Tacos	0	1	1
Apple cider	0	1	1
Game meat	0	1	1
Cake	1	0	1
Cheese curd	1	0	1
Fruit salad	1	0	1
Macaroni salad	1	0	1
Multiple	1	0	1
Unknown	0	2	2
Total	21	21	42

Sources: ([38]; [39]).

Table 2.
Food vehicles implicated in outbreaks of E. coli *O157:H7, United States, 1998–1999.*

In the United States since 2008, 37 outbreaks of *E. coli* O157:H7 infection linked to leafy greens were identified: 5 linked to iceberg (13.5%), 11 linked to romaine lettuce (29.7%), and 21 linked to other or unspecified types of leafy greens (56.8%). These 37 outbreaks resulted in 1070 illness cases: 491 linked romaine lettuce (45.9%), 144 linked to iceberg lettuce (13.5%), and 435 linked to other or unspecified types of leafy greens (40.7%) [45].

Information on which month the outbreaks occurred is available for 17 of the 18 outbreaks linked to romaine lettuce in Canada and the United States from 2008 to 2018. The majority of these outbreaks happened during two seasons: eight occurred in the spring (March to June) and eight occurred in the fall (September to December) [46].

A possible theory for the distribution of *E. coli* O157:H7 illness cases observed in Canada could be related to the commercial distribution of lettuce. Lettuce imported from U. S. lettuce-growing regions can travel long distances to reach Canada and even distances farther t the eastern part of the country [47].

In the spring 2018 outbreaks in U.S., trace back investigation identified 36 growing fields on 23 farms in the Yuma, AZ, growing region as potential sources of contaminated lettuce. Growers reported the following common elements: romaine lettuce was grown under conventional agricultural practices; Colorado River water via an open irrigation canal was used to irrigate the romaine lettuce and to dilute agricultural chemicals; and overhead sprinkler irrigation was used during the germination of romaine lettuce followed in most fields by furrow irrigation [48].

In November 5, 2019 to November 16, 2019, CDC and FDA investigated a multistate outbreak of Shiga toxin-producing *E. coli* O157:H7 infection started in Georgia, Illinois, Minnesota, North Dakota, and Wisconsin. A total of 10 people ranged from 21 to 91, with a median age of 33, infected with the outbreak strain of *E. coli* O157:H7 were reported and 60% were female. Four of 10 ill people were hospitalized, including one person who developed hemolytic uremic syndrome. No deaths were reported. Information collected during the investigation showed that Fresh Express Sunflower Crisp chopped salad kits were the likely source of this outbreak. Of the 10 ill people with information available, all 10 reported eating any leafy greens in the week before their illness started. Eight ill people specifically reported eating a Fresh Express Sunflower Crisp chopped salad kit.

As of January 15, 2020, this outbreak appears to be over [49].

In December 2020, CDC, U.S. Food and Drug Administration, and public health regulatory officials reported an outbreak of *E. coli* O157:H7 in several states.

Public health investigators used the national PulseNet system to identify illness that may have been included in this outbreak. PulseNet system is the subtyping network of public health and food regulatory agency laboratories conducted by CDC, and with the help of the whole genome sequencing (WGS) method, analyzing the DNA fingerprinting is being done by official investigators. Molecular Investigations showed that people in this outbreak were more likely to share a common source of infection. As of December 16, 2020, a total of 32 people infected with the outbreak strain of *E. coli* O15:H7 were reported from 12 states. Illnesses started on dates ranging from June 6, 2020 to October 25, 2020. Ill people ranged in age from 2 to 75 years, with a median age of 27 years, and 72% were female. Of 29 ill people with information available, 15 were hospitalized and one developed hemolytic uremic syndrome (HUS). One death was reported from Michigan.

The officials interviewed ill people to determine what they ate, they reported variety of food items. Several ill people also reported eating at the same restaurant with common foods. As of December 18, 2020, this outbreak is over and that ended unknown, before enough information was available for investigators [50].

In November, 2021 CDC and FDA collected data to investigate a multistate outbreak of *E. coli* O157:H7 infections. Epidemiologic and laboratory data show that Josie's Organics prepackaged baby spinach may be contaminated with *E. coli* and may be making people sick. As of the November 15, a total of 10 people infected with the outbreak strain of *E. coli* O157:H7 have been reported from seven states. Illnesses started on dates ranging from October 15 to October 27, 2021. Sick people range in age from 2 to 71 years, with a median age of 26, and 70% are female. Of eight people with information available, two have been hospitalized and no deaths have been reported.

Public health investigators are using the PulseNet system and WGS method to identify illnesses that could be part of the outbreak, which showed that bacteria from sick people samples are closely related genetically, that suggests that people in the outbreak got sick from the same food. This outbreak is not over yet and CDC is advising people not to eat, sell or serve Josie's Organics prepackaged with best by date of October 23, 2021 [51].

On December 11, 2017 the Public Health Agency of Canada (PHAC) announced an outbreak of 21 STEC O157:H7 infections in three provinces linked to romaine lettuce.

This outbreak appears to be over as of January 25, 2018, and 25 people infected with the outbreak strain of STEC O157:H7 were reported [52].

On January 4, 2017 to April 18, 2018 *E. coli* O157:H7 infections linked to I.M. Healthy Brand SoyNut Butter was started and there were 32 people infected in Arizona, California, Florida, Illinois, Oregon, and Virginia and 9 people developed HUS. The source of the infection was a nut-free substitute for peanut butter and this outbreak seems to be over [53].

On June 27, 2016 to September 10, 2016, there were 11 reports based on *E. coli* O157:H7 infections in Connecticut, West Virginia, Pennsylvania, and Massachusetts. Epidemiologic and laboratory evidence indicated that beef products produced by Adams Farm Slaughterhouse in Athol, Massachusetts, were likely source of this outbreak and one ill person developed HUS. On October 19, 2016 officials declared this outbreak over [54].

On October 6, 2015 to November 3, 2015, there were a total of 19 people infected with the outbreak strain of Shiga toxin-producing STEC O157:H7 in Missouri, Colorado, Utah, Virginia, Washington, and Montana, ranged in age from 5 to 84 years, with a median age of 18. Two ill people developed HUS. The epidemiologic evidence collected during this investigation suggested that rotisserie chicken salad made and sold in Costco stores was the likely source of this outbreak. This outbreak seems to be over reported to CDC [55].

On May 19, 2014, a total number of 12 persons were infected with the outbreak strains of Shiga toxin-producing *E. coli* O157:H7 were reported from Michigan, Missouri, Ohio, and Massachusetts. No ill people developed HUS. Federal officials indicated that contaminated ground beef produced by Wolverine Packing Company was the likely source of this outbreak of STEC O157:H7 infections. On June 20, 2014 this outbreak was over reported by CDC [56].

A total of 33 ill persons infected with the outbreak strain of STEC O157:H7 were reported from Arizona, California, Texas, and Washington, on November 10, 2013. Federal officials indicated that consumption of two ready-to-eat salads, Field Fresh Chopped Salad with Grilled Chicken and Mexicali Salad with Chili Lime Chicken, produced by Glass Onion Catering and sold at Trader Joe's grocery store locations, was the source of this outbreak of STEC O157:H7 infections and even two ill people developed HUS. This outbreak appears to be over [57].

From October 18, 2012 to November 12, 2012, a total of 33 people infected with the outbreak strain of STEC O157:H7 were reported from Connecticut, Massachusetts, New

York, Pennsylvania, and Virginia and two ill persons developed HUS. Epidemiologic investigation conducted by officials in local linked this outbreak to prepackaged leafy greens, produced by State Garden of Chelsea, Massachusetts. Testing conducted by the New York Department Health Wadsworth Center Laboratories isolated the outbreak strain of STEC O157:H7 from four leftover packages of Wegmans brand Organic Spinach and Spring Mix blend collected from four ill person's homes [58].

As of March 22, 2011, 14 persons infected with the outbreak strain of *E. coli* serotype O157:H7 have been reported from Maryland (three cases), New Jersey (two cases), North Carolina (one case), Ohio (two cases) and Pennsylvania (six cases) and none have reported HUS. Reported dates of illness onset range from January 10, 2011 to February 15, 2011. Collaborative investigative efforts of local, state, and federal officials have associated this outbreak with eating Lebanon bologna, produced by Palmyra Bologna Company, which is a fermented, semi-dry sausage. In an epidemiologic study conducted during March 15–18, a total of 13 ill persons reported about the common disease [59].

From October 16, 2010 through October 27, 2010, 38 persons infected with the outbreak strain of *E. coli* O157:H7 have been reported from New Mexico (3 cases), Arizona (19 cases), California (3 cases), Colorado (11 cases), and Nevada (2 cases) with one case of HUS. The officials linked this outbreak to Bravo Farms Dutch Style Gouda Cheese purchased from a Costco stores [60].

As of Tuesday, June 30, 2009, 72 persons infected with *E. coli* O157:H7 with a particular DNA fingerprint have been reported. The number of ill persons in each state is as follows: Arizona (2 cases), California (3 cases), Georgia (1 cases), Illinois (5 cases), Kentucky (2 cases), Montana (1 cases), Maine (3 cases), Maryland (2 cases), Nevada (2 cases), Ohio (3 cases), Oklahoma (1 cases), Utah (4 cases), Texas (3 cases), New Jersey (1 case), and Wisconsin (1 case). Ill persons range in age from 2 to 65 years, Thirty-four persons have been hospitalized and 10 people have developed HUS. The epidemiological study indicated a strong association with eating raw prepackaged cookie dough. Most patients reported eating refrigerated prepackaged Nestle Toll House cookie dough products raw. This outbreak is over [61].

As of July 17, 2008, 49 confirmed cases have been linked both epidemiologically and molecular fingerprinting to *E. coli* O157:H7. The number of cases in each state is as follows: Georgia (4 cases), Indiana (1 case), Kentucky (1 case), Michigan (20 cases), Ohio (21 cases), and Utah (1 case). Their illnesses began between May 27 and July 1, 2008. Twenty-seven persons have been hospitalized and one person developed HUS. The officials reported the ground beef sold at Kroger Co was the main source of the infection. This outbreak is over [62].

On December 14, 2006, 71 persons with illness associated with the Taco Bell restaurant outbreak have been reported to CDC from five states: New Jersey (33 cases), New York (22 cases), Pennsylvania (13 cases), Delaware (2 cases), and South Carolina (1 case). A total of 71 ill persons, 53 were hospitalized and 8 developed HUS. This outbreak has ended [63].

7. Hemolytic uremic syndrome

Hemolytic uremic syndrome (HUS) is complication of Shiga-toxin producing *E. coli* (STEC) infection. Prompted by the finding of an STEC in the stool of a patient who died from HUS Karmali et al. examined the samples of different cases of HUS and found evidence of STEC infection in 11–15 patients [64].

HUS actually develops 1 week or more after diarrhea begins. Due to the use of the immunomagnetic separation (IMS) the isolation of O157:H7 in the stools from patients with HUS has been increased dramatically [65].

You can see the timeframe of the development and evolution of STEC-HUS (**Figure 2**).

The pathogenesis is related to the endothelial cell damage caused by Shiga-toxin which is produced in the gastrointestinal tract. These damaged cells become separated from the basement membranes of the glomeruli and in blood vessels to other organs. Clinically HUS from *E. coli* O157:H7 first appears similar to other common severe gastroenteritis, stools may be bloody, fever is absent, and thrombocytopenia as anemia is a hallmark. Edema is common in later stages and also blood loss via gastrointestinal tract and small blood vessels due to active hemolysis is common too. The *E. coli* O157:H7 can easily be cultured from feces. For more information please see (**Table 3**) which shows the distribution of children and adults in studies of STEC-HUS [67].

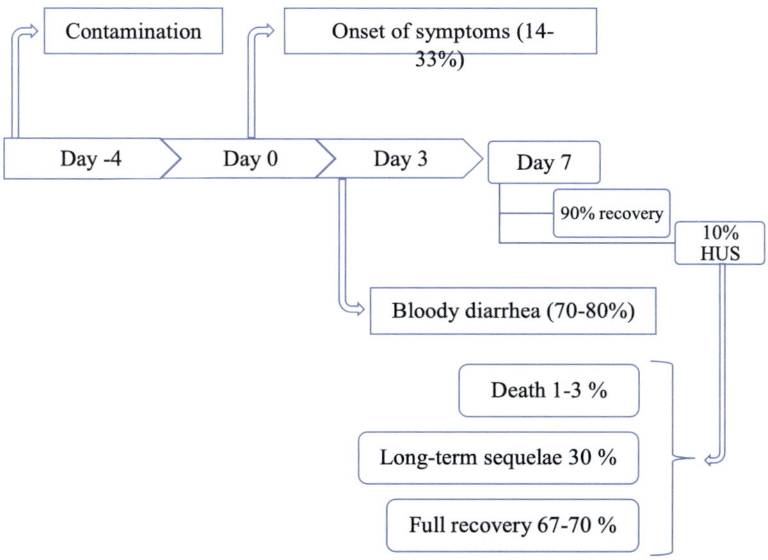

Figure 2.
Timeframe of the development and evolution of STEC-HUS. The timeframe and proportion represented are based on median values and are highly variable, depending on strain, epidemiological, and individual patient characteristics [66].

Year or time period and geographical region	STEC-HUS cases (children)	STEC-HUS cases (adult)
1979–1983 Washington and Baltimore	20	0
2000–2006 USA (8 states)	190	28
1992–2012 Norway	24	1
2017 Switzerland	4	3
1989–2006 Oklahoma	—	21
2009–2016 Alberta, Canada	33	6
2009–2013 England	66	20
2009–2017 France	—	96
2014 USA (10 states)	42	0

Table 3.
Distribution of children and adults in STEC-HUS cases [67].

Year	HUS cases	Incidence
2003	8	5.00
2004	3	1.88
2005	8	5.00
2006	11	6.88
2007	14	8.75
2008	7	4.38
2009	10	6.25
2010	15	9.38
2011	6	3.75
2012	19	11.88
Total	101	6.31

Table 4.
Annual HUS in U.S. incidence per 1,000,000 children [67].

HUS primarily affects the kidneys, however it can also lead to sepsis and neurological damages [66]. In (**Table 4**) there are some information based on Annual HUS incidence per 1,000,000 children in U.S. [67]. All the data based on HUS cases in last 10 years, are given in Section 6.

8. Prevention

As *E. coli* O157:H7 is an emerging cause of foodborne illness associated with undercooked all ground beef or hamburger, avoiding any undercooked hamburgers in the restaurant, practicing proper hygiene especially good handwashing, and consuming only pasteurized milk, can prevent *E. coli* O157:H7 infection [54].

9. Conclusion

Through decades several outbreaks related to *E. coli* O157:H7 have been occurred. According to *E. coli*'s genetic traits, the number of patients in outbreaks, *E. coli* O157:H7 may be considered a potentially deadly biological weapon agent, which anytime can be used for mass destruction.

Author details

Mojtaba Mohseni[1], Benyamin Djawadi[1*] and Noushin Khazaei[2]

1 Department of Microbiology, University of Mazandaran, Babolsar, Iran

2 Department of Medical Science, Zahedan University of Medical Sciences, Zahedan, Iran

*Address all correspondence to: benyaaminjavadi1377@gmail.com

References

[1] Karmali MA. Infection by verocytotoxin-producing *Escherichia coli*. Clinical Microbiology Reviews. 1989;**2**(1):15-38

[2] Hayashi T, Makino M, Ohnishi K, Kurokawa K, Ishii K, Yokoyama C, Nakayama K. Complete genome sequence of enterohemorrhagic *Escherichia coli* O157:H7 and genomic comparison with a laboratory strain K-12. 2001

[3] Fontaine R, Arnon S, Martin W. Raw hamburger: An interstate common source of human salmonellosis. 1978

[4] Armstrong G, Hollingsworth J, Morris J. Emerging foodborne pathogens: *Escherichia coli* O157:H7 as a model of entry of a new pathogen into the food supply of the developed world. 1996

[5] Ostroff S, Kobayashi J, Lewis J. Infections with *Escherichia coli* O157:H7 in Washington state: The first year of statewide disease surveillance. JAMA. 1989;**262**(3):355-359

[6] Stephenson J. New approaches for detecting and curtailing foodborne microbial infections. 1997

[7] Blanco M, Blanco J, Mora A, Dahbi G, Alonso M, González E. Serotypes, virulence genes, and intimin types of Shiga toxin (verotoxin)-producing *Escherichia coli* isolates from cattle in Spain and identification of a new intimin variant gene (eae-xi). 2004

[8] Bustamante A, Sanso A, Lucchesi P, Parma A. Multiplex PCR assay for the detection of five putative virulence genes encoded in verotoxigenic *Escherichia coli* plasmids. 2011

[9] Guth B, Prado V, Rivas M. Shiga toxin-producing *Escherichia coli*. In: Torres AG, editor. Pathogenic *Escherichia coli* in Latin America. Galveston, TX, USA: Department of Microbiology and Immunology, Sealy Institute for Vaccine Sciences, University of Texas Medical Branch; 2010

[10] Gyles C. Shiga toxin-producing *Escherichia coli*. An overview. Journal of Animal Science. 2007;**85**(13 Suppl):E45-E62

[11] Kaper J, Mobley H, Nataro J. Pathogenic *Escherichia coli*. Nature Reviews. 2004;**2**(2):123-140

[12] Paton JC. Pathogenesis and diagnosis of Shiga toxin-producing *Escherichia coli* infections. Clinical Microbiology Reviews. 1998;**11**(3):450-479

[13] Schmidt H, Karch H, Herold S. Shiga toxin-encoding bacteriophages—Genomes. International Journal of Medical Microbiology. 2004

[14] McIngvale SC. Optimization of reverse transcriptase PCR to detect viable Shiga-producing *Escherichia coli*. American Society Microbiology. 2002

[15] Muhldorfer I. Pathogenicity islands of virulent bacteria: Structure, function and impact on microbial evolution. Wiley Online Library. 1997

[16] Nataro J, Kaper J. Diarrheagenic *Escherichia coli*. Clinical Microbiology Reviews. 1998;**11**(1):142-201

[17] Davis M, Carpenter J, Hovde C. *Escherichia coli* O157:H7 in environments of culture-positive cattle. 2005

[18] Hancock D, Besser T, Herriott D, Rice D. Multiple sources of *Escherichia coli* O157 in feedlots and dairy farms in the Northwestern USA. 1998

[19] Schets F, During M, Italiaander R, Heijnen L, Rutjes S, Van Der Zwaluw W. *Escherichia coli* O157:H7 in drinking water from private water supplies in the Netherlands. 2005

[20] Lynch M, Nguyen T, Painter J. Outbreaks of Shiga toxin producing *Escherichia coli* (STEC) infections in the United Sates, 1998-2005: Trends in foodborne STEC outbreaks. 2006

[21] Besser T, Richards B, Rice D, Hancock D. *Escherichia coli* O157:H7 infection of calves: Infectious dose and direct contact transmission. 2001

[22] Hancock D, Besser T, Rice D. Ecology of *Escherichia coli* O157:H7 in cattle and impact of management practices. *Escherichia coli* O157:H7 and other Shiga toxin producing *Escherichia coli*. 1998

[23] Callaway T, Anderson R, Edrington T, Genovese K, Bischoff K. What are we doing about *Escherichia coli* O157:H7 in cattle? 2004

[24] Center for Veterinary Medicine. A proposed framework for evaluating and assuring the human safety of the microbial effects of antimicrobial new animal drugs intended for use in food-producing 6 animals. n.d.

[25] Klapproth J, Scaletsky I, McNamara B, Malstrom C. A large toxin from pathogenic *Escherichia coli* strains that inhibits lymphocyte activation. 2000

[26] Church D, Emshey D, Semeniuk H, Lloyd T, Pitout J. Evaluation of BBL CHROMagar O157. 2007

[27] Gould L, Bopp C, Strockbine N, Atkinson R, Baselski V. Recommendations for diagnosis of Shiga toxin-producing *Escherichia coli* infections by clinical laboratories. 2009

[28] Thompson J, Hodge D, Borczyk A. Rapid biochemical test to identify verocytotoxin-positive strains of *Escherichia coli* serotype O157. 1990

[29] Atkinson M. Tracking media reports on the Shiga toxin-producing *Escherichia coli* O104: H4 outbreak in Germany. 2012

[30] Scheutz F, Teel L, Beutin L, Piérard D, Buvens G, Karch H. Multicenter evaluation of a sequence-based protocol for subtyping Shiga toxins and standardizing Stx nomenclature. 2012

[31] Wang X, Kim Y, Hong S, Pokusaeva K, Sturino J, Wood T. Cryptic prophages help bacteria cope with adverse environments. 2010

[32] Skinner C, Patfield S, Hernlem B. New Stx2e monoclonal antibodies for immunological detection and distinction of Stx2 subtypes. 2015

[33] He X, Kong Q, Patfield S, Rasooly R. A new immunoassay for detecting all subtypes of Shiga toxins produced by Shiga toxin-producing *E. coli* in ground beef. 2016

[34] Atkinson R, Besser J, Bopp C, Carlson C, Crandall C, George K, Gladbach S. Guidance for public health laboratories on the isolation and characterization of Shiga toxin-producing *Escherichia coli* (STEC) from clinical specimens. 2012

[35] Mead P, Griffin P. *Escherichia coli* O157:H7. 1998

[36] Banatvala N. The United States national prospective hemolytic uremic syndrome study: Microbiologic, serologic, clinical, and epidemiologic findings. 2001

[37] Griffin P, Tauxe R. The epidemiology of infections caused by

Escherichia coli O157:H7, other enterohemorrhagic *E. coli*, and the associated hemolytic uremic syndrome. Epidemiologic Reviews. 1991;**13**: 60-98

[38] Outbreak of *Escherichia coli* O157:H7 among attendees of the Washington Country Fair-New York. 1999. Retrieved from Centers for Disease Control and Prevention: https://www.cdc.gov

[39] Outbreak of *Escherichia coli* O157:H7 infections among patients. 2001. Retrieved from Centers for Disease Control and Prevention: https://www.cdc.gov

[40] Rangel J. Epidemiology of *Escherichia coli* O157:H7 outbreaks, United States. 1982-2002

[41] Hoff C, Higa J, Patel K. Notes from the field: An Outbreak of *Escherichia coli* O157:H7. Centers for Disease Control and Prevention. 2021

[42] Retrieved from Centers for Disease Control and Prevention. 2020

[43] Public Health Agency of Canada. Outbreak summaries. 2018

[44] Retrieved from Centers for Disease Control and Prevention. 2018

[45] Foodborne viral outbreaks associated with *Escherichia coli* reported by PAIFOD. 2018

[46] Catford A, Martinez A. Outbreaks of Escherichia coli O157:H7 infections linked to Romaine lettuce. 2020

[47] Global Food Safety Initiative. 2020

[48] FDA. 2020. Retrieved from FDA: www.fda.gov

[49] CDC. 2021. *E. coli* outbreaks. Retrieved from CDC.gov: https://www.cdc.gov/ecoli/outbreaks.html

[50] CDC. 2021. Outbreak of *E. coli* infections. Retrieved from CDC.gov: https://www.cdc.gov/ecoli/2020/o157h7-10-20a/index.html

[51] CDC. 2021. Outbreak of *E. coli* infections linked to Fresh Express Sunflower Crisp chopped salad kits. Retrieved from CDC.gov: https://www.cdc.gov/ecoli/2019/o157h7-12-19/index.html

[52] CDC. 2018. O157:H7 infections linked to leafy greens. Retrieved from CDC.gov: https://www.cdc.gov/ecoli/2017/o157h7-12-17/index.html

[53] CDC. 2017. *E. coli* infections linked to I.M. Healthy Brand SoyNut Butter. Retrieved from CDC.gov: https://www.cdc.gov/ecoli/2017/o157h7-03-17/index.html

[54] CDC. 2016. O157:H7 infections linked to beef products produced by Adams farm. Retrieved from CDC.gov: https://www.cdc.gov/ecoli/2016/o157h7-09-16/index.html

[55] CDC. 2015. O157:H7 infections linked to Costco rotisserie chicken salad. Retrieved from CDC.gov: https://www.cdc.gov/ecoli/2015/o157h7-11-15/index.html

[56] CDC. 2014. O157:H7 infections linked to ground beef. Retrieved from CDC.gov: https://www.cdc.gov/ecoli/2014/O157H7-05-14/index.html

[57] CDC. 2013. O157:H7 infections linked to ready-to-eat salads. Retrieved from CDC.gov: https://www.cdc.gov/ecoli/2013/o157h7-11-13/index.html

[58] CDC. 2012. O157:H7 infections linked to organic spinach and spring mix blend. Retrieved from CDC.gov: https://www.cdc.gov/ecoli/2012/O157H7-11-12/index.html

[59] CDC. 2011. O157:H7 infections linked to Lebanon bologna. Retrieved

from CDC.gov: https://www.cdc.gov/ecoli/2011/lebanon-bologna-3-23-11.html

[60] CDC. 2010. O157:H7 infections linked to bravo farms cheese. Retrieved from CDC.gov: https://www.cdc.gov/ecoli/2010/bravo-farms-cheese-11-24-10.html

[61] CDC. 2009. O157:H7 linked to prepackaged cookie dough. Retrieved from CDC.gov: https://www.cdc.gov/ecoli/2009/cookie-dough-6-30-2009.html

[62] CDC. 2008. O157:H7 infections linked to beef from Kroger. Retrieved from CDC.gov: https://www.cdc.gov/ecoli/2008/ground-beef-kroger-7-18-2008.html

[63] CDC. 2006. O157:H7 infections linked to Taco Bell. Retrieved from CDC.gov: https://www.cdc.gov/ecoli/2006/taco-bell-12-2006.html

[64] Karmali M. Sporadic cases of haemolytic-uraemic syndrome associated with faecal cytotoxin-producing Escherichia coli in stools. 1983

[65] Karmali M. The association between idiopathic hemolytic uremic syndrome and infection by verotoxin-producing *Escherichia coli*. 1985

[66] Roberts J. HUS and TTP associated with *Escherichia coli* O157:H7. Emergency Medicine News. 2004

[67] Ardissino G. Early volume expansion and outcomes of hemolytic uremic syndrome. 2016

Chapter 7

Virulence Factors of Uropathogenic *Escherichia coli*

Etefia Etefia

Abstract

Uropathogenic *Escherichia coli* (UPEC) strains are those that cause infections in the urinary tract. They acquired virulence factors which enable them to survive in the urinary tract and elicit pathogenicity. The virulence factors are classified into two categories: (i) bacterial cell surface virulence factors and (ii) bacteria secreted virulence factors. Adhesins, toxins and iron up-take systems are major groups of virulence factors. The variety of virulence factors of UPEC is presented in this chapter.

Keywords: extraintestinal *E. coli*, uropathogenic *Escherichia coli*, urinary tract infection

1. Introduction

Uropathogenic *Escherichia coli* (UPEC) is a type of extraintestinal pathogenic *E. coli* (ExPEC) responsible for urinary tract infection (UTI). It is reported to be the ExPEC with the greatest medical importance. This is so because UPEC is responsible for most of the UTIs and humans of all ages are affected [1, 2]. These bacteria are associated with both asymptomatic bacteriuria and symptomatic UTIs. UTIs are categorized based on the parts of the body which the infections occur. These are cystitis which occurs in the bladder and pyelonephritis which occurs in the kidney [3–6]. UPEC strains have a lot of virulence factors which are responsible for the pathogenicity associated with symptomatic UTIs [7, 8]. The virulence factors are classified into two categories: (i) bacterial cell surface virulence factors and (ii) bacteria secreted virulence factors [9–11]. Many of virulence-associated genes can be found on pathogenicity islands (PAIs) [12, 13]. Though the mechanisms of asymptomatic bacteriuria are still not clear, studies have reported that UPEC becomes nonadherent and nonhemolytic resulting to asymptomatic bacteriuria [14–16]. Thus, this chapter will elucidate on the important UPEC virulence factors which are responsible for UTIs.

2. Adhesins of uropathogenic *Escherichia coli*

Adhesins are adhesive organelles, notably fimbriae, that promote bacterial colonization. Some adhesins also promote bacterial invasion of the host cell. Adhesins are thought to be the most important virulence-associated molecules which function in UPEC pathogenicity. The adhesins can also directly trigger host and bacterial cell signaling pathways. They can also facilitate the delivery of other

IntechOpen

bacterial products to the host tissues [17]. Prominent bacterial cell surface virulence factors, which play significant roles in UPEC pathogenicity include type 1 fimbriae [11]; Class I, Class II, and Class III of P-fimbriae [18–20]; Dr. family of adhesins for binding to the decay-accelerating factor (DAF) [21]; Curli fimbriae which functions as binding factor and biofilm producer [22]; and S-fimbriae [14, 23, 24]. Type 1 fimbriae have the most significant effects in UTIs as they enhance bacterial survival and growth, enhance inflammatory reaction at the mucosa, bacterial invasion, and control biofilm production [7]. P-fimbriae have the second most prominent role in UPEC-associated pathogenesis of human ascending UTIs and pyelonephritis. They promote UPEC adherence to the matrix of the mucosa and tissues and trigger cytokine production [25–30].

3. Toxins of uropathogenic *Escherichia coli*

UPEC secrete several virulence toxins which are responsible for the damage of the host cells and host inflammatory response. α-hemolysin (HlyA) is the most virulent toxin produced by UPEC. The effects of HlyA in UTIs are dependent on its dosage produced by UPEC. At high concentration, HlyA destroys the erythrocytes and allow UPEC to break through the mucosal barriers, damage immune system, and depletes iron stores of the host [31–34]. At low concentration, HlyA induces cell death in the bladder using proinflammatorycaspase-1/caspase-4. This causes kidney damage and scarring; oscillations of $Ca2+$; ascension and colonization of ureters and kidney parenchyma in the renal tubule epithelia resulting in the disruption of normal flow of urine [35–38]. The stimulation of *in vitro* production of actin stress fibers and membrane ruffle in a Rho GTPase-dependent manner is enhanced by cytotoxic necrotizing factor 1 (CNF1) produced by many strains of UPEC. This also facilitates the invasion of UPEC into the kidney cells [39, 40]. However, the extensiveness of CNF1 activities in causing invasion-associated pyelonephritis is not well understood and it has different schools of thoughts [41]. CNF1 also causes polymorphonuclear phagocytosis to trigger apoptosis and scarring of the epithelia of the bladder [42]. The uropathogenic specific protein (Usp) is important in the movement of UPEC from the urinary tract to the bloodstream. High prevalence of Usp has been reported UPEC isolated in cystitis, pyelonephritis, and prostatitis [43]. Serine-autotransporter toxin (Sat) secreted by UPEC is toxic to the cell lines of bladder or kidney origin thereby enhancing pathogenesis of UTI [44, 45]. Also, cytolethal distending toxin (CDT) is another toxin secreted by UPEC which is virulent in UTIs [46, 47].

4. Iron uptake systems of uropathogenic *Escherichia coli*

Urinary tract has limited iron. However, UPEC are able to produce small iron chelator molecules, known as siderophores, to scavenge ferric iron ($Fe3+$) in the host. The most prominent ones are yersiniabactin, salmochelin, and aerobactin [48, 49]. The yersiniabactin and its receptor, FyuA, are encoded in a PAI [50, 51]. It has also been reported that for efficient biofilm formation by UPEC, FyuA is required [52]. UPEC also secretes another important hydroxamate siderophore called aerobactin. This is produced from the condensation of two lysine and a citrate molecules. During UPEC invasion, the bacterium secretes salmochelin. Its outer membrane siderophore receptor (IroN) transports different catechol siderophores, including N-(2,3-dihydroxybenzoy)-L-serine and enterochelin also called enterobactin [53]. Enterobactin has less solubility and stability than

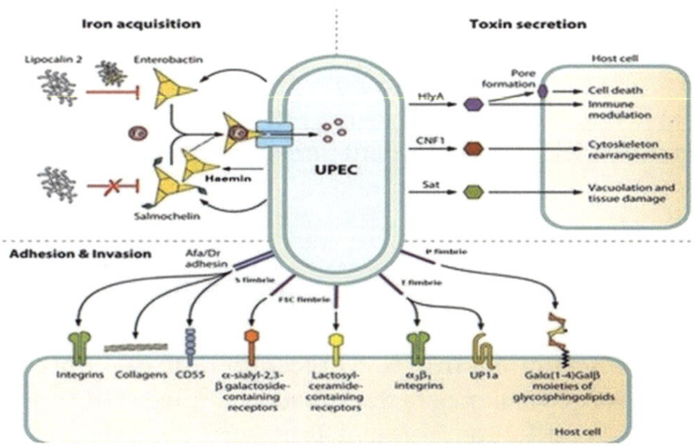

Figure 1.
UPEC-associated fitness and virulence. Adapted from the work by Servin [64].

aerobactin [54–56] but has higher iron affinity than aerobactin in aqueous [55, 57]. UPEC also uses enterobactin for Fe3+ scavenging in the urinary tract [9]. However, enterobactin can be inactivated by the host proteins such as serum albumin and siderocalin thereby preventing its uptake [58]. UPEC overcomes this instability by modifying the enterobactin to salmochelin by glucosylation through the enzymatic action of glucosyltransferase and prevents it from being recognized by the host proteins [9]. Also, UPEC has another iron acquisition system called haemin uptake system consisting of Ton-B dependent receptor (ChuA) and heavy metal associated (*Hma*) receptor that takes part in direct upregulation of haem receptors from free iron during UPEC infection. This system has also been reported to play significant role in the formation of biofilm [59–61]. The expression of ChuA is controlled by other regulatory proteins. It has been reported that the production of ChuA is triggered as RfaH increases [62]. However, Hma does not depend on ChuA and it is controlled by Tyr-126. Both Hma and ChuA are associated with haem uptake for optimal kidney utilization [63]. **Figure 1** shows the diagram of UPEC-associated fitness and virulence factors.

5. Lipopolysaccarides of uropathogenic *Escherichia coli*

Lipopolysaccharide (LPS) is a major part of the cell wall which has highly conserved lipid A-core and repeating O-antigen subunits which vary in different strains of *E. coli* depending on the sugar residues and their linkage patterns within the repeating subunits [41, 65]. LPS is very prominent in activating the host immune response and the stimulation of nitric oxide and cytokine (IL-1, TNF-α) for inflammatory response [11, 66]. Also, it triggers the production of specific antibodies to the somatic antigen and the humoral immune response to other antigens of the pathogen [31]. Several antigenic types of LPS help UPEC to escape being killed by the host serum [31]. A study on animal models has reported that LPS-associated acute renal failure is due to the response of the host to the LPS and not based on the expression of TLR4 (LPS receptor) in the kidney [66].

6. Capsule of uropathogenic *Escherichia coli*

Capsule is made up of polysaccharides and it covers and protects UPEC from various harsh environmental conditions [66]. The capsule helps UPEC to resist phagocytosis and bactericidal effects of complements in the host. It also confers antimicrobial resistance and antiserum activity to UPEC [54, 61]. Capsules like K1 and K5 interfere with the proper response of the humoral immunity of the infected host [66]. The K1 polysaccharide plays a significant role in intracellular bacterial community (IBC) development and the pathogenesis of several UTI stages [54, 67].

7. Other virulence factors of uropathogenic *Escherichia coli*

Toll receptor (TIR)/interleukin1 (IL-1) receptor domain-containing protein (TcpC) is a novel class of virulence factors that destabilize TIR signaling for UPEC to survive during UTIs [68]. Interaction of TcpC with myeloid differentiation primary response 88 (MyD88) found in the host ends the downstream signaling pathways mediated by TLRs [69].

UPEC produces outer membrane protease T (OmpT) that catalyzes plasminogen activation to plasmin [70]. OmpT helps UPEC to persist in the urinary tract when protamine and other cation peptides cleave with antibiotic activity [71, 72]. UPEC also decreases cytokines production by blocking nuclear factor kappa-light-chain-enhancer of activated B cells (NF-κB) [68]. In **Table 1**, prominent UPEC virulence factors, their role and genetic markers are presented.

Virulence factor	Role	Genetic markers/gene name	References
Afimbrial adhesions	Binding factor	*afa*	[23, 24, 54]
Cytotoxic necrotizing factor 1	Toxin	*cnf1*	[38, 39]
Curlifimbriae	Binding factor	*csgA-G*	[22]
Dr family of adhesions	Binding factor	*drb*	[21]
Haemin	Iron uptake and biofilm formation	*hmn, chuA*	[59–61]
Type 1 fimbriae	Binding factor	*fimH*	[8]
Ferric yersiniabactin uptake receptor	Iron uptake and biofilm formation	*fyuA*	[62]
α-hemolysin	Lyses red blood cells	*hlyA*	[33]
Salmochelin	Siderophore receptor	$iroN_{E.\ coli}$	[51]
Aerobactin	Iron chelation and uptake	*iucD, iutA*	[50]
Outer membrane protease T	Outer membrane protease production to degrade protamine peptides	*ompT*	[73, 74]
Uropathogen specific protein	Movement of UPEC from the urinary tract to the bloodstream	*usp*	[42]
Class I, Class II, and Class III P-fimbriae	For binding to the uroepithelial cells	*papGJ96, papGAD/IA2, and prsGJ96*	[18, 20, 21]
Serine-protease autotransporter toxin	Vacuolation and tissue damage	*sat*	[73, 74]
S-fimbrial family	Binding factor	*sfa*	[8, 23, 24]

Table 1.
Virulence factors of uropathogenic Escherichia coli *and their functions.*

DOI: http://dx.doi.org/10.5772/intechopen.99891

8. Conclusion

Apart from possessing virulence factors, for the medical importance of *E. coli* strains the ability to form biofilms is also significant. Biofilms play a major role in urology. Biofilms are namely usually associated with pyelonephritis and chronic or recurrent infections [75]. Biofilm formation is a complex process that may involve multiples adhesins and factors [76]. Biofilm contributes to bacterial resistance [60, 77–81]. Studies have reported that biofilm production mediated by co-expression of curli and cellulose facilitates in *E. coli* helps UPEC to survive in the urinary tract for a long time through the production of an inert, hydrophobic extracellular matrix which surrounds the organism [60, 77, 78].

Author details

Etefia Etefia
Department of Medical Laboratory Science, University of Calabar, Calabar, Nigeria

*Address all correspondence to: etefiaetefia1@gmail.com

References

[1] Bergsten G, Wullt B, Svanborg C. *Escherichia coli*, fimbriae, bacterial persistence and host response induction in the human urinary tract. Int J Med Microbiol 2005;295:487-502.

[2] Lloyd AL, Rasko DA, Mobley HL. Defining genomic islands and uropathogen-specific genes in uropathogenic *Escherichia coli*. J Bacteriol 2007;189:3532-3546.

[3] Foxman B. Epidemiology of urinary tract infections: Incidence, morbidity, and economic costs. Dis Mon 2003;49:53-70.

[4] Hooton TM, Stamm WE. Diagnosis and treatment of uncomplicated urinary tract infection. Infect Dis Clin North Am 1997;11:551-581.

[5] Svanborg C, Godaly G. Bacterial virulence in urinary tract infection. Infect Dis Clin North Am 1997;11: 513-529.

[6] Sadler I, Chiang A, Kurihara T, Rothblatt J, Way J, Silver P, et al. A yeast gene important for protein assembly into the endoplasmic reticulum and the nucleus has homology to DnaJ, an *Escherichia coli* heat shock protein. J Cell Biol 1989;109:2665-2675.

[7] Mulvey MA, Lopez-Boado YS, Wilson CL, Roth R, Parks WC, Heuser J, et al. Induction and evasion of host defenses by Type 1-piliated uropathogenic *Escherichia coli*. Science 1998;282:1494-1497.

[8] Martinez JJ, Mulvey MA, Schilling JD, Pinkner JS, Hultgren SJ. Type 1 pilus-mediated bacterial invasion of bladder epithelial cells. EMBO J 2000;19:2803-2812.

[9] Wiles TJ, Kulesus RR, Mulvey MA. Origins and virulence mechanisms of uropathogenic *Escherichia coli*. ExpMolPathol 2008;85:11-19.

[10] uropathogenic*E. coli* and the evolution of virulence. Int J Antimicrob Agents 2002;19:517-521.

[11] Emody L, Kerényi M, Nagy G. Virulence factors of uropathogenic *Escherichia coli*. Int J Antimicrob Agents 2003;22Suppl 2:29-33.

[12] Etefia EU, Ben SA. Virulence markers, phylogenetic evolution, and molecular techniques of uropathogenic *Escherichia coli*. J Nat Sci Med 2020;3:13-22.

[13] Starčič Erjavec M, Žgur-Bertok D. Virulence potential for extraintestinal infections among commensal *Escherichia coli* isolated from healthy humans—the Trojan horse within our gut. FEMS Microbiology Letters 2015, 362, fnu061.

[14] Kaijser B, Ahlstedt S. Protective capacity of antibodies against *Escherichia coli* and K antigens. Infect Immun 1977;17:286-289.

[15] Edén CS, Hanson LA, Jodal U, Lindberg U, Akerlund AS. Variable adherence to normal human urinary-tract epithelial cells of *Escherichia coli* strains associated with various forms of urinary-tract infection. Lancet 1976;1:490-492.

[16] Lindberg U, Hanson LA, Jodal U, Lidin-Janson G, Lincoln K, Olling S, et al. Asymptomatic bacteriuria in schoolgirls. II. Differences in *Escherichia coli* causing asymptomatic bacteriuria. ActaPaediatrScand 1975;64: 432-436.

[17] Mulvey MA. Adhesion and entry of uropathogenic *Escherichia coli*. Cell Microbiol 2002;4:257-271.

[18] Kaper JB, Nataro JP, Mobley HL. Pathogenic *Escherichia coli*. Nat Rev Microbiol 2004;2:123-140.

[19] Hull RA, Gill RE, Hsu P, Minshew BH, Falkow S. Construction and expression of recombinant plasmids encoding type 1 or D-mannose-resistant pili from a urinary tract infection *Escherichia coli* isolate. Infect Immun 1981;33:933-938.

[20] Wullt B, Bergsten G, Connell H, Röllano P, Gebretsadik N, Hull R, *et al.* P fimbriae enhance the early establishment of *Escherichia coli* in the human urinary tract. MolMicrobiol 2000;38:456-464.

[21] Nowicki B, Selvarangan R, Nowicki S. Family of *Escherichia coli* Dr Adhesins: Decay-accelerating factor receptor recognition and invasiveness. J Infect Dis 2001;183Suppl 1:S24-7.

[22] Gophna U, Barlev M, Seijffers R, Oelschlager TA, Hacker J, Ron EZ, *et al.* Curlifibers mediate internalization of *Escherichia coli* by eukaryotic cells. Infect Immun 2001;69:2659-2665.

[23] Hacker JS, Morschhauser J. In: Klemm P, editor. Fimbriae, Adhesion, Genetics, Biogenesis, and Vaccines. Boca Raton, Fla, USA: CRC Press; 1994. p. 27-36.

[24] Marre R, Kreft B, Hacker J. Genetically engineered S and F1C fimbriae differ in their contribution to adherence of *Escherichia coli* to cultured renal tubular cells. Infect Immun 1990;58:3434-3437.

[25] Godaly G, Bergsten G, Frendéus B, Hang L, Hedlund M, Karpman D, *et al.* Innate defences and resistance to gram negative mucosal infection. AdvExp Med Biol 2000;485:9-24.

[26] Hedlund M, Wachtler C, Johansson E, Hang L, Somerville JE, Darveau RP, *et al.* P fimbriae-dependent, lipopolysaccharide-independent activation of epithelial cytokine responses. MolMicrobiol 1999;33:693-703.

[27] Leffler H, Eden CS. Chemical identification of a glycosphingolipid receptor for *Escherichia coli* attaching to human urinary tract epithelial cells and agglutinating human erythrocytes. FEMS Microbiol Lett 1980;8:127-134.

[28] Leffler H, Svanborg-Edén C. Glycolipid receptors for uropathogenic *Escherichia coli* on human erythrocytes and uroepithelial cells. Infect Immun 1981;34:920-929.

[29] Väisänen V, Elo J, Tallgren LG, Siitonen A, Mäkelä PH, Svanborg-Eden C, *et al.* Mannose-resistant haemagglutination and *P* antigen recognition are characteristic of *Escherichia coli* causing primary pyelonephritis. Lancet 1981;2:1366-1369.

[30] Plos K, Connell H, Jodal U, Marklund BI, Mårild S, Wettergren B, *et al.* Intestinal carriage of *P* fimbriated*Escherichia coli* and the susceptibility to urinary tract infection in young children. J Infect Dis 1995;171:625-631.

[31] Cirl C, Wieser A, Yadav M, Duerr S, Schubert S, Fischer H, *et al.* Subversion of toll-like receptor signaling by a unique family of bacterial toll/interleukin-1 receptor domain-containing proteins. Nat Med 2008;14:399-406.

[32] Laestadius A, Richter-Dahlfors A, Aperia A. Dual effects of *Escherichia coli* alpha-hemolysin on rat renal proximal tubule cells. Kidney Int 2002;62: 2035-2042.

[33] Keane WF, Welch R, Gekker G, Peterson PK. Mechanism of *Escherichia coli* alpha-hemolysin-induced injury to isolated renal tubular cells. Am J Pathol 1987;126:350-357.

[34] Uhlén P, Laestadius A, Jahnukainen T, Söderblom T, Bäckhed F, Celsi G, *et al.* Alpha-haemolysin of uropathogenic *E. coli* induces ca2+

oscillations in renal epithelial cells. Nature 2000;405:694-697.

[35] Kohan DE. Role of endothelin and tumour necrosis factor in the renal response to sepsis. Nephrol Dial Transplant 1994;9Suppl 4:73-7. Jakobsson B, Berg U, Svensson L. Renal scarring after acute pyelonephritis. Arch Dis Child 1994;70:111-115.

[36] Ditchfield MR, de Campo JF, Nolan TM, Cook DJ, Grimwood K, Powell HR, *et al.* Risk factors in the development of early renal cortical defects in children with urinary tract infection. AJR Am J Roentgenol 1994;162:1393-1397.

[37] Mobley HL, Green DM, Trifillis AL, Johnson DE, Chippendale GR, Lockatell CV, *et al.* Pyelonephritogenic *Escherichia coli* and killing of cultured human renal proximal tubular epithelial cells: Role of hemolysin in some strains. Infect Immun 1990;58:1281-1289.

[38] Landraud L, Gauthier M, Fosse T, Boquet P. Frequency of *Escherichia coli* strains producing the cytotoxic necrotizing factor (CNF1) in nosocomial urinary tract infections. Lett ApplMicrobiol 2000;30:213-216.

[39] De Rycke J, Milon A, Oswald E. Necrotoxic*Escherichia coli* (NTEC): Two emerging categories of human and animal pathogens. Vet Res 1999;30: 221-233.

[40] Chen M, Tofighi R, Bao W, Aspevall O, Jahnukainen T, Gustafsson LE, *et al.* Carbon monoxide prevents apoptosis induced by uropathogenic *Escherichia coli* toxins. PediatrNephrol 2006;21:382-389.

[41] Bower JM, Eto DS, Mulvey MA. Covert operations of uropathogenic *Escherichia coli* within the urinary tract. Traffic 2005;6:18-31.

[42] Kanamaru S, Kurazono H, Nakano M, Terai A, Ogawa O, Yamamoto S, *et al.* Subtyping of uropathogenic *Escherichia coli* according to the pathogenicity island encoding uropathogenic-specific protein: Comparison with phylogenetic groups. Int J Urol 2006;13:754-760.

[43] Hui CY, Guo Y, He QS, Peng L, Wu SC, Cao H, *et al. Escherichia coli* outer membrane protease OmpT confers resistance to urinary cationic peptides. MicrobiolImmunol 2010;54:452-459.

[44] Guyer DM, Henderson IR, Nataro JP, Mobley HL. Identification of sat, an autotransporter toxin produced by uropathogenic *Escherichia coli*. MolMicrobiol 2000;38:53-66.

[45] Féria CP, Correia JD, Gonçalves J, Machado J. Detection of virulence factors in uropathogenic *Escherichia coli* isolated from humans, dogs and cats in Portugal. AdvExp Med Biol 2000;485: 305-308.

[46] Tóth I, Hérault F, Beutin L, Oswald E. Production of cytolethal distending toxins by pathogenic *Escherichia coli* strains isolated from human and animal sources: Establishment of the existence of a new cdt variant (Type IV). J ClinMicrobiol 2003;41:4285-4291.

[47] Chen M, Jahnukainen T, Bao W, Daré E, Ceccatelli S, Celsi G, *et al.* Uropathogenic *Escherichia coli* toxins induce caspase-independent apoptosis in renal proximal tubular cells via ERK signaling. Am J Nephrol 2003;23: 140-151.

[48] Skaar EP. The battle for iron between bacterial pathogens and their vertebrate hosts. PLoSPathog 2010;6:e1000949.

[49] O'Brien VP, Hannan TJ, Nielsen HV, Hultgren SJ. Drug and vaccine development for the treatment and prevention of urinary tract infections. MicrobiolSpectr 2016;4:42.

[50] Schubert S, Picard B, Gouriou S, Heesemann J, Denamur E. Yersinia high-pathogenicity island contributes to virulence in *Escherichia coli* causing extraintestinal infections. Infect Immun 2002;70:5335-5337.

[51] Hancock V, Klemm P. Global gene expression profiling of asymptomatic bacteriuria *Escherichia coli* during biofilm growth in human urine. Infect Immun 2007;75:966-976.

[52] Carbonetti NH, Boonchai S, Parry SH, Vaisanen-Rhen V, Korhonen TK, Williams PH, *et al.* Aerobactin-mediated iron uptake by *Escherichia coli* isolates from human extraintestinal infections. Infect Immun 1986;51:966-968.

[53] Barber AE, Norton JP, Wiles TJ, Mulvey MA. Strengths and limitations of model systems for the study of urinary tract infections and related pathologies. MicrobiolMolBiol Rev 2016;80:351-367.

[54] Hagberg L, Jodal U, Korhonen TK, Lidin-Janson G, Lindberg U, Eden CS. Adhesion, hemagglutination, and virulence of *Escherichia coli* causing urinary tract infections. Infect Immun 1981;31(2):564-570.

[55] Warner PJ, Williams PH, Bindereif A, Neilands JB. ColV plasmid-specific aerobactin synthesis by invasive strains of *Escherichia coli*. Infect Immun 1981;33(2):540-545.

[56] De Lorenzo V, Martinez JL. Aerobactin production as a virulence factor: A reevaluation. Eur J Clin. Microbiol Infect Dis 1988;7(5):621-629.

[57] Johnson JR. Virulence factors in *Escherichia coli* urinary tract infection. ClinMicrobiol Rev 1991;4(1):80-128.

[58] Nowicki B, Labigne A, Moseley S, Hull R, Hull S, Moulds J. The Dr hemagglutinin, afimbrial adhesins AFA-I and AFA-III, and F1845 fimbriae of uropathogenic and diarrhea-associated *Escherichia coli* belong to a family of hemagglutinins with Dr receptor recognition. Infect Immun 1990;58(1):279-281

[59] Jahandeh N, Ranjbar R, Behzadi P, Behzadi E. Uropathogenic *Escherichia coli* virulence genes: Invaluable approaches for designing DNA microarray probes. Cent European J Urol 2015;68(4):452.

[60] Reigstad CS, Hultgren SJ, Gordon JI. Functional genomic studies of uropathogenic *Escherichia coli* and host urothelial cells when intracellular bacterial communities are assembled. J BiolChem 2007;282:21259-21267.

[61] Garcia EC, Brumbaugh AR, Mobley HL. Redundancy and specificity of *Escherichia coli* iron acquisition systems during urinary tract infection. Infect and Immun 2011;79(3):1225-1235.

[62] Nagy G, Dobrindt U, Emody L, Karch H, Hacker J. Expression of hemin receptor molecule ChuA is influenced by RfaH in Uropathogenic *Escherichia coli* strain 536. Infect Immun 2001;69(3):1924-1928.

[63] Hagan EC, Mobley HLT. Haem acquisition is facilitated by a novel receptor Hma and required by uropathogenic *Escherichia coli* for kidney infection. MolMicrobiol 2008;71(1): 79-91

[64] Servin AL. Pathogenesis of human diffusely adhering *Escherichia coli* expressing afa/Dr adhesins (Afa/Dr DAEC): Current insights and future challenges. ClinMicrobiol Rev 2014;27:823-869.

[65] Sarkar S, Ulett GC, Totsika M, Phan MD, Schembri MA. Role of capsule and O antigen in the virulence of uropathogenic *Escherichia coli*. PLoS One 2014;9(4):e94786.

[66] Bien J, Sokolova O, Bozko P. Role of uropathogenic *Escherichia coli* virulence factors in development of urinary tract infection and kidney damage. Int J Nephrol 2012:e681473.

[67] Agarwal J, Srivastava S, Singh M. Pathogenomics of uropathogenic *Escherichia coli*. Indian J Med Microbiol 2012;30(2):141.

[68] Guyer DM, Radulovic S, Jones FE, Mobley HL. Sat, the secreted autotransporter toxin of uropathogenic *Escherichia coli*, is a vacuolating cytotoxin for bladder and kidney epithelial cells. Infect Immun 2002;70:4539-4546.

[69] Parvez SA and Rahman D. Virulence Factors of Uropathogenic *E. coli*. Microbiology of Urinary Tract Infections-Microbial Agents and Predisposing Factors. Behzadi P. IntechOpen, 2018.

[70] Olson PD, Justice SS, Hunstad D. *Escherichia coli* in urinary tract infections. Molecular Medical Microbiology, 2nd ed. London: Academic Press; 2015;76(7):1373-87.

[71] Guina T, Yi EC, Wang H, Hackett M, Miller SI. A phoP-regulated outer membrane protease of *Salmonella enteric* serovar Typhimurium promotes resistance to alpha-helical antimicrobial peptides. J Bacteriol 2000;182: 4077-4086.

[72] Stumpe S, Schmid R, Stephens DL, Georgiou G, Bakker EP. Identification of OmpT as the protease that hydrolyzes the antimicrobial peptide protamine before it enters growing cells of *Escherichia coli*. J Bacteriol 1998;180:4002-4006.

[73] Jacobson SH, Hammarlind M, Lidefeldt KJ, Osterberg E, Tullus K, Brauner A, *et al.* Incidence of aerobactin-positive *Escherichia coli* strains in patients with symptomatic urinary tract infection. Eur J ClinMicrobiol Infect Dis 1988;7:630-634.

[74] Colonna B, Nicoletti M, Visca P, Casalino M, Valenti P, Maimone F, *et al.* Composite IS1 elements encoding hydroxamate-mediated iron uptake in FIme plasmids from epidemic *Salmonella*spp. J Bacteriol 1985;162: 307-316.

[75] Soto SM. Importance of biofilms in urinary tract infections: New therapeutic approaches. AdvBiol 2014;2014:1-13.

[76] Stærk K, Khandige S, Kolmos HJ, Møller-Jensen J, Andersen TE. Uropathogenic *Escherichia coli* express Type 1 fimbriae only in surface adherent populations under physiological growth conditions. J Infect Dis 2016;213:386-394.

[77] Kai-Larsen Y, Lüthje P, Chromek M, Peters V, Wang X, Holm A, *et al.* Uropathogenic *Escherichia coli* modulates immune responses and its curli fimbriae interact with the antimicrobial peptide LL-37. PLoSPathog 2010;6:e1001010.

[78] Kostakioti M, Hadjifrangiskou M, Hultgren SJ. Bacterial bio films: Development, dispersal, and therapeutic strategies in the dawn of the postantibiotic era. Cold Spring HarbPerspect Med 2013;3:a10306.

[79] de la Fuente-Nunez C, Reffuveille F, Fernández L, Hancock RE. Bacterial biofilm development as a multicellular adaptation: Antibiotic resistance and new therapeutic strategies. CurrOpinMicrobiol 2013;16:580-589.

[80] Lüthje P, Brauner A. Ag43 promotes persistence of uropathogenic *Escherichia coli* isolates in the urinary tract. J ClinMicrobiol 2010;48:2316-2317.

[81] Schroeder M, Brooks BD, Brooks AE. The complex relationship between virulence and antibiotic resistance. Genes (Basel) 2017;8:39.

Chapter 8

Pathogenic *Escherichia coli*: An Overview on Pre-Harvest Factors That Impact the Microbial Safety of Leafy Greens

Aura Darabă

Abstract

Consumption of fresh leafy greens has been repeatedly reported and linked to pathogenic *Escherichia coli*-associated foodborne illnesses outbreaks. Leafy greens are mostly eaten raw, based on the increased consumers' preferences for natural, nutritious diets. Recent studies indicate the incidence of infections caused by pathogenic *Escherichia coli* remained almost unchanged or even increased. In this context, fresh produces increased the awareness about their primary contamination level, namely the pre-harvest phase. Fully eliminating pathogenic *Escherichia coli* from pre-harvest environment proved to be impossible. Emphasis must be placed on the pre-harvest factors that affect the food safety and, subsequently, on the identification of possible mitigation strategies that can be used on-farm for reducing the risk of leafy greens contamination with pathogenic *Escherichia coli*.

Keywords: pathogenic *Escherichia coli*, leafy greens, foodborne illnesses outbreaks, pre-harvest on-farm contamination factors, pre-harvest microbial safety mitigation strategies

1. Introduction

Leafy greens are mostly eaten raw, based on the increased consumers' preferences for natural, nutritious diets. The consumption of leafy greens is recommended to reduce the risk of malnutrition, diet-related chronic diseases such as cardiovascular diseases, cancer, metabolic disorders, and may help to slow down the cognitive decline with aging [1–4]. For preserving their bioactive compounds, leafy greens are commonly consumed raw, and the lack of a kill-step to inactivate the potentially present pathogens leads to greater risk to the health of consumers. Among other fresh produce, leafy greens are more exposed to pathogen contamination because they grow low to the ground and can be easily contaminated in open fields. The increased consumption of fresh, ready-to-eat leafy greens has been repeatedly, reported worldwide and linked to pathogenic *Escherichia coli* (herein *E. coli or E. coli* O157:H7) associated foodborne illnesses outbreaks.

Pathogenic *Escherichia coli* group is comprised of six pathotypes out of which Shiga toxin-producing *E. coli* (STEC)—STEC (also be referred to as Verocytotoxin-producing *E. coli* (VTEC) or enterohemorrhagic *E. coli* (EHEC)) is the one most associated with foodborne outbreaks [5]. Some other STEC *E. coli* strains, namely *E. coli* O145, *E. coli* O26, and *E. coli* O104:H4 were involved in rare foodborne outbreaks due to consumption of shredded lettuce, raw clover sprouts, and raw sprouted seeds [6–8].

According to World Health Organization (WHO) pre-harvest food safety is an important element in creating sustainable food safety policies and must be considered in the context of *farm-to-fork* for the protection of human health [9]. Despite the existing food safety regulations and the undertaken on-farm food safety measures, according to recent studies performed by the United States Center of Diseases Control (CDC), European Centre for Disease Prevention and Control (ECDC) in collaboration with the European Food Safety Authority (EFSA) the incidence of infections caused by pathogenic *Escherichia coli*, between 2016 and 2019, remained almost unchanged or even increased (**Table 1**) [10–14]. In this context, in the past years, these fresh produces emerged as a food safety concern that, ultimately, increased the awareness about their primary contamination stage, namely pre-harvest. It has been established that once contaminated leafy greens leave the farm's site it will be difficult to prevent further transmission of *E. coli* to consumers. Usually, large quantities of contaminated leafy greens are recalled from the markets, a fact which pose a great economic burden on leafy greens growers but also on public health [15]. The on-farm contamination with pathogenic *Escherichia coli* largely depends on agricultural and environmental factors, unsafe on-site agronomic practices including the harvesting stage, and ineffective or missing post-harvest decontamination steps. However, eliminating completely the presence of the *E. coli* from the on-farm, the natural growing environment of leafy greens, during pre-harvest stage, proves to be impossible due to the high number of factors which are involved in the harboring and transmission of this pathogen. Based on the vast number of on-field and experimental results it was unanimously agreed that it is more feasible to first understand the main agricultural factors affecting the

Year	Type of vehicle	Total reported cases (hospitalizations/deaths)	Pathogen involved in the outbreak	Sources of contamination
2020	Spinach; romaine lettuce	40 (20/0)	*E. coli* O157:H7	• Cattle feces
2019	Romaine lettuce	167 (85/0)	*E. coli* O157:H7/ strains of Shiga toxin-producing *E.coli* (STEC)	• Farm in proximity to cattle grazing land; • On-farm water drainage basins
2018	Red leaf lettuce; green leaf lettuce	62 (25/0)	*E. coli* O157:H7	• Agricultural water reservoir
2017	Romaine lettuce and other leafy greens	25 (9/1)	*E. coli* O157:H7	• Source of contamination not identified
2016	Alfalfa sprouts	11 (2/0)	*E. coli* O157 (STEC O157)	• Farming practices
2014	Raw clover sprouts	19 (44%/0)	*E. coli* O121 (STEC O121)	• Farming practices

[a]*Reported by the CDC, between 2014 and 2020.*
[b]*Source: https://www.cdc.gov/ecoli/outbreaks.html.*

Table 1.
Selected pathogenic E. coli outbreaks associated with fresh leafy greens[a,b].

prevalence, incidence, and survival of this pathogen, as well as pathogen contamination of leafy greens, which ultimately negatively impacts the microbial safety of the produce. In turn, this will assist leafy greens producers to improve their on-farm pre-harvest agronomic practices for reducing the pathogen contamination to levels that will be a lesser hazard to public health.

Subsequent to the reported pathogenic *Escherichia coli*-related foodborne outbreaks, epidemiological trace-back studies identified the following as main contamination factors: (a) the use of manure, as a soil organic fertilizer; (b) irrigation water; (c) the domestic and wild animals which either can be found in the proximity of the growing sites or as free-roaming animals; and (d) on-farm human activity [15–17].

Similarly, in European countries, over the years, the consumption of fresh leafy greens led to multiple foodborne outbreaks. For example, in Germany (2011) the consumption of sprouts led to 3816 total illnesses (810 hospitalizations and 54 deaths) due to *E.coli* O104:H4. Between 2010 and 2011, in England, Wales and Scotland, 252 fell ill and one died following the consumption of raw leeks; the identified pathogen being *E.coli* O157 PT8. In Denmark (2010), the consumption of lettuce resulted in 264 illnesses due to *E. coli* ETEC O6:K15:H16. The consumption of fresh basil, provoked in Denmark (2006) about 200 illnesses due to *E. coli* ETEC O92:H- and O153:H2 [18].

2. Leafy greens are an easy target for contamination with pathogens: mechanisms of microbial contamination

2.1 General considerations

Leafy greens are known as an important vector for microbial hazards responsible for foodborne outbreak illnesses and almost 20% of leafy greens contamination with pathogens takes place on-farm [19, 20]. In leafy greens, *E. coli* O157:H7 is found to be more frequent than other pathogens due to its ability to contaminate mostly via biofilm formation on the produce surface which could explain the large number *E. coli* O157:H7 related outbreaks [21]. The on-farm fate of enteric pathogens on leafy greens depends on multiple conditions that the pathogenic bacteria are facing in the soil-produce environment, and on the pathogen's ability and strategies to survive and contaminate the fresh produce, such as biofilm formation or internalization. In the preharvest stage, due to the pathogen-produce interaction pathways and mechanisms, some of the pathogens could become endopathogenic in leafy greens—a stage which raises serious food safety concerns since the post-harvest decontamination treatments have almost a null chance to reduce the numbers of viable cells to a harmless level [22]. The "*points-of-entry*" used by pathogens to contaminate the leafy greens are the plant's rhizosphere and/or phyllosphere. Due to its richness in nutrients (root exudates including compounds released as a consequence of root cell metabolism or after lysis of plant cells), the root zone (rhizosphere) is an excellent environment for pathogens and it could support the presence of 10^6 to 10^9 bacteria per gram of roots [23]. On the phylloshere, *E. coli* O157:H7 is capable of attaching on these plant's parts, and can remain viable for weeks to months, and even multiply if the environmental conditions are favorable (i.e., warm temperatures, high humidity, adequate nutrients, plant's leaves' characteristics and integrity) [21]. Compared to the rhizosphere, the leaves surface nutrients are scarce. The nutrients found on leaves, probably originating from mesophyll and epidermal cell exudates that leak onto the leaves surface from wounds and broken trichomes, are not distributed homogenously. Since the phyllosphere is subjected to many stress factors which can have rapid fluctuations will affect the bacterial survival: temperature, solar radiation and humidity, phyllospheres typically could support fewer than 10^3 to 10^7 pathogen per gram of leaf [19]. Therefore, understanding the pathogen contamination pathways and mechanisms will provide important information to fresh produce growers for either adopting preventive actions or protecting their produce during the pre-harvest stage.

2.2 Biofilm formation

Leafy greens, as pathogenic biofilm carriers, pose a great threat to produce microbial safety since the biofilms poses a great resiliency towards decontamination methods applied during post-harvest processing (i.e., chemical washing solutions) [24, 25]. The general mechanisms of leafy greens contamination by pathogens' colonization takes place in phases: (a) attachment to phylloshere and/or to rhizoplane, and (b) pathogens' adaptation to environmental factors followed by survival and multiplication on the plant parts. The whole general bacterial attachment-colonization mechanism takes pace in a similar manner for human enteric pathogens that are either environmentally shed by domestic animals and/or wildlife: cattle, sheep, pigs, goats, wild birds, deer, mice, insects, or can originate from other sources: soil, manure, irrigation waters, etc. [26]. The leafy greens' structure (leaves' roughness, leaves' surface degree of porousness, crests etc.), influences the pathogen's attachment phenomenon that results in biofilm formation. When leaves are damaged (i.e., cuts, wounds) the pathogen may further become internalized due to pathogen's multiplication in these areas where damaged plant tissue exudes inner nutrients [27]. In addition, the amount of contaminating bacteria is a factor which can affect the degree of pathogen's attachment to the leafy greens. The colonization of leafy greens, as the first stage of biofilm formation, could take place through multiple routes, such as: contaminated soil (i.e., via dust or splashes), roots, seeds, or by wetting of produce leaves (i.e., via irrigation waters) and depends on the pathogens' ability to adapt to the new environment following the attachment phase. Once colonization takes place, biofilm formation is initiated. According to Ximenes and Tarver biofilm formation on leafy greens (i.e., lettuce, spinach, basil, cilantro, green onions, and parsley) by enteric pathogens involves in several stages: (a) initial contact of *E. coli* with the leafy greens and pathogen's subsequent attachment to the produce; (b) *E.coli* cells' proliferation and cells' aggregation by the excretion of the extracellular polymeric substances – which helps the formation of the initial "matrix" where the pathogen will grow and multiply; (c) *E. coli* biofilm maturation, and (d) sporadic *E. coli* cells' dispersion or detachment into the environment and contaminate other produce from the vicinity of the "infected" produce [26, 28]. According to Beattie and Lindow, bacteria found on leaves possess two major strategies which they can apply for their attachment, growth and survival, and biofilm formation on the plant surface: (a) "*tolerance strategy*" that requires the bacteria's ability to resist exposure to environmental stresses on leaf surfaces; or (b) "*avoidance strategy*", in this case bacteria seek plant sites that are protected from those stresses. Using these bacterial strategies, a general step-by step-model of leaf colonization and biofilm formation was developed: 1. the landed bacteria on the leaf surface are randomly distributed; 2. some of bacteria will enter into the leaf via openings such as stomata while some will stay on the surface of the plant leaves and modify the local environment to fit their needs; 3. surface adhered bacteria start to multiply and to form aggregates or micro-colonies, which subsequently will develop into biofilms [29]. Subsequent to the tight adhesion on favorable sites found on plants (niches), the biofilm formation process is facing environmental factors, plant properties, and the innate plant microbiota [20]. Nevertheless, once the biofilm is formed it has the capability to protects the rest of attached bacteria against environmental stressors (i.e., desiccation, UV radiation etc.), from the plant immune response, and from endogenous (plant-origin) or exogenous (indigenous microorganisms-origin) antimicrobial compounds. Studies on the attachment of human enteric *E. coli* indicate that it can rapidly adhere to a variety of growing plant tissues such leaves and roots. Surface attachment is possible due to the presence of the plant's cuticles and the plant's surface characteristics. The cuticle present on the plant surfaces favors attachment of hydrophobic molecules and any breaks in the cuticle may expose the hydrophilic structures for further attachment [30]. The characteristics of the plant's surface is also important in the microbial adhesion process. For example, the surface roughness of the plant parts depends on the nature and age of the plant, and it is important not only for adherence but also for the pathogen's survival on the produce as demonstrated for *E. coli* O157:H7 adhesion on leaves of different spinach cultivars [31]. The microbiota found on the plants is not homogenously distributed on the leaf surface, bacterial cells predominantly attaching and colonizing on specific sites of leaf surfaces such as epidermal cell wall junctions, in grooves along veins and depressions, or beneath in the cuticle [29]. Under certain factors (on-field circumstances, bacterial unspecific binding based on hydrophobic and electrostatic interactions), attachment phenomenon could be reversible. However, when the pathogen cells form the exopolymeric material, are able to fix themselves more strongly on the leafy greens, the attachment is irreversible and the pathogen cannot be removed by washing treatments [20, 32, 33].

Studies showed that both, produce and bacterial properties, are factors involved in attachment of pathogenic *E. coli*. Leafy greens surface properties (i.e., cuticles, roughness) is favoring the pathogen attachment and colonization at specific sites of leaves: base of trichomes, stomata, epidermal cell wall junctions, or in grooves existing along the produce veins and depressions [29]. The study by Takeuchi *et al.* indicated specific attachment and colonization sites the cut surfaces of lettuce are rich in water and nutrients and offer *E. coli* O157:H7 stress-protection [34]. *E. coli* strains possess an attachment-adhesion system due to its ability to produce a diversity of pili and fimbriae and non-fimbrial adhesins, that function as 'professional' adhesion systems, and flagella; these compounds could play alternative functions in attachment and adhesion stages [35, 36]. An earlier study led by Torres *et al.* showed that *E. coli* O157:H7 possesses several redundant protein adhesins and the overexpression of each adhesin alone is sufficient to promote binding to alfalfa sprouts [37]. Ximenes *et al.* indicated the importance of some bacterial hydrolytic enzymes, such as: pectinases, cellulases, proteinases, and amylases which can further enhance the ability of pathogens to invade and spread on plant tissues [26]. Several experimental studies showed that *E. coli* ability to adhere and attach varies in time and some influence factors could be the initial number of viable pathogenic cells contaminating the plant and the type of leafy green. For arugula leaves, 2 $\log_{10}$ CFU/g of pathogen attached after 60 min, for lettuce leaves attachment time varied between 25 and 120 min (final level of pathogenic viable cells being 1–2.5 $\log_{10}$ CFU/cm^2) and for spinach approximately 3 $\log_{10}$/spinach leaf attached in less than 60 min [31, 38, 39].

2.3 Internalization

Experimental studies indicated that there are many mechanisms used by *E. coli* O157:H7 to contaminate and internalize both the leafy greens root and leaves tissues [40–42]. From the roots, the pathogen can pass to the leaves by using the produce's vascular system or can penetrate the produce internal tissues using the existing wounds or other natural "openings" of the leaf system [42–44]. While due to the difficulty to study the pathogen internalization in the natural growing plant environment, internalization has been extensively studied in systems that mimic the natural environment, many factors which can promote produce contamination are yet to be clarified. Although the leafy greens possess physical and chemical defense mechanisms to restrict the internalization of pathogens under certain circumstances the produce defense can be disrupted either by biological or mechanical means and *E. coli* O157:H7 access to produce inner tissues is favored [45–47]. Once the pathogen penetrates the produce inner tissues it can potentially evade the produce defense systems by adapting to the plant environment and becomes internalized [48].

Generally, it has been accepted that the internalization of pathogens depends on several factors, such as: (a) plant type, age, and exposure time to pathogen; (b) produce growing system (i.e., soil, hydroponic, aquaponic), (c) the level of contamination of produce with the pathogen, (d) the type and the degree of roots or leaves injury, (e) length of time given to the pathogen to spread from injured roots to the mature leaves etc. (**Table 2**) [40].

For soil-grown plants, internalization was observed as a sporadic phenomenon and with low incidence. Usually, the contaminated soil, could have a little to no influence on the noted internalization, soil presenting a relatively low risk of internalization as compared to other produce growing systems (i.e., hydroponic or aquaponic systems). Generally, the soil-grown produces are protected by environmental stressful conditions which are not favoring the pathogen internalization [40, 49]. The pathogen internalization in soil-grown leafy greens remains controversial: while several studies on leafy greens (lettuce or spinach) grown on contaminated soil have shown that internalization of *E. coli* O157:H7 could occur [50, 51] other researchers found little to no pathogen internalization in soil-grown produce [52]. When pathogen internalization in leafy greens grown in soil was observed, the incriminated factors were either the root damage during growth or soil's microbial profile lacking the microorganisms that may compete with the pathogen [53]. Despite the extensive experimental studies, there are many possible factors which can interact together in promoting the pathogen internalization, and it remains controversial whether *E. coli* O157:H7, when introduced through soil or irrigation water, could internalize the edible parts of the mature produce. For example, the specific role of produce type in bacterial internalization is very difficult to assess in detail given the multiple existing interfering variables. In this regard, it was found that *E. coli* O157:H7 was able to internalize into inoculated seeds of cress, spinach, and lettuce [54]. In spinach plants, internalization was observed in the root tissue or seedlings but not in mature leaves [55]. Plant roots appear to be preferred by the pathogen as attachment and entrance site, and the roots contamination was reported to be dependent on roots health status (healthy, non-damaged roots versus damaged roots) and on the degree of pathogen contamination level [40]. While produce roots are getting mature, the differences in the produce developmental stages may also influence the ability of *E. coli* O157:H7 to interact with the produce, the pathogen could be enabled to enter the produce leaves by traveling through the root system [56]. Hora *et al.* [55] found that the degree of *E. coli* O157:H7 internalization of the spinach roots depends on the type of roots damage and produce age but it does not favor the internalization of leaves (**Table 3**).

When the produce contamination occurs, produce age, produce exposure to pathogen, and contact length of time with the pathogen can result in possible internalization of the pathogen [40]. Produce leaf's age has been shown to influence the growth and survival of *E. coli* O157:H7; young lettuce leaves were found to be associated with a greater risk of pathogen contamination and internalization [21].

Type of produce	Pathogen	Growth system	Inoculation method and level	Plant age	Internalization status
Spinach lettuce, parsley	*E.coli* O517:H7 (Shiga toxin negative)	Field-grown	Drip irrigation or compost (2, 4, or 6 $\log_{10}$ CFU/mL)	Transplanted to field 8 weeks, inoculated 1, 8, and 10 weeks after transplant	No internalization was detected in any leaf tissues; detection in root occurred at one sampling time
Whole romaine lettuce and iceberg lettuce	*E. coli* O157:H7	Soil	4 $\log_{10}$and 6 $\log_{10}$ CFU/g	Seedlings with 4–5 leaves were transplanted and inoculated 30 days after transplant	Heat and drought stress applied individually or in combination did not promote internalization
Leaves of romaine lettuce and iceberg lettuce	*E. coli* O157:H7 (5 strains cocktail)	Soil	Inoculated 3 and 6 $\log_{10}$ CFU/mL by manure, soil, and water	Inoculated when 3–4 leaves present, analyzed on days 26 and 60 post-inoculation	All samples were negative for bacterial at all inoculums, routes of inoculation, and times post-inoculation
Green ice lettuce	*E. coli* O157:H7	Manure amended soil	Inoculated manure with 8, 6, and 4 $\log_{10}$ CFU/g and added to lettuce flats	Days 3, 6, and 9 post-planting seedlings were cut 1 cm above the soil surface	*E. coli* O157:H7 was visualized at depths of up to 45 μm below the tissue surface; edible portions can become contaminated through transport by the root system
Spinach	*E. coli* O157:H7	Soil (drench)	6 $\log_{10}$ CFU/mL	Inoculated at 4-leaf stage, analysis on days 0, 7, and 14	No internalization into leaf observed by direct plating and qPCR; bacterial presence on roots observed by confocal laser microscopy on day 7 and 14
Spinach	*E. coli* O157:H7	Hydro-ponic system	Low (5 logs) or high (8.5 logs)	Plants (12-day) inoculated, allowed to grow for 21 days	*E. coli* O157:H7 was recovered from shoot tissue from 3 replicates on days 14 and 21
Cress, spinach, lettuce	*E. coli* O157:H7	Hydro-ponic system	Seeds were soaked in bacterial cell suspension (2 $\log_{10}$ CFU/mL)	Plants surface sterilized, seedlings analyzed on day 9	*E. coli* O157:H7 was recovered from external and internal tissues of all plants

**Adapted from [40].*

Table 2.
Examples of Escherichia coli O157:H7 internalization status in leafy greens grown in different environments.*

As an example, in spinach grown under greenhouse conditions, the internalization of *E. coli* O157:H7 in the leaves is rare and mostly is taking place from outside of the produce to the inside if the plant surface is exposed to a heavy contamination with the pathogen [57]. Some studies show that hydroponic systems favor a greater

Root treatment (damage) type	Number of sampled spinach plants	Number of *E. coli* O157:H7 positive samples	
		Roots	Leaves
Control[a]	9	9	0
Cutting of seminal root[b]	8	8	0
Removal of root hairs[c]	8	8	0
Biological damage[d]	8	8	0

**Source: [55].*
[a]Roots without damage; spinach plant was not removed from soil.
[b]Seminal root was severed from 5-week-old spinach plants; plants were repotted.
[c]After the removal of root hairs transferred to soil.
[d]Roots inoculated with nematodes (Meloidogyne hapla); plant age-14 days.

Table 3.
The degree of E. coli O157:H7 internalization of spinach roots and leaves following different types of root damages.*

internalization of leafy greens compared to soil-growing system [58–60] and the water, as a growing environment, is indicated as the main source of produce contamination via pathogen uptake by roots [61, 62]. In aquaponic growing systems, under certain circumstances, STEC *E. coli* can internalize both roots and the plant leaves. STEC *E. coli* first internalizes the roots which are mechanically injured due to manipulation during transplanting. Subsequently internalization into the leaves occurs when the pathogen is given sufficient time to spread into the plant shoots and into mature leaves. Internalization of STEC *E. coli* into the whole plant grown in aquaponic system seems to be dependent on the plant age at the time of root injury: if the infection takes place during the early stage of plant development the STEC *E. coli* internalization in the whole young plant is favored [63].

Although the variability of the published experimental results is great, several conclusions can be reached in relationship with leafy greens pathogen internalization: (a) the produce growth environment plays an important role in pathogen internalization; (b) internalization is a plant-pathogen specific interaction; (c) health status of the roots does not enable the uptake of pathogen into produce, and (d) the presence of internalized pathogens into roots of plants is cannot be used as an indicator for pathogen internalization in leaves and does not directly correlate with internalized pathogens in the produce leaves.

3. Leafy greens pre-harvest pathogen contamination: risk factors and management strategies

3.1 General considerations

For the leafy greens grown in open fields, during pre-harvest stage, there is a constant and concomitant exposure to factors which favor the produce contamination with pathogens. While manure (i.e., improperly stored raw manure, improperly treated or composted manure) deposited nearby plating fields or without using any protective barriers, agricultural soil (manure amended or non-amended), and irrigation water, are considered main risks for the microbial safety of the leafy greens. Other factors such as the presence of domestic or wild animals, which are usually shedding the enteric pathogens via feces, and field workers are involved in leafy greens pathogen contamination [64, 65]. Proper identification and management of these factors are paramount for reducing the contamination of leafy greens in the pre-harvest stage [16, 66, 67].

3.2 Non-amended agricultural soil: the soil-substrate management

The non-amended agricultural soil ("soil" herein) represents a habitat for pathogenic E. *coli*, as well as for other microorganisms (pathogenic or not), and it is recognized as a potential environmental factor which could contribute to pre-harvest leafy greens contamination [68, 69]. In soil, the fate of *E. coli* (i.e., survival or die-off rates) depends on a myriad of soil properties, such as: abiotic (physico-chemical composition) and biotic properties (inherent existing microbiota), the growing soil localization (i.e., nearby unprotected animal farming operations, sewage etc.), and soil type (i.e., sandy, clay etc.) [68, 70, 71]. From an experimental standpoint, due to the difficulty to study and predict the effect of a combination of factors of influence, as well as their importance against the pathogenic *E. coli*, focus has usually been placed upon a single factor or soil component [71, 72]. Based on the experimental results, the soil-related factors which can influence the pathogenic *E. coli* survival have been divided into soil's biotic and abiotic characteristics [68, 70, 73]. Soil's biotic profile is very complex and experimental targeted studies indicated a high die-off of *E. coli* O157:H7 rates in soils containing rich microbial communities (i.e., bacteria and fungi), especially those characterized by a high metabolic diversity, and an increased *E. coli* O157:H7 concentration in sterile soils due the absence of competitive and/or predatory interactions [74]. Competition for existing nutrients, the release of secondary metabolites, such bacteriocins [75], by the microbial community, or direct antagonism could determine the fate of *E. coli* O157:H7 in soil [72, 75]. Zhang *et al.* and Majeed *et al.* confirm that the Gram-negative bacteria exhibit a greater antagonism against *E. coli* O157:H7 than the Gram-positive bacteria and are known to out-compete Gram-positive bacteria for nutrients in soil [70, 76]. Soil temperature can affect the activity of microbial communities against *E. coli* O157:H7. At 18°C the decrease of the pathogen was likely caused by enhanced antagonistic activity among soil microorganisms [74]. Also, Vidovic *et al.* confirms that *E. coli* O157:H7 declined more rapidly at 22°C compared to 4°C in autoclaved soil [77].

Since the survivability of *E. coli* O157:H7 is considered a huge risk for contaminating the leafy greens or other fresh produce, the determination of essential nutrients availability including carbon, nitrogen, trace elements, salinity, soil's pH and temperature are paramount prior to planting [78]. Zhang *et al.* found that the soil's pH influences the survival of *E. coli* O157:H7. While low pH soil values could shorten the *E. coli* O157:H7 survival to 6–7 days, in a more neutral pH *E. coli* O157:H7 could survive between 32 and 33 days. In addition, the association of an acidic soil with the richness in organic carbon could result in a prolonged survival of *E. coli* O157:H7. This experimental study indicates the fact that the soil pH influences the adsorption and desorption of soil minerals by the pathogen, nutritional availability of soil components, and heavy metal toxicity [70]. Similarly, Li and Stevens showed that the soil with low pH reduces the risk of contamination regardless the virulence of *E. coli* O157:H7 strains [79]; however, it was noted that the virulent *E. coli* O157:H7 strains survived less than the nonvirulent ones [68]. Cools *et al.* indicated that the soil's content in organic matter can be more influential on the pathogen survival than soil type [80, 81]. In this context, Brennan *et al.* found that clay loam soil has a greater nutrient availability and a fine texture which is favoring the long-term survivability of pathogenic *E. coli* [82]. In addition, the clay soil has more available micropores that favors the nutrient adsorption by the pathogen [83, 84]. For example, Fenlon *et al.* were able to isolate inoculated pathogenic *E. coli* over 4 months from clay and loam soils, and for 8 weeks from sandy soils [85]. For minimizing the long-term persistence of pathogenic *E. coli* in soil before planting, regulators and researchers are proposing several mitigation strategies (**Table 4**).

Recommendation	References
In agricultural areas where the risk of pathogen presence is high and the pathogen could be transferred to fresh produce crops without having in place a validated kill step process, planting should not be carried out.	[86]
Stoppage of soil amendment for a period of time prior to harvesting of fresh produce. After the use of manure, the "*90 to 120 days rule*" of not harvesting farm produce 90 days (for farm produce whose edible parts touch the soil) or 120 days (for farm produce whose edible parts do not touch the soil) must be applied.	[87, 88]
For either reducing the level of pathogens in the soil or applying the time rule to reduce pathogen to acceptable levels, the assessment of the planting land history, and of the adjacent land activities is required. Reducing the human (anthropogenic) activities which could disturb the nutrient resources and modifying the competition between native microbial communities and invasive species.	[79, 89, 90]
Similar hazards raise concern for proximity to waste stockpiling and management, composting operations, and run-off from areas of concentrated wildlife populations and urban environments.	[91]
Topographical features of the growing fields and adjacent land should also be considered in a hazard analysis to identify potential contamination sources.	[18]
Implementation of Good Agricultural Practices (GAP) and intervention strategies focused on the construction of ditches, establishment of buffer areas/physical barriers, setting up of fences around the farms to prevent animal intrusion, to re-direct or reduce runoff from animal production or other waste management operations.	[87, 92–94]
Before planting, the soil's acidity must be tested.	[95]
Encouraging growers to apply HACCP system in their primary production stage.	[24, 96, 97]

Table 4.
Examples of mitigation strategies recommended to be applied to growing soils and adjacent lands.

3.3 Manure and manure amended soils

In the fresh produce pre-planting and pre-harvest stages, amending the soil with organic fertilizers, such as manure, or biosolid fertilizers is a cost-effective alternative to chemical fertilizers, the later posing a great threat to humans and environment due to their potential toxicity. In farming, the use of manure is of paramount importance to enhancing the soil's fertility by primarily increasing its content of nutrients and other organic compounds required for improved production yields and agricultural sustainability. From a practical standpoint, manure is the solid part resulted after the segregation of the solid and liquid portions of the organic residual compounds from different origins (i.e., cattle, poultry, pigs etc.). Since manure has been used as an old, traditional farming practice, the advantages of using manure for soil replenishing with nutrients is well known. The studies performed over the last decades are scientifically validating the additional, multiple benefits of amending soils with manure: improving the soil's microbial diversity along with soils' agricultural properties such as soil density and structure (i.e., loosening up/breaking down the heavy soils), increment of water holding capacity [98], soil erosion, and to maintain the quality of "exhausted" soils due to the repeated use of agricultural lands—by application at the beginning of each growing season [99].

The addition of manure is performed before planting the soil and at different time periods during the fresh produce growth stages but not immediately before the harvesting stage. Manure can be applied as: solid manure (i.e., aged manure, compost, manure slurries, or manure tea). Among the identified pitfalls of soil manuring, the most important aspect is that the manure contains high levels of pathogens which can contaminate the leafy greens due to its ability to harbor and spread both animal- and human-origin pathogens including the *E. coli* O157:H7 in the farming environment. For minimizing to reduce this microbial risk, as a thumb rule, the manure must be generally added into the soil after being processed (aging or composting) and not at a stage near the produce harvesting time.

As a route of produce contamination, internalization of *E. coli* O157:H7 has been found highly prevalent on leafy greens, including lettuce leaves, when the soil has been fertilized with contaminated manure possibly due to the intake of the pathogen up through the leaves via the produce's root system [50]. Ekman *et al.* found that the *E. coli* can survive in manure amended soils and the viable *E. coli* O157:H7 numbers were declining by at least 3 logs after 50 days of manure application to the soil, regardless of the season of application [100]. Maximizing the time between manure application and harvest stage of the leafy greens is one avenue to allow the natural reductions of the target pathogen into the soil. Additionally, in manure amended soils, existing pathogens can colonize the seedlings during germination, or transfer from the manure amended soil to the leafy greens through water splashing (during irrigation or rain) or through soil dust [101, 102]. Islam *et al.* found more than 10 CFU/g *E. coli* O157:H7 on parsley and lettuce even when these produces were harvested after 160 and 70 days, respectively, when soil was amended with manure containing $\log_{10}$ 7 CFU/g *E. coli* [49]. In an experimental transfer "soil-to-crops" of *E.coli* O157:H7-inoculated manure, Suslow predicted that, once the contaminated manure was incorporated into the soil, a 99% reduction of *E. coli* O157:H7 viable population could take place after 60–120 days depending on soil type but also on other factors yet to be determined [103]. Later, other several other factors responsible for leafy greens contamination with manure pathogens were indicated by Baker and were based on the high variation of farming practices, from site to site: the use of untreated manure; the differences in manure storage methods, type of manure applied treatment including the time of manure piles resting undisturbed; the manure-handling equipment cleaning, sanitation, and segregation practices; lack of protection against wild animals of the manure sitting piles' [104]. The type of manure, aged (dried and compact) or manure slurry, and temperature could also influence the survivability of *E. coli* O157:H7. Under experimental conditions, Himathongkham found that the *E. coli* O157:H7 survival in aged cattle manure was higher at 20°C, while in fresh cattle manure slurry (1-part aged manure and 2-parts water) survivability was higher at 4°C and slightly reduced at 20°C [105]. Jiang *et al.* observed a more rapid decline of *E. coli* O157 in manure-amended unautoclaved soil at 21°C than at 5°C. This was attributed to an increase in microbial activity with temperature and consequently, greater competition for nutrients. These findings are important for elucidating the influence of temperature on *E. coli* O157:H7 survival in different types of manures used for soil fertilization [106].

It is established that the contaminated, untreated, or improperly treated manure has been implicated, worldwide, as a major source of pathogenic *E. coli* O157:H7-related foodborne outbreaks due to the consumption of leafy greens and fresh produce. Therefore, efficient manure management strategies and policies are required to be established and used on-farm. The public health is the ultimate, main objective of the manure management strategies which, for being successful, require multi-pronged approaches. Once adopted, these management strategies and policies should efficiently mitigate the negative impact of manure on the environment and on the leafy greens. Epidemiological and experimental studies conducted by CDC and FDA indicated manure as a major factor in the outbreaks due to *E. coli* O157:H7 and Shiga toxin-producing *E. coli* (STEC) [107–109].

To reduce the target pathogen and minimize the risk of leafy greens contamination via use of manure, FDA established a set of Good Agricultural Practices (GAPs),

Manure treatment status	Type of requirement for the application of manure.	Minimum application interval
Untreated	• Manure does not have contact with the leafy greens during application or the potential for contact with the leafy greens is minimized during manure application.	"Reserved"[a]
Treated[b,c]	• Manure does not have contact with the leafy greens during or after application.	0 days
	• Manure is applied in a manner that minimizes any potential contact with the leafy greens during or after application.	0 days

**Adapted from [110].*
[a]FDA is conducting additional research, working with other researchers, and working to conduct a formal risk assessment [111].
[b]A scientifically valid controlled physical, chemical, or biological process, or a combination of scientifically valid controlled physical, chemical, and/or biological processes to meet the requirements of microbial standard for E. coli O157:H7.
[c]Relevant national standard or E. coli O157:H7 is not detected using a method that has a detection limit of 0.3 MPN (Most Probable Numbers) per 1 gram or per 1 mL if liquid (i.e., agricultural manure tea) is being sampled as analytical portion.

Table 5.
Application requirements and the minimum application intervals of the manure depending on their treatment status and the potential on-field contact with leafy greens.*

as mitigation strategies to reduce the pathogen hazard. These strategies are related to the application and use of animal-origin manure and to the minimum application intervals for leafy greens according to manure treatment types (**Tables 5** and **6**) [110, 111]. In addition, minimizing direct or indirect contact between manure and the leafy greens especially at a stage closer to the harvesting time could be used as a method to reduce the contamination with *E. coli* O157:H7 [111].

Accordingly, there are other treatments on which fresh produce growers can rely on for minimizing the pathogens hazards, such as: allowing enough passage of time in conjunction with the action of other environmental factors (i.e., environmental temperature, moisture fluctuations, and solar ultraviolet irradiation) to ensure the manure is properly aged and decomposed before first application to fields. These type of manure treatments, are known as "*passive treatments*". Its disadvantage is that the treatments are time consuming compared to the "*active treatments*" because they depend on the type and source of manure, and on the climatic factors (regional and/or seasonal). When manure aging is used as a passive treatment U.S. FDA cautioned on not confusing this process with composting process, the latter being solely applied as an active treatment [111]. In addition, produce contamination with pathogens occurs if the manure is not treated before use, or if the untreated manure does not respect the recommended application method during produce growing (**Table 5**) [110].

The accepted manure active treatments consist in the application of a scientifically controlled processes such as: physical (i.e., thermal treatment), chemical (i.e., highly alkaline digestion), biological (i.e., composting), or a combination of those so that *E. coli* O157:H7 levels satisfy the microbial accepted standard levels (**Table 6**).

Similarly, in 2017, the European Union Commission in collaboration with the European Food Safety Agency (EFSA) established Good Agricultural Practices for the application of animal manures and the minimum pre-harvest intervals that should be followed when growers use organic fertilizers for leafy greens based on manure treatment types and manure microbial quality [112].

Examples of good agricultural practices (GAPs) for reducing the pathogen levels	
1. Manure treatments	
	1. Passive treatments: • Relying primarily on the passage of time, in conjunction with the influence of environmental factors: temperature, moisture fluctuations and natural ultraviolet (UV) irradiation; holding time for passive treatments will vary depending on regional and seasonal climatic factors and on the type and source of manure. • Growers should ensure the passive treated manure is well aged and decomposed before applying to fields.
	2. Active treatments: • Pasteurization, heat drying, anaerobic digestion, alkali stabilization, aerobic digestion, or combinations of these.
2. Manure handling and application	
	1. General: • Manure storage and treatment sites should be situated as far as practicable from fresh produce production and handling areas. • Consider barriers or physical containment to secure manure storage or treatment areas where contamination from runoff, leaching, or wind spread is a concern. • Consider good agricultural practices to minimize leachate resulting from manure storage or treatment areas contaminating produce. • Consider practices to minimize the recontamination potential of the treated manure.
	2. Untreated manure: • Consider incorporating manure into the soil prior to planting. • Applying raw manure, or leachate from raw manure, to produce fields during the growing season prior to harvest is not recommended. • Maximize the time between application of manure to produce production areas and harvest. • Where it is not possible to maximize the time between application and harvest, such as for fresh produce crops which are harvested throughout most of the year, raw manure should not be used.
	3. Treated manure[a]: • Avoid contamination of fresh produce from manure that is in the process of being composted or otherwise treated. • Apply good agricultural practices that ensure that all materials receive an adequate treatment.

**Source: [146].*
[a]If the manure is not treated on-farm then: (i) Growers purchasing manure should obtain a specification sheet from the manure supplier for each shipment of manure containing information about the method of treatment, (ii) Growers should contact state or local manure handling experts for advice specific to their individual operations and regions.

Table 6.
Control measures for minimizing E. coli O157:H7 and other microbial hazards.*

3.4 Use of irrigation waters

When grown in open fields, leafy greens can become contaminated inside roots and leaves with *E. coli* O157:H7 when irrigation is performed with contaminated water and by the irrigation method [49, 113] and could become the source of many outbreaks [114, 115]. The transmission of the pathogens from contaminated irrigation waters has been elucidated [50, 116], and secondary vehicles by which *E. coli* O157:H7 may contaminate the leafy greens were identified: flood irrigation with water contaminated either with animal feces or by contact with surface runoff [117, 118]. Experimental and on-field studies indicated the ability of the pathogen to survive for extended periods in water [119, 120].

Many different sources of water and methods are used for irrigation of fresh produce [121]. As water sources are identified two main groups: (a) surface water or treated wastewater (more prone to contamination and presents variables in water quality parameters); and (b) ground water reserves or collected rainfall water (which is less prone to contamination and more controlled from microbial quality standpoint if stored properly). Using drip or subsurface irrigation limits direct contact between edible plant tissue and irrigation water (splashes) and thus is less likely to introduce pathogens than furrow or sprinkler/overhead irrigation. Drip irrigation (subsurface irrigation) has less impact on leafy greens' contact with the pathogen and pathogen survival compared to other irrigation methods such as spraying, surface irrigation, and furrow which favor the subsequent survival of the pathogen up to 56 days [49, 122–124].

On the farm, to ensure the leafy greens protection from pathogen contamination, checking the water source history, application of preventive control measures to prevent contamination or to eliminate the pathogen (i.e., frequent sanitary surveys of water reservoirs and distribution systems, identification, and surveillance of drainages at the confluence points of water sources) are aspects of importance [14, 93, 125]. The preventive control measures are usually combined with different water treatments: filtration, disinfection, or solar irradiation (UV natural treatment) [89, 126, 127]. Similarly, FDA issued a set of GAPs for produce growers which includes: (a) identification of the source and distribution of water used and check its relative potential for being a source of pathogens; (b) maintain water wells in good working condition; (c) revision of existing practices and conditions to identify potential sources of contamination (direct or indirect contamination, contamination from human or animal waste); (d) check the current and historical use of land since agricultural water is frequently a shared resource with other operations or affected by human activity); and (e) test the irrigation water microbial quality [128]. Regarding the GAPs implementation, the Canadian Ministry of Agriculture, Food and Rural Affairs, proposed to farmers additional management practices to avoid or reduce the risk of contamination: (a) choose a different irrigation method (i.e., use drip or trickle irrigation systems rather than overhead sprinklers); (b) choose a different water source, or (c) for some irrigation systems and applications, water treatment is required to improve its quality (**Table** 7) [129].

3.5 Other factors which can contribute to pathogen contamination of leafy greens

3.5.1 Domestic and wild animals

An extensive number of post-foodborne outbreaks epidemiological surveys recognize the interconnection between animal activity on or in the proximity of growing fields and, leafy greens contamination with pathogenic *E. coli* [49, 130]. Regardless the leafy greens production phase, animals, domesticated (i.e., nearby livestock and on-farm working animals) and wildlife, can shed and transfer *E. coli* O:157:H7 to the produce, even the animals do not display any signs of illness. Among animals themselves, a zoonotic vicious cycle can take place. In many instances, cross contamination via fecal matter between domestic and wild animals have been identified, and approximately 77% of the pathogens

Irrigation water source	Best management practices
Streams	a. Use an off stream settling pond-allows large particles that may contain pathogens to settle out of the water and reduce the potential contaminant load.
	b. Work with neighbors (animal farms, industrial parks, households etc.) to reduce livestock access to water sources.
	c. Establish vegetative buffer zones to filter water and slow down run-off.
Ponds	a. Fence pond to prevent animals, both wildlife and domestic, from defecating in or near water.
	b. Re-direct runoff so that it flows around the pond and avoids contaminants entering pond through runoff.
	c. Establish grassed waterways or vegetative buffer strips to filter water before it enters the pond.
	d. Install steep sides or rocky berms to discourage geese from nesting.
Stream-fed ponds	a. Avoid harvesting water during the peak flows after a rainfall—this water carries most of the sediment (and possibly pathogens) washed by the rainfall.
	b. Establish vegetative buffer zones to filter water and slow down run-off.
Wells	a. Mound up the ground around the outside of the well or well pit with clean earth to provide drainage for surface water so that runoff flows away from the well.
	b. Maintain well casing above grade.
	c. Ensure that well casing is intact and there are no cracks or openings.
	d. Don't allow any space between the well casing and the surrounding soil (this could act as a pathway for surface water to contaminate the well).

**Source: [129].*

Table 7.
Best management practices for different irrigation water sources.*

that infect livestock can also infect wildlife (deer, geese, rodents, foxes etc.) which, in turn, can re-infect the livestock populations [131, 132]. For the past 10 years, FDA investigation findings on previous foodborne illness outbreak indicate the proximity of cattle operations as a main contributing factor for pathogenic *E. coli* contamination of leafy greens, cattle being repeatedly demonstrated to be a persistent source of E. coli O157:H7 [133]. In addition, leafy greens can become contaminated with antibiotic-resistant *E. coli* which can represent a real danger for public health. This fact was discovered when the *E. coli* isolates from lettuce production sites were compared with the animal-derived *E. coli* strains, and it was determined that these antimicrobial-resistant strains was prevalent in cattle [134]. Due to this high risk, a relatively recent report was issued by the European Food Safety Authority (EFSA) which attempted to ascertain to what extent fresh produce represents a vehicle for the acquisition by humans of antimicrobial-resistant bacteria and to identify possible control options [135–137]. Since food safety is a shared responsibility among all sectors ample animal management guidelines and mitigation strategies were proposed for protecting leafy greens but also fresh produces from being contaminated with pathogens at any stage of production [96, 138]. The on-field protection of the produce against pathogen cross-contamination from the existing multiple sources, regardless if the contamination sources are placed on the farm's premises or outside the farm, several practical protection strategies can used without disturbing the production chain (**Table 8**).

Location and type of pathogen contamination sources	Examples of management strategies
On-farm:	
a. Working animals	When working animals are needed to be used during harvest, minimizing animal contact with the produce must be reduced by: • Establishing "*no harvest*" buffer zones since working animals defecate in the field. • For animal and manure handling, development of standard operating procedure (SOPs) regarding hand washing, cleaning and sanitizing tools, and other sanitary practices to be completed after animal handling.
b. Mixed farming (i.e., animal farming and leafy greens production)	When both livestock and fresh produce production facilities are located on the same farm, implement farm policies, such as: • Require workers from animal holding areas to change their shoes or boots and clothing before entering fresh produce fields to prevent cross-contamination. • Proceed to train the employees to identify contaminants and determine when to not harvest produce that is likely to be contaminated with a known or reasonably foreseeable hazard. • Train the workers to washing hands after touching animals or any waste of animal origin before handling the produce.
Outside of the farm/produce fields:	
a. Dairy, livestock, or poultry nearby production facilities	Avoid locating produce fields and packing areas adjacent to dairy, livestock, or poultry production facilities unless adequate physical barriers are put in place, such as: ditches, mounds, grass/sod waterways, diversion berms, and vegetative buffer areas. The physical barriers will help to re-direct or reduce runoff from animal production or waste management operations, and to exclude free-roaming livestock from fresh produce fields.
b. Other animals (such as livestock from nearby farms)	Monitor for any signs of animal entry such as the presence of feces or damage to the crop and consider adding barriers to prevent animal waste from adjacent fields from contaminating produce fields during heavy rains, especially if fresh produce is grown in low-lying fields or orchards. If any animal holding areas are nearby, assure that the produce fields are uphill and the manure or urine runoffs are away, downhill, from produce fields.
c. Wild animals, pests, birds	Addition of distress machines (i.e., sonic fences), scarecrows, reflective strips, or gunshots to ward off birds and pests from crops, and repellent substances.

**Adapted from [138].*

Table 8.
Examples of using on-farm management strategies to avoiding leafy greens pathogen contamination.*

3.5.2 Workers, on-farm activities, and farming equipment and tools

3.5.2.1 Farm workers

Authorized or unauthorized human activity, regardless the status, farm worker or trespassers, could take place on the farm premises and on growing fields and could result in the produce contamination with pathogens. However, due to the daily, continuous type of work, the farmers and farm workers are playing an important role in maintaining uncompromised the microbial safety of the fresh produce, leafy greens, respectively, while performing their duties. A survey of Midwestern United States farms brought up an important aspect: while the most farmers were familiar with GAPs, the GAPs were not fully implemented on farms because they did not believe that the fresh produce contamination with pathogens were the direct result of their on-farm practices and there are several factors which farmers view as obstacles in GAPs implementation (**Table 9**) [139, 140].

Also, multiple studies indicated that the workers' clothing, hands, feet, and training are involved in fresh produce contamination with pathogens [141–143]. The survey data obtained by Antwi-Agyei *et al.* supports the fact that on-farm workers' hygiene practices could favor the on-field produce contamination with pathogen via hands- and feet-to-soil contact: (a) 73% of workers are practicing open field defecation, while only 25% use a public toilet, and 2.4% other toilets; (b) the percentage of farmers' prior contact to fecal contamination was 69 (as hand-to-soil contact) and 74 (as feet-to-soil contact) [144]. In this context, the data from **Table 7** correlates with other findings related to farm workers hygiene and on-field activities, and, more important with the willingness and the ability of farmers to provide proper conditions for avoiding fresh produce contamination by field workers. Practically, the hygiene interventions specifically designed for produce farm workers and workers' hygiene behavior is affected by farmers on-site policies and offered food safety training. Surveys performed by Bartz *et al.*, Antwi-Agyei *et al.*, and Fabiszewski *et al.* support the fact that the farm workers' hands are the main contamination vehicle of leafy greens during pre-harvest activities, such as: bed preparation, transplanting, soil tilling, weed removal, irrigation, due to the lack of field workers' accessibility of toilets, handwashing posts, eating and resting posts, training, and facilitative work policies to encourage workers to respect and practice the on-field hygiene [141, 142, 144, 145]. According to common guide issued in 1998 by FDA in collaboration with U.S. Department of Agriculture (USDA) and CDC, both farmers and workers must be reminded: (a) that anything that comes in contact with fresh produce has the potential of contaminating it and, for most foodborne pathogens associated with the fresh produce, the major source of contamination is human or animal feces, and (b) worker hygiene and sanitation practices during production, harvesting, sorting, packing, and transportation play a critical role in minimizing the potential for microbial contamination of fresh produce [146]. The multiple survey-based research on farm management and on farm workers indicate several mitigation strategies that could be implemented concomitantly (**Table 10**).

Type of factors	(%)
High costs of workers' training and GAPs implementation	67
Lack of time for implementing GAPs	68
Non-existent on-farm technical solutions (i.e., water and soil testing, testing the health status of workers etc.)	26
Lack of knowledge of GAPs	17
Lack of knowledge how to prioritize and implement the GAPs	27
Lack of personnel training opportunities	35
On-farm implementation of GAPs minimizes the growers' profit	40

**Adapted from: [140].*
[a]Number of survey respondents, n = 143.

Table 9.
Examples of factors that farmers consider obstacles in implementing GAPs[a] on their farms[*,a].*

Strategies to avoid on-field pathogen contamination by workers activities or via workers hands	References
Encourage workers to taking time away from their activities to use sanitary facilities (toilets, hand washing posts), especially if these facilities are not placed within a reasonable distance from the growing fields to avoiding the pathogen transfer from hands to leafy greens.	[147, 148]
Farmers should consider that workers, who are conditioned by the payment of hourly or daily pre-established quantity of harvested fresh produce and by the fragile nature of leafy greens, may be strongly discouraged from taking time away from their activities to use distant sanitary facilities and, as a consequence they will use the actual growing fields as a "*sanitary facility*". Therefore, the position and the number of handwashing posts and toilets must be well established prior to start growing the leafy greens.	[144, 147]
Farmers must establish labor policies and food safety training to encourage workers to adopt hygiene behavior even when the temporary on-farm workforce represents a unique challenge to farmers, and although these activities could affect financially the farmers.	[141, 142, 145]

Table 10.
Examples of mitigation strategies for being applied concomitantly for ensuring on-farm food safety during pre-harvest stage.

3.5.2.2 Farming equipment and tools

Farming is labor intensive and require a variety of working equipment and tools for land preparation (i.e., primary and secondar tilling, primary and secondary applications of manure/fertilizers or pesticides and insecticides, sowing the seeds or transplanting the seedlings, folding, irrigation etc.) and management, manure management, and workers protective equipment. Focusing on the pre-harvest phase of the farm-to-fork chain, the manipulation of leafy greens in the field is of particular concern due the risk of cross-contamination of the produce from unsanitary, soiled farm equipment. Little or no cleaning and sanitation between activities, lack of equipment and tools segregation, and lack of proper storage represent major causes of concern since these can become a direct source of produce contamination with pathogens [66, 93, 148]. In addition, processing of produce in the field such manual practices, and mechanical activities should be performed in ways that reduce the contamination of produce from soil, workers, or equipment surfaces [66]. Since on-farm surveys indicated that some farms do not clean and sanitize properly their equipment, or the equipment was most commonly cleaned by using only water without applying detergents and sanitizers. However, if water alone is used for cleaning the equipment and tools, famers should use only water with high microbiological quality (comparable with drinking water). Several management strategies (standard operation procedures and good hygiene practices, good agricultural practices) were designed to assist farmers and farm workers to reduce the microbial hazard and the microbial cross-contamination between equipment and tools and the fresh produce (**Table 11**) [149–151].

4. Conclusions

A better understanding of the pathogens' behavior in pre-harvest environments will support the developing of effective on-farm food safety management strategies (GAPs, HACCP) and interventions that will ensure the delivery of a safe produce to the consumer. Leafy greens should be given a high food safety priority since they are an important vehicle for pathogenic *E. coli* and are playing an important role in the emergence of new foodborne outbreaks. There are many possible sources of contamination of leafy greens due to their exposure to many different environmental factors, and multiple handling phases until reaching the consumers. Moreover,

What	How
• Equipment and tools that may contact raw produce should be sanitized and maintained clean to reduce the risk of cross-contamination. • Produce contact implements should be cleaned using adequate washing, sanitizing, and rinsing protocols, and the frequency of these operations should be determined, and the schedule maintained. • Cleaning of implements should be performed in a separate area and at appropriate times to prevent contamination of growing produce. • Storage of these implements should be in a clean area separate from that of manure/compost to avoid contact and cross-contamination.	*Step 1*: The surface should be rinsed so any obvious dirt and debris are removed. *Step 2*: Apply an appropriate detergent[a] and scrub the surface. *Step 3*: Rinse the surface with water that is the microbial equivalent of drinking water (potable). *Step 4*: Apply an appropriate sanitizer[a]. *Step 5*: If the sanitizer requires a final rinse, this will require an extra step, namely surface air dry.

[a]*Detergents and sanitizers must be food grade.*

Table 11.
Examples of on-farm cleaning and sanitation procedures for equipment and tools.

pathogenic *E. coli* could survive in leafy greens for commercially relevant periods even multiple disinfection procedures are applied. Therefore, pre-harvest stage must be viewed and approached as an important process which favors the contamination with pathogenic *E. coli*. The improvement of leafy greens microbial safety can be achieved by embracing the farming management strategies which will help growers to re-examine their own farming processes for reducing or eliminating the food safety risks. Comprehensive surveys, risk assessments, and scientific research on pre-harvest factors are needed to continue to identify risks, mitigation priorities, and the efficacy of different intervention strategies. Because of the frequent growers' failure to implement food safety rules and guidelines on their production premises, the existing mitigation strategies are not a "silver bullet" for minimizing the risk of leafy greens pathogen contamination. Therefore, both regulators and researchers should use the existing and the new incoming information for proposing and continuously designing potential mitigation strategies to be implemented by farmers for reducing the risk of leafy greens contamination with pathogenic *Escherichia coli* to harmless levels. These mitigation strategies have to undergo changes and be re-designed to address newly identified and reported on-farm deviations or violations of the food safety guidelines or of the Good Agricultural Practices (GAPs). Accordingly, the farmers and farm managers should be persuaded and helped to undergo more training sessions. National and international organizations and agencies, and researchers must support farmers to maximize their understanding and adherence to food safety guidelines for increasing their awareness on their role in the assurance of food safety throughout the leafy greens *farm-to-fork* continuum.

Author details

Aura Darabă
Department of Food Science, Food Engineering and Food Biotechnology, Faculty of Food Science and Engineering, University "Dunarea de Jos" of Galati, Galati, Romania

*Address all correspondence to: aura.daraba@ugal.ro

References

[1] Stratton AE, Finley JW, Gustafson DI, Mitcham EJ, Myers SS, Naylor RL, et al. Mitigating sustainability tradeoffs as global fruit and vegetable systems expand to meet dietary recommendations. Environmental Research Letters. 2021;**16**:055010. DOI: 10.1088/1748-9326/abe25a

[2] Morris MC, Wang Y, Barnes LL, Bennett DA, Dawson-Hughes B, Booth SL. Nutrients and bioactives in green leafy vegetables and cognitive decline: Prospective study. Neurology. 2018;**90**(3):214-e222. DOI: 10.1212/WNL.0000000000004815

[3] Lyons GH, Goebel RG, Tikai P, Stanley K-J, Taylor M. Promoting nutritious leafy vegetables in the Pacific and northern Australia. Acta Horticulturae. 2015;**1102**:253-260. DOI: 10.17660/ActaHortic.2015.1102.31

[4] Ross MW, Robin DG. A new paradigm for world agriculture: Meeting human needs: Productive, sustainable, nutritious. Field Crops Research. 1999;**60**(1-2):1-10. DOI: 10.1016/S0378-4290(98)00129-4

[5] US Center of Disease Control and Prevention (CDC). *Escherichia coli*. 2014. Available from: https://www.cdc.gov/ecoli/general/index.html [Accessed: July 25, 2021]

[6] Linda L. New *E. coli* O145:H28 Outbreak Sickens 15. Little Information Provided. Food Poisoning Bulletin. 2021. Available from: https://foodpoisoningbulletin.com/2021/new-e-coli-o145h28-outbreak-sickens-15-little-information-provided/ [Accessed: July 25, 2021]

[7] US Center of Disease Control and Prevention (CDC). Multistate Outbreak of Shiga Toxin-producing *Escherichia coli* O26 Infections Linked to Raw Clover Sprouts at Jimmy John's Restaurants (Final Update). 2012. Available from: https://www.cdc.gov/ecoli/2012/o26-02-12/index.html [Accessed: July 21, 2021]

[8] Soon JM, Seaman P, Baines RN. *Escherichia coli* O104:H4 outbreak from sprouted seeds. International Journal of Hygiene and Environmental Health. 2013;**216**(3):346-354. DOI: 10.1016/j.ijheh.2012.07.005

[9] World Health Organization (WHO). Pre-harvest food safety: Report of a WHO consultation with the participation of the Food and Agriculture Organization of the United Nations and the Office of International des Epizooties. 2001. Available from: https://apps.who.int/iris/bitstream/handle/10665/68889/WHO_CDS_CSR_EPH_2002.9.pdf?sequence=1&isAllowed=y [Accessed: May 21, 2021]

[10] European Food Safety Authority and European Centre for Disease Prevention and Control (EFSA and ECDC). The European Union one health 2019 zoonoses report. EFSA Journal. 2021;**19**(2):6406. DOI: 10.2903/j.efsa.2021.6406

[11] US Center of Disease Control and Prevention (CDC). Reports of Selected *E. coli* Outbreak Investigations. 2020. Available from: https://www.cdc.gov/ecoli/2020/o157h7-10-20b/index.html [Accessed: July 1, 2021]

[12] Luna-Guevara JJ, Arenas-Hernandez MMP, Martínez de la Peña C, Silva JL, Luna-Guevara ML. The role of pathogenic *E. coli* in fresh vegetables: Behavior, contamination factors, and preventive measures. International Journal of Microbiology. 2019;**2019**:2894328. DOI: 10.1155/2019/2894328

[13] European Union Commission. Commission notice on guidance document on addressing microbiological

risks in fresh fruits and vegetables at primary production through good hygiene, 2017/C 163/01. Official Journal of the European Union. 2017;**60**:1-40

[14] EFSA Panel on Biological Hazards (BIOHAZ) Panel. Scientific Opinion on the risk posed by pathogens in food of non-animal origin. Part 1 (outbreak data analysis and risk ranking of food/ pathogen combinations). EFSA Journal. 2013;**11**(1):3025. DOI: 10.2903/j. efsa.2013.3025, . Available from: https:// efsa.onlinelibrary.wiley.com/doi/ epdf/10.2903/j.efsa.2013.3025 [Accessed: July 1, 2021]

[15] US Center of Disease Control and Prevention (CDC). Reports of Selected *E. coli* Outbreak Investigations. 2021. Available from: https://www.cdc. gov/ecoli/outbreaks.html [Accessed: July 1, 2021]

[16] Iwu CD, Okoh AI. Preharvest transmission routes of fresh produce associated bacterial pathogens with outbreak potentials: A review. International Journal of Environmental Research and Public Health. 2019;**16**:4407. DOI: 10.3390/ijerph16224407

[17] Food and Agriculture Organization (FAO). Food safety and quality: Preventing *Escherichia coli* in food. 2011. Available from: http://www.fao.org/ fileadmin/user_upload/agns/pdf/ FAO_E.Coli_FCC_2011.06.231.pdf [Accessed: May 22, 2021]

[18] Yang S-C, Lin C-H, Aljuffali IA, Fang J-Y. Current pathogenic *Escherichia coli* foodborne outbreak cases and therapy development. Archives of Microbiology. 2017;**199**:811-825. DOI: 10.1007/s00203-017-1393-y

[19] FAO/WHO. Microbiological hazards in fresh leafy vegetables and herbs: Meeting report. Microbiological Risk Assessment Series No. 14. 2008. ISBN 978-92-4-156378-9 (WHO), ISBN 978-92-5-106118-3 (FAO). Available from: http://www.fao.org/3/i0452e/ i0452e.pdf [Accessed: July 11, 2021]

[20] Yaron S, Römling U. Biofilm formation by enteric pathogens and its role in plant colonization and persistence. Microbial Biotechnology. 2014;7(6):496-516. DOI: 10.1111/ 1751-7915.12186

[21] Brandl MT, Amundson R. Leaf age as a risk factor in the contamination of lettuce with *Escherichia coli* O157:H7 and *Salmonella enterica*. Applied and Environmental Microbiology. 2008;**74**(8): 2298-2306. DOI: 10.1128/AEM.02459-07

[22] Lynch MF, Tauxe RV, Hedberg CW. The growing burden of foodborne outbreaks due to contaminated fresh produce: Risks and opportunities. Epidemiology and Infection. 2009; **137**(3):307-315

[23] Hallmann J, Quadt-Hallmann A, Miller WG, Sikora RA, Lindow SE. Endophytic colonization of plants by the biocontrol agent *Rhizobium etli* G12 in relation to *Meloidogyne incognita* infection. Phytopathology. 2001;**91**:415-422. DOI: 10.1094/phyto.2001.91.4.415

[24] Gil MI, Selma MV, López-Gálvez F, Allende A. Fresh-cut product sanitation and wash water disinfection: Problems and solutions. International Journal of Food Microbiology. 2009;**134**(1-2):37-45. DOI: 10.1016/j.ijfoodmicro. 2009.05.021

[25] Parish ME, Beuchat LR, Suslow TV, Harris LJ, Garrett EH, Farber JN, et al. Methods to reduce/eliminate pathogens from fresh and fresh-cut produce. Comprehensive Reviews in Food Science and Food Safety. 2003;**2**(Supplement): 161-173. DOI: 10.1111/j.1541-4337.2003. tb00033.x

[26] Ximenes E, Hoagland L, Ku S, Li X, Ladisch M. Human pathogens in plant biofilms: Formation, physiology, and detection. Biotechnology and

Bioengineering. 2017;**114**:1403-1418. DOI: 10.1002/bit.26247

[27] Rowaida KK, Joseph FF. Behavior of *Escherichia coli* O157:H7 on damaged leaves of spinach, lettuce, cilantro, and parsley stored at abusive temperatures. Journal of Food Protection. 2010;**73**(2): 212-220. DOI: 10.4315/0362-028X-73. 2.212

[28] Tarver T. Biofilms: A threat to food safety. Food Technology Magazine. 2009;**63**(2):46-52. Available from: https://www.ift.org/news-and-publications/food-technology-magazine/issues/2009/february/features/biofilms-a-threat-to-food-safety [Accessed July 6, 2021]

[29] Beattie GA, Lindow SE. Bacterial colonization of leaves: A spectrum of strategies. Phytopathology. 1999;**89**:353-359. DOI: 10.1094/PHYTO.1999.89.5.353

[30] Patel J, Sharma M. Differences in attachment of *Salmonella enterica* serovars to cabbage and lettuce leaves. International Journal of Food Microbiology. 2010;**139**:41-47. DOI: 10.1016/j.ijfoodmicro.2010.02.005

[31] Macarisin D, Patel J, Bauchan G, Giron JA, Sharma VK. Role of curli and cellulose expression in adherence of *Escherichia coli* O157:H7 to spinach leaves. Foodborne Pathogens and Disease. 2012;**9**:160-167. DOI: 10.1089/fpd.2011.1020

[32] Laus MC, van Brussel AA, Kijne JW. Role of cellulose fibrils and exopolysaccharides of *Rhizobium leguminosarum* in attachment to and infection of *Vicia sativa* root hairs. Molecular Plant-Microbe Interactions. 2005;**18**:533-538. DOI: 10.1094/MPMI-18-0533

[33] Dunne WM Jr. Bacterial adhesion: Seen any good biofilms lately? Clinical Microbiology Reviews. 2002;**15**: 155-166. DOI: 10.1128/CMR.15.2.155-166.2002

[34] Takeuchi K, Matute CM, Hassan AN, Frank JF. Comparison of the attachment of *Escherichia coli* O157:H7, *Listeria monocytogenes*, *Salmonella Typhimurium*, and *Pseudomonas fluorescens* to lettuce leaves. Journal of Food Protection. 2000;**63**:1433-1437. DOI: 10.4315/0362-028x-63.10.1433

[35] Hernandez-Reyes C, Schikora A. *Salmonella*, a cross-kingdom pathogen infecting humans and plants. FEMS Microbiology Letters. 2013;**343**:1-7. DOI: 10.1111/1574-6968.12127

[36] Yaron S. Microbial attachment and persistence on plants. In: Matthews KR, Sapers GM, Gerba CP, editors. The Produce Contamination Problem. 2nd ed. Cambridge, Massachusetts, USA: Academic Press; 2014. pp. 21-57. DOI: 10.1016/B978-0-12-404611-5.00002-6. ISBN 9780124046115

[37] Torres AG, Jeter C, Langley W, Matthysse AG. Differential binding of *Escherichia coli* O157:H7 to alfalfa, human epithelial cells, and plastic is mediated by a variety of surface structures. Applied and Environmental Microbiology. 2005;**71**:8008-8015.DOI: 10.1128/AEM.71.12.8008-8015.2005

[38] Shaw RK, Berger CN, Feys B, Knutton S, Pallen MJ, Frankel G. Enterohemorrhagic *Escherichia coli* exploits EspA filaments for attachment to salad leaves. Applied and Environmental Microbiology. 2008;**74**:2908-2914. DOI: 10.1128/AEM.02704-07

[39] Fink RC, Black EP, Hou Z, Sugawara M, Sadowsky MJ, Diez-Gonzalez F. Transcriptional responses of *Escherichia coli* K-12 and O157:H7 associated with lettuce leaves. Applied and Environmental Microbiology. 2012;**78**:1752-1764. DOI: 10.1128/AEM.00787-12

[40] Hirneisen KA, Sharma M, Kniel KE. Human enteric pathogen internalization

by root uptake into food crops. Foodborne Pathogens and Disease. 2012;**9**:396-405. DOI: 10.1089/fpd.2011.1044

[41] Pinto R, Ioffe M. Evidence for internalization of *Escherichia coli* into the aerial parts of maize via the root system. Journal of Food Protection. 2007;**70**:471-475. DOI: 10.1016/j.foodres.2011.06.058

[42] Melotto M, Underwood W, Koczan J, Nomura K, He SY. Plant stomata function in innate immunity against bacterial invasion. Cell. 2006;**126**(5):969-980. DOI: 10.1016/j.cell.2006.06.054

[43] Warriner K, Ibrahim F, Dickinson M, Wright C, Waites WM. Internalization of human pathogens within growing salad vegetables. Biotechnology and Genetic Engineering Reviews. 2003;**20**:117. DOI: 10.1080/02648725.2003.10648040

[44] Hirano SS, Upper CD. Bacteria in the leaf ecosystem with emphasis on *Pseudomonas syringae*-a pathogen, ice nucleus, and epiphyte. Microbiology and Molecular Biology Reviews. 2000;**64**:624-653. DOI: 10.1128/MMBR.64.3.624-653.2000

[45] Thordal-Christensen H. Fresh insight into processes of non-host resistance. Current Opinion in Plant Biology. 2003;**6**:351-335. DOI: 10.1016/S1369-5266(03)00063-3

[46] Guo X, Van Iersel MW, Chen J, Bracketand RE, Beuchat LR. Evidence of association of salmonellae with tomato plants grown hydroponically in inoculated nutrient solution. Applied and Environmental Microbiology. 2002;**68**:3639-3643. DOI: 10.1128/AEM.68.7.3639-3643.2002

[47] Abawi GS, Chen J. Concomitant pathogen and pest interactions. In: Barker KR, Pederson GA, Windham GL, Bartels J, Hatfield JM, Baenziger P, Bigham J, editors. Plant and Nematode Interactions. Series: Agronomy Monograph. Madison, USA: American Society of Agronomy. Vol. 36. 1998. DOI: 10.2134/agronmonogr36.c7

[48] Warriner K, Namvar A. The tricks learnt by human enteric pathogens from phytopathogens to persist within the plant environment. Current Opinion in Biotechnology. 2010;**21**(2):131-136. DOI: 10.1016/j.copbio.2010.01.004. ISSN 0958-1669. Available from: https://www.sciencedirect.com/science/article/pii/S0958166910000054

[49] Islam M, Doyle MP, Phatak SC, Millner P, Jiang X. Persistence of enterohemorrhagic *Escherichia coli* O157:H7 in soil and on leaf lettuce and parsley grown in fields treated with contaminated manure composts or irrigation water. Journal of Food Protection. 2004;**67**:1365-1370. DOI: 10.4315/0362-028X-67.7.1365

[50] Solomon EB, Yaron S, Matthews KR. Transmission of *Escherichia coli* O157:H7 from contaminated manure and irrigation water to lettuce plant tissue and its subsequent internalization. Applied and Environmental Microbiology. 2002;**68**:397-400. DOI: 10.1128/AEM.68.1.397-400.2002

[51] Warriner K, Ibrahim F, Dickinson M, Wright C, Waites WM. Interaction of *Escherichia coli* with growing salad spinach plants. Journal of Food Protection. 2003;**66**:1790-1797. DOI: 10.4315/0362-028X-66.10.1790

[52] Johannessen GS, Bengtsson GB, Heier BT, Bredholt S, Wasteson Y, Rorvik LM. Potential uptake of *Escherichia coli* O157:H7 from organic manure into crisp head lettuce. Applied and Environmental Microbiology. 2005;**71**:2221-2225. DOI: 10.1128/AEM.71.5.2221-2225.2005

[53] Franz E, Visser AA, Van Diepeningen AD, Klerks MM,

Termorshuizen AJ, van Bruggen AHC. Quantification of contamination of lettuce by GFP-expressing *Escherichia coli* O157:H7 and *Salmonella enterica* serovar Typhimurium. Food Microbiology. 2007;**24**:106-112. DOI: 10.1016/j.fm.2006.03.002

[54] Jablasone J, Warriner K. Griffiths M. Interactions of *Escherichia coli* O157:H7, *Salmonella typhimurium* and *Listeria monocytogenes* plants cultivated in a gnotobiotic system. International Journal of Food Microbiology. 2005;**99**(1):7-18. DOI: 10.1016/j.ijfoodmicro.2004.06.011

[55] Hora R, Warriner K, Shelp BJ, Griffiths MW. Internalization of *Escherichia coli* O157:H7 following Biological and Mechanical Disruption of Growing Spinach Plants. Journal of Food Protection. 2005;**68**(12):2506-2509. DOI: 10.4315/0362-028X-68.12.2506

[56] Mootian G, Wu WH, Matthews KR. Transfer of *Escherichia coli* O157:H7 from soil, water, and manure contaminated with low numbers of the pathogen to lettuce plants. Journal of Food Protection. 2009;**72**:2308-2312

[57] Pu S, Beaulieu JC, Prinyawiwatkul W, Ge B. Effects of plant maturity and growth media bacterial inoculum level on the surface contamination and internalization of *Escherichia coli* O157:H7 in growing spinach leaves. Journal of Food Protection. 2009;**72**(11):2313-2320. DOI: 10.4315/0362-028x-72.11.2313

[58] Erickson MC. Internalization of fresh produce by foodborne pathogens. Annual Review of Food Science and Technology. 2012;**3**:283-310. DOI: 10.1146/annurev-food-022811-101211

[59] Sharma M, Ingram DT, Patel JR, Millner PD, Wang X, Hull AE, et al. A novel approach to investigate the uptake and internalization of *Escherichia coli* O157:H7 in spinach cultivated in soil and hydroponic medium. Journal of Food Protection. 2009;**72**:1513-1520. DOI: 10.4315/0362-028X-72.7.1513

[60] Deering AJ, Mauer LJ, Pruitt RE. Internalization of *E. coli* O157:H7 and *Salmonella* spp. in plants: A review. Food Research International. 2012;**45**:567-575. DOI: 10.1016/j.foodres.2011.06.058

[61] Riggio GM, Jones SL, Gibson KE. Risk of human pathogen internalization in leafy vegetables during lab-scale hydroponic cultivation. Horticulturae. 2019;**5**(1):25. DOI: 10.3390/horticulturae5010025

[62] Wang Y-J, Deering J, Kim H-J. Effects of plant age and root damage on internalization of shiga toxin-Producing *Escherichia coli* in leafy vegetables and herbs. Horticulturae. 2021;**7**(4):68. DOI: 10.3390/horticulturae7040068

[63] Beuchat LR. Vectors and conditions for preharvest contamination of fruits and vegetables with pathogens capable of causing enteric diseases. British Food Journal. 2006;**108**:38-53. DOI: 10.1108/00070700610637625

[64] Park S, Navratil S, Gregory A, Bauer A, Srinath I, Szonyi B, et al. Multifactorial effects of ambient temperature, precipitation, farm management, and environmental factors determine the level of generic *Escherichia coli* contamination on preharvested spinach. Applied and Environmental Microbiology. 2015;**81**:2635-2650

[65] Jung Y, Jang H, Matthews KR. Effect of the food production chain from farm practices to vegetable processing on outbreak incidence. Microbial Biotechnology. 2014;**7**:517-527. DOI: 10.1111/1751-7915.12178

[66] Machado-Moreira B, Richards K, Brennan F, Abram F, Burgess CM. Microbial contamination of fresh produce: What, where, and how?

Comprehensive Reviews in Food Science and Food Safety. 2019;**18**:1727-1750. DOI: 10.1111/1541-4337.12487

[67] Marine SC, Pagadala S, Wang F, Pahl DM, Melendez MV, Kline WL, et al. The growing season, but not the farming system, is a food safety risk determinant for leafy greens in the mid-Atlantic region of the United States. Applied and Environmental Microbiology. 2015;**81**:2395-2407. DOI: 10.1128/AEM.00051-15

[68] Xing J, Wang H, Brookes PC, Salles JF, Xu J. Soil pH and microbial diversity constrain the survival of *E. coli* in soil. Soil Biology and Biochemistry. 2019;**128**:139-149. DOI: 10.1016/j.soilbio.2018.10.01. ISSN 0038-0717

[69] Mallon CA, Poly F, Le Roux X, Marring I, van Elsas JD, Salles JF. Resource pulses can alleviate the biodiversity–invasion relationship in soil microbial communities. Ecology. 2015;**96**:915-926. DOI: 10.1890/14-1001.1

[70] Zhang T, Wang H, Wu L, Lou J, Wu J, Brookes PC, et al. Survival of *Escherichia coli* O157:H7 in soils from Jiangsu Province, China. PLoS One. 2013;**8**(12):e81178. DOI: 10.1371/journal.pone.0081178

[71] Mubiru DN, Coyne MS, Grove JH. Mortality of *Escherichia coli* O157:H7 in two soils with different physical and chemical properties. Journal of Environmental Quality. 2000;**29**: 1821-1825. DOI: 10.2134/jeq2000.00472425002900060012x

[72] van Elsas JD, Chiurazzi M, Mallon CA, Elhottovā D, Krištůfek V, Salles JF. Microbial diversity determines the invasion of soil by a bacterial pathogen. Proceedings of the National Academy of Sciences. 2012;**109**(4):1159-1164. DOI: 10.1073/pnas.1109326109

[73] Gagliardi JV, Karns JS. Persistence of *Escherichia coli* O157:H7 in soil and on plant roots. Environmental Microbiology. 2002;**4**:89-86. DOI: 10.1046/j.1462-2920.2002.00273.x

[74] Moynihan EL, Richards KG, Ritz K, Tyrrel SF, Brennan FP. Impact of soil type, biology and temperature on the survival of nontoxigenic *Escherichia coli* O157. Biology and Environment: Proceedings of the Royal Irish Academy. 2013;**113B**:1-6. DOI: 10.3318/BIOE.2013.05

[75] Westphal A, Williams ML, Baysal-Gurel F, LeJeune JT, McSpadden Gardener BB. General suppression of Escherichia coli O157:H7 in sand-based dairy livestock bedding. Applied and Environmental Microbiology. 2011; **77**(6):2113-2121. DOI: 10.1128/AEM.01655-10

[76] Majeed H, Gillor O, Kerr B, et al. Competitive interactions in *Escherichia coli* populations: The role of bacteriocins. The ISME Journal. 2011;**5**:71-81. DOI: 10.1038/ismej.2010.90

[77] Vidovic S, Block HC, Korber DR. Effect of soil composition, temperature, indigenous microflora, and environmental conditions on the survival of *Escherichia coli* O157:H7. Canadian Journal of Microbiology. 2007;**53**:822-829. DOI: 10.1139/W07-041

[78] Jincai M, Ibekwe AM, Crowley DE, Yang C-H. Persistence of *Escherichia coli* O157:H7 in major leafy green producing soils. Environmental Science & Technology. 2012;**46**(21):12154-12161. DOI: 10.1021/es302738z

[79] Li W, Stevens MHH. Fluctuating resource availability increases invasibility in microbial microcosms. Oikos. 2011;**121**:435-441. DOI: 10.1111/j.1600-0706.2011.19762.x

[80] Cools D, Merckx R, Vlassak K, Verhaegen J. Survival of *E. coli* and *Enterococcus* spp. derived from pig slurry in soils of different texture. Applied Soil

Ecology. 2001;**17**:53-62. DOI: 10.1016/S0929-1393(00)00133-5

[81] Lang NL, Smith SR. Influence of soil type, moisture content and biosolids application on the fate of *Escherichia coli* in agricultural soil under controlled laboratory conditions. Journal of Applied Microbiology. 2007;**103**:2122-2131. DOI: 10.1111/j.1365-2672.2007.03490.x

[82] Brennan FP, O'Flaherty V, Kramers G, Grant J, Richards KG. Long-term persistence and leaching of *Escherichia coli* in temperate maritime soils. Applied and Environmental Microbiology. 2010;**76**:1449-1455. DOI: 10.1128/AEM.02335-09

[83] Postma J, Hok-Hin ACH, Van Veen JA. Role of microniches in protecting introduced *Rhizobium leguminosarum* biovar trifolii against competition and predation in soil. Applied and Environmental Microbiology. 1990;**56**:495-502. ISSN 0099-2240

[84] Coleman DC, Crossley DA, editors. Fundamentals of soil ecology. London, UK: Academic Press; 2003. p. 205. ISBN: 0121797279

[85] Fenlon DR, Ogden ID, Vinten A, Svoboda I. The fate of *Escherichia coli* and *E. coli* O157 in cattle slurry after application to land. Symposium series (Society for Applied Microbiology). 2000;**29**:149S-156S

[86] Codex Alimentarius Commission Code of Hygienic Practice for Fresh Fruits and Vegetables. CAC/RCP 53. 2003. Revised in 2010 for Fresh Leafy Greens. Available from: http://www.fao.org/fao-who-codexalimentarius/sh-proxy/en/?lnk=1&url=https%253A%252F%252Fworkspace.fao.org%252Fsites%252Fcodex%252FStandards%252FCXC%2B53-2003%252FCXC_053e.pdf [Accessed: July 21, 2021]

[87] Food and Agriculture Organization of the United Nations FAO. Good Agricultural Practices for Greenhouse Vegetable Production in the South East European Countries. FAO Plant Production and Protection Paper. Rome, Italy: FAO; 2017. Available from: https://cris.unibo.it/retrieve/handle/11585/650492/471838/Irrigation%20management%20challenges%20and%20opportunities.pdf [Accessed July 20, 2021]

[88] Oliver DM, Haygarth PM, Clegg CD, Heathwaite AL. Differential, *E. coli* die-off patterns associated with agricultural matrices. Environmental Science & Technology. 2006;**40**:5710-5716

[89] Suslow TV, Oria MP, Beuchat LR, Garrett EH, Parish ME, Harris LJ, et al. Production practices as risk factors in microbial food safety of fresh and fresh cut produce. Comprehensive Reviews in Food Science and Food Safety. 2003;**2**:38-77. DOI: 10.1128/microbiolspec.PFS-0022-2018

[90] James J. Overview of microbial hazards in fresh fruit and vegetables operations. In: James J, editor. Microbial Hazards Identification in Fresh Fruit and Vegetables. New York: Wiley; 2006. pp. 1-36. DOI: 10.1002/0470007761.ch1

[91] Keraita B, Drechsel P, Amoah P. Influence of urban wastewater on stream water quality and agriculture in and around Kumasi, Ghana. Environment and Urbanization. 2003;**15**:171-178. DOI: 10.1177/095624780301500207

[92] Abu-Ashour J, Lee H. Transport of bacteria on sloping soil surfaces by runoff. Environmental Toxicology. 2000;**15**:149-153. DOI: 10.1002/(SICI)1522-7278(2000)15:2<149::AID-TOX11>3.0.CO;2-O

[93] Gil MI, Selma MV, Suslow T, Jacxsens L, Allende A. Pre and postharvest preventive measures and intervention strategies to microbial food safety hazards of fresh leafy vegetables.

Critical Reviews in Food Science and Nutrition. 2015;**55**:453-468. DOI: 10.1080/10408398.2012.657808

[94] Stier RF, Nagle NE. Growers Beware: Adopt GAPs or Else. Available from: https://www.foodsafetymagazine.com/magazine-archive1/octobernovember-2001/growers-beware-adopt-gaps-or-else/ [Accessed: May 7, 2019]

[95] Zhang T, Wang H, Wu L, Lou J, Wu J, Brookes PC, et al. Survival of *Escherichia coli* O157:H7 in soils from Jiangsu Province, China. PLoS One. 2013;**8**(12):e81178. DOI: 10.1371/journal.pone.0081178

[96] Aber H, Mulindwa J, Lung'aho M, Nyakundi F, Wambui J, Jager M, et al. Hazard analysis and critical control point plan for hazards in Ugandan Amaranth vegetable value chain. African Journal of Food Agriculture Nutrition and Development. 2019;**19**(2):14458-14482. DOI: 10.18697/ajfand.85.17425

[97] Suslow T, Jacxsens L, Uyttendaele M, Allende A. Pre- and postharvest preventive measures and intervention strategies to control microbial food safety hazards of fresh leafy vegetables. Critical Reviews in Food Science and Nutrition. 2015;**55**(4):453-468. DOI: 10.1080/10408398.2012.657808

[98] Sharma M, Millner PD, HashemF VBT, East C, Handy ET, White K, et al. Survival of *Escherichia coli* in manure-amended soils is affected by spatiotemporal, agricultural, and weather factors. Applied and Environmental Microbiology. 2019;**85**:e02392-e02318. DOI: 10.1128/aem.02392-18

[99] Larney FJ, Angers DA. The role of organic amendments in soil reclamation: A review. Canadian Journal of Soil Science. 2012;**92**:19-38. DOI: 10.4141/cjss2010-064

[100] Ekman J, Goldwater A, Bradbury M, Matthews J, Rogers G. Persistence of human pathogens in manure-amended Australian soils used for production of leafy vegetables. Agriculture. 2021;**11**:14. DOI: 10.3390/agriculture11010014

[101] Ingham SC, Fanslau MA, Engel RA, Breuer JR, Breuer JE, Wright TH, et al. Evaluation of fertilization-to-planting and fertilization-to-harvest intervals for safe use of noncomposted bovine manure in Wisconsin vegetable production. Journal of Food Protection. 2005;**68**:1134-1142. DOI: 10.4315/0362-028X-68.6.1134

[102] Jacobsen CS, Bech TB. Soil survival of *Salmonella* and transfer to freshwater and fresh produce. Food Research International. 2012;**45**:557-566. DOI: 10.1016/j.foodres.2011.07.026

[103] Suslow T. Addressing animal manure management issues for fresh vegetable production. Perishables Handling Quarterly. 1999;**98**:7-9. Available from: http://ucce.ucdavis.edu/files/datastore/234-207.pdf [Accessed: June 27, 2021]

[104] Baker CA, De J, Bertoldi B, Dunn L, Chapin T, Jay-Russell M, et al. Prevalence and concentration of stx+ *E. coli* and *E. coli* O157 in bovine manure from Florida farms. PLoS One. 2019;**14**(5):e0217445. DOI: 10.1371/journal. pone.0217445

[105] Himathongkham S, Bahari S, Riemann H, Cliver D. Survival of *Escherichia coli* O157:H7 and *Salmonella typhimurium* in cow manure and cow manure slurry. FEMS Microbiology Letters. 1999;**178**(2):251-257. DOI: 10.1111/j.1574-6968.1999.tb08684.x

[106] Jiang X, Morgan J, Doyle MP. Fate of *Escherichia coli* O157:H7 in manure-amended soil. Applied and Environmental Microbiology. 2002; **68**:2605-2609. DOI: 10.1128/AEM.68.5.2605-2609.2002

[107] US Center of Disease Control and Prevention CDC. Multistate outbreak of

shiga toxin-producing *Escherichia coli* O157:H7 infections linked to leafy greens (final update). 2018. Available from: https://www.cdc.gov/ecoli/2017/o157h7-12-17/index.html [Accessed: June 27, 2021]

[108] Food and Drug Administration (FDA). Outbreak Investigation of *E. coli*: Romaine from Salinas, California. 2019. Available from: https://www.fda.gov/food/outbreaks-foodborne-illness/outbreak-investigation-e-coli-romaine-salinas-california-november-2019 [Accessed: June 30, 2021]

[109] US Center of Disease Control and Prevention CDC. Reports of Selected *E. coli* Outbreak Investigations. 2021. Available from: https://www.cdc.gov/ecoli/outbreaks.html [Accessed: June 23, 2021]

[110] Food and Drug Administration FDA 21 CFR 112 (Subpart F). 2020. Available from: https://www.accessdata.fda.gov/scripts/cdrh/cfdocs/cfCFR/CFRSearch.cfm?CFRPart=112&showFR=1&subpartNode=21:2.0.1.1.12.6 [Accessed: June 27, 2021]

[111] Cornell University, USA. Produce Safety Alliance. Soil Summits. Available from: https://producesafetyalliance.cornell.edu/resources/soil-summits/ [Accessed: July 24, 2021]

[112] Commission notice on guidance document on addressing microbiological risks in fresh fruits and vegetables at primary production through good hygiene (2017/C 163/01). Official Journal of the European Union. 2017;**60**. ISSN 1977-091X. Available from: https://eur-lex.europa.eu/legal-content/EN/TXT/PDF/?uri=CELEX:52017XC0523(03)&from=EN [Accessed: June 30, 2021]

[113] Kljujev I, Raicevic V, Andrews S, Jackson R, Lalevic B, Dorati F. Transmission of *E. coli* from contaminated irrigation water and soil to plant tissue. Journal of Hygienic Engineering and Design. 2012;**1**:83-87. Available from: https://www.cabdirect.org/cabdirect/abstract/20133175028

[114] Hillborn ED, Mermin JH, Mshar PA, Hadler JL, Voetsch A, Wojtkunski C, et al. A multistate outbreak of *Escherichia coli* O157:H7 infections associated with consumption of mesclun lettuce. Archives of Internal Medicine. 1999;**159**:1758-1764. DOI: 10.1001/archinte.159.15.1758

[115] Centers for Disease Control and Prevention (CDC). Outbreak of *Escherichia coli* O157:H7 and Campylobacter among attendees of the Washington County Fair-New York. Morbidity and Mortality Weekly Report. 1999;**48**(36):803-805

[116] Ibekwe M, Shouse PJ, Grieve CM. Quantification of survival of *Escherichia coli* O157:H7 on plants affected by contaminated irrigation water. Engineering in Life Sciences. 2006;**6**(6): 566-572. DOI: 10.1002/elsc.200620157

[117] Ackers M-L, Mahon BE, Leahy E, Goode B, Damrow T, Hayes PS, et al. An outbreak of *Escherichia coli* O157:H7 infections associated with leaf lettuce consumption. The Journal of Infectious Diseases, June. 1998;**177**(6):1588-1593. DOI: 10.1086/515323

[118] Hilborn ED, Mermin JH, Mshar PA, Hadler JL, Voetsch A, Wojtkunski C, et al. A multistate outbreak of *Escherichia coli* O157:H7 infections associated with consumption of mesclun lettuce. Archives of Internal Medicine. 1999; **159**(15):1758-1764. DOI: 10.1001/archinte.159.15.1758

[119] Chalmers RM, Aird H, Bolton FJ. Waterborne *Escherichia coli* O157. Journal of Applied Microbiology. 2000;**88**:124S-132S. DOI: 10.1046/j.1365-2672.2000.00927.x

[120] Wang G, Doyle MP. Survival of enterohemorrhagic *Escherichia coli*

O157:H7 in water. Journal of Food Protection. 1998;**61**:662-667. DOI: 10.4315/0362-028X-61.6.662

[121] Uyttendaele M, Jaykus L-A, Amoah P, Chiodini A, Cunliffe D, Jacxsens L, et al. Microbial hazards in irrigation water: Standards, norms, and testing to manage use of water in fresh produce primary production. Comprehensive Reviews In Food Science And Food Safety. 2015;**14**: 336-356. DOI: 10.1111/1541-4337.12133

[122] Erickson MC, Webb CC, az-Perez JC, Phatak SC, Silvoy JJ, Davey L, et al. Surface and internalized *Escherichia coli* O157: H7 on field-grown spinach and lettuce treated with spray-contaminated irrigation water. Journal of Food Protection. 2010;**73**:1023-1029. DOI: 10.4315/0362-028X-73.6.1023

[123] Fonseca JM, Fallon SD, Sanchez CA, Nolte KD. *Escherichia coli* survival in lettuce fields following its introduction through different irrigation systems. Journal of Applied Microbiology. 2011;**110**:893-902. DOI: 10.4315/0362-028X-73.6.1023

[124] Mootian G, Wu WH, Matthews KR. Transfer of *Escherichia coli* O157:H7 from soil, water, and manure contaminated with low numbers of the pathogen to lettuce plants. Journal of Food Protection. 2009;**72**:2308-2312. DOI: 10.4315/0362-028X-72.11.2308

[125] EFSA BIOHAZ Panel (EFSA Panel on Biological Hazards). Scientific opinion on the risk posed by pathogens in food of non-animal origin. Part 2 (*Salmonella* and Norovirus in leafy greens eaten raw as salads). EFSA Journal. 2014;**12**(3):3600. DOI: 10.2903/j.efsa.2014.3600, . Available from: https://orbit.dtu.dk/en/publications/efsa-biohaz-panel-efsa-panel-on-biological-hazards-2014-scientifi-9 [Accessed: July 21, 2021]

[126] Carr R, Blumenthal U. Guidelines for the safe use of wastewater in agriculture: Revisiting WHO guidelines. Water Science & Technology. 2004;**50**:31-38. DOI: 10.2166/wst.2004.0081

[127] Bichai F, Polo-López MI, Ibáñez PF. Solar disinfection of wastewater to reduce contamination of lettuce crops by *Escherichia coli* in reclaimed water irrigation. Water Science and Technology. 2012;**46**(18):6040-6050. DOI: 10.1016/j.watres.2012.08.024

[128] Bourquin LD. Good Agricultural Practices—Summary of Guidance on Irrigation Water Quality. Department of Food Science and Human Nutrition, and Michigan State University Extension; 2000. Available from: https://www.ksre.k-state.edu/foodsafety/produce/guidance/docs/irrigation.pdf#:~:text=Good%20agricultural%20practices%20may%20include%20protecting%20surface%20waters%2C,from%20contaminating%20agricultural%20water%20sources%20and%20produce%20crops [Accessed: July 20, 2021]

[129] Jones S, Shortt R. Improving On-Farm Food Safety through Good Irrigation Practices. Canada: Ministry of Agriculture, Food and Rural Affairs; 2010. Available from: http://omafra.gov.on.ca/english/crops/facts/10-037.htm [Accessed: July 23, 2021]

[130] Wang G, Zhao T, Doyle MP. Fate of enterohemorrhagic *Escherichia coli* O157:H7 in bovine feces. Applied and Environmental Microbiology. 1996;**62**:2567-2570. DOI: 10.1128/aem.62.7.2567-2570.1996

[131] Brackett RE. Incidence, contributing factors, and control of bacterial pathogens in produce. Postharvest Biology and Technology. 1999;**15**(3):305-311. DOI: 10.1016/S0925-5214(98)00096-9

[132] McAllister TA, Topp E. Role of livestock in microbiological contamination of water: Commonly the

blame, but not always the source. Animal Frontiers. 2012;**2**(2):17-27. DOI: 10.2527/af.2012-0039

[133] Food and Drug Administration (FDA). Investigation Report: Factors Potentially Contributing to the Contamination of Leafy Greens Implicated in the Fall 2020 Outbreak of *E. coli* O157:H7. The challenge of successful coexistence of various types of agricultural industries to ensure food safety and protect consumers against foodborne illnesses. 2020. Available from: https://www.fda.gov/media/147349/download [Accessed: July 6, 2021]

[134] Holvoet K, Sampers I, Callens B, Dewulf J, Uyttendaele M. Moderate prevalence of antimicrobial resistance in *Escherichia coli* isolates from lettuce, irrigation water, and soil. Applied and Environmental Microbiology. 2013; **79**(21):6677-6683. DOI: 10.1128/AEM.01995-13

[135] EFSA BIOHAZ Panel (EFSA Panel on Biological Hazards). Scientific Opinion on the risk posed by pathogens in food of non-animal origin. Part 2 (Salmonella and Norovirus in leafy greens eaten raw as salads). EFSA Journal. 2014;**12**(3):3600. DOI: 10.2903/j.efsa.2014.3600. Available from: https://orbit.dtu.dk/en/publications/efsa-biohaz-panel-efsa-panel-on-biological-hazards-2014-scientifi-9 [Accessed: July 21, 2021]

[136] EFSA Publication. Scientific opinion of the panel on biological hazards on a request from the European Food Safety Authority on foodborne antimicrobial resistance as a biological hazard: Question No EFSA-Q-2007-089. EFSA Journal. 2008;**765**:87. DOI: 10.2903/j.efsa.2008.765. Available from: https://www.efsa.europa.eu/en/efsajournal/pub/765 [Accessed: July 26, 2021]

[137] EFSA Publication. Report of the Public Consultation on the EFSA Draft Opinion on "Foodborne antimicrobial resistance as a biological hazard". EFSA Journal. 2008;**749**. DOI: 10.2903/j.efsa.2008.749. Available from: https://www.efsa.europa.eu/en/efsajournal/pub/749 [Accessed: July 26, 2021]

[138] Bhullar M. Domesticated Animals Factsheet—FSMA Produce Safety Rule. Available from: https://www.ksre.k-state.edu/foodsafety/produce/docs/NCR-Domesticated-Animals-Factsheet.pdf [Accessed July 27, 2021]

[139] Ivey ML, LeJeune JT, Miller SA. Vegetable producers perceptions of food safety hazard in the Midwestern USA. Food Control. 2012;**26**:453-465. DOI: 10.1016/j.foodcont.2012.01.065

[140] Sinkel D, Khouryieh H, Daday JK, Stone M, Shen C. Knowledge and implementation of good agricultural practices among Kentucky fresh produce farmers, Food. Protection Trends. 2018;**38**(2):111-121. Available from: http://www.foodprotection.org/files/food-protection-trends/mar-apr-18-sinkel.pdf [Accessed: July 27, 2021]

[141] Fabiszewski de Aceituno A, Heredia N, Stern A, Bartz FE, Venegas F, Solis-Soto L, et al. Efficacy of two hygiene methods to reduce soil and microbial contamination on farmworker hands during harvest. Food Control. 2016;**59**:787-792. DOI: 10.1016/j.foodcont.2015.06.059

[142] Fabiszewski de Aceituno A, Bartz FE, Hodge DW, Shumaker DJ, Grubb JE, Arbogast JW, et al. Ability of hand hygiene interventions using alcohol-based hand sanitizers and soap to reduce microbial load on farmworker hands soiled during harvest. Journal of Food Protection. 2015;**78**:2024-2032. DOI: 10.4315/0362-028X.JFP-15-102

[143] Todd ECD, Greig JD, Bartleson CA, Michaels BS. Outbreaks where food workers have been implicated in the spread of foodborne disease. Part 5.

Sources of contamination and pathogen excretion from infected persons. Journal of Food Protection. 2008;**71**:2582-2595. DOI: 10.4315/0362-028X-71.12.2582

[144] Antwi-Agyei P, Biran A, Peasey A, et al. A faecal exposure assessment of farm workers in Accra, Ghana: A cross sectional study. BMC Public Health. 2016;**16**:587. DOI: 10.1186/s12889-016-3266-8

[145] Bartz FE, Lickness JS, Heredia N, Fabiszewski de Aceituno A, Newman KL, Hodge DW, et al. Contamination of fresh produce by microbial indicators on farms and in packing facilities: Elucidation of environmental routes. Applied and Environmental Microbiology. 2017;**83**:e02984-16. DOI: 10.1128/AEM.02984-16

[146] Food and Drug Administration FDA. Guidance for Industry: Guide to Minimize Microbial Food Safety Hazards for Fresh Fruits and Vegetables. 1998. Available from: https://www.fda.gov/regulatory-information/search-fda-guidance-documents/guidance-industry-guide-minimize-microbial-food-safety-hazards-fresh-fruits-and-vegetables [Accessed: June 25, 2021]

[147] Barak JD, Schroeder BK. Interrelationships of food safety and plant pathology: The life cycle of human pathogens on plants. Annual Review of Phytopathology. 2012;**50**:241-266. DOI: 10.1146/annurev-phyto-081211-172936

[148] Mehta K, Gabbard SM, Barratt V, Lewis M, Carroll D, Mines R. Findings from the National Agricultural Workers Survey (NAWS) 1997-1998: A demographic and employment profile of United States farmworkers. Washington, DC: US Department of Labor, Office of the Assistant Secretary for Policy, Office of Program Economics; 2000. Available from: https://www.doleta.gov/agworker/report_8.pdf

[149] Yang Y, Luo Y, Millner P, Turner E, Feng H. Assessment of *Escherichia coli* O157:H7 transference from soil to iceberg lettuce via a contaminated field coring harvesting knife. International Journal of Food Microbiology. 2012;**153**(3):345-350. DOI: 10.1016/j.ijfoodmicro.2011.11.024

[150] Stopforth JD, Kottapalliand B, Sofos JN. Interventions for hazard control in foods preharvest. In: Juneja VK, Sofos JN, editors. Pathogens and Toxins in Foods: Challenges and Interventions. Washington, DC: American Society for Microbiology Press; 2010. pp. 357-378. ISBN 978-1-55581-459-5

[151] Morris W. 2.00 Standard Operating Procedures. Available from: https://www.agconfoodsafety.com/uploads/2/5/5/7/2557880/sop_2.18_cleaning_and_maintenance_of_field_harvest_equipment.pdf [Accessed: July 28, 2021]

Section 3

Escherichia coli and Antimicrobial Resistance

Chapter 9

Antimicrobial Resistance in *Escherichia coli*

Waheed Ullah and Shandana Ali

Abstract

The ability of microbes to resist or neutralize the action of drugs that have been used against microbes is considered as antimicrobial resistance (AMR). AMR among different strains of *Escherichia coli* is considered as a major threat to public health. Drug-resistant in *E. coli* is found predominantly in the hospital sittings, in the community, and surrounding environment. It has adopted different defensive strategies to minimize the effects of drugs. Extended-spectrum β-lactamase (ESBL), fluoroquinolones, and carbapenemases have been considered as strong resistance strategies being present in most of resistant bacterial strains. Mobile genetic elements (MGEs) have the major contribution in the transfer of resistance genes in between or among bacterial cells. Plasmids are normally present in most of resistant strains, helping in the transfer of genetic material between bacterial cells. Transposons another MGEs, are being considered as one of the major sources of resistance transmission. Collectively, MGEs play an important role in facilitating in exchange, acquisition, and dissemination of resistance genes. Resistance in *E. coli* has been reported worldwide and there is variation in its resistance pattern. *CTX-M* ESBLs, carbapenems, colistin-resistant, and ST-131 *E. coli* resistant clones are considered the most dominant phenotypes. The aforesaid resistant variants are predominantly found in densely populated regions, Sub-Saharan Africa, China, and South Asian countries.

Keywords: antibiotics resistance, trends, mobile genetic elements, epidemiology

1. Introduction

Antimicrobial resistance is the capability of bacterial pathogens to neutralize the bactericidal effects of antibiotics. Antibiotic resistance arises due to the changes that take place in bacteria in a way that decreases the efficiency of antibiotics, chemicals, or other mediators that are used for infections control [1]. Globally, antimicrobial resistance is the main problem associated with humans' and animals' health. With the emergence of resistance clones, those antibiotics that were previously considered as broad-spectrum lost their efficacy, this increasing trend in resistant clone posture serious problem for the clinicians to deal with such pathogens. As we know that antibiotics are categorized according to the type of bactericidal activity, their mode of action, their chemical nature, and their origin. Further, these drugs can be characterized on the basis of their mode of action like their involvement in bringing complexity in the synthesis of the bacterial cell wall, depolarization of cell membrane, inhibiting microbial key protein synthesis, and altering nucleic acid synthesis. In early era, microbial secondary metabolites were considered the main treatment option for microbial infection, but later due to increasing resistance issues, synthetic derivatives of these natural products were being searched. There is a different reason that has pushed microbes to adopt drug resistance strategies. The use and misuse of antimicrobial agents have led to the emergence of resistance [2] Similarly the usage of low-standard antibiotics in some parts of the world particularly in underdeveloped countries may be the source of the emergence of drug resistance [3]. *Escherichia coli* strains are resistant because they are part of the natural microbiota of animals, humans and are found in the ecosystem [4]. *E. coli* is the most prevalent facultative bacteria found in humans and animals some strains being responsible for initiating infections. The foremost concern is their probable transmission of resistant *E. coli* strains among humans and animals. It uses different routes for their transmission such as direct contact, through food chains, or contact with animal excretions. *E. coli* strains that are considered as multidrug or extreme drug-resistant responsible for enteropathogenic and uropathogenic clones are a specific concern for world health. World Health Organization (WHO) have shown serious concern over the freely spread of resistant clone in the community and environment as it will pose threat to human health and the economy [5]. Although, it is one of the main reservoirs of resistance genes that might be responsible for treatment failures in both human and animal medicine. An increasing trend of resistant genes has been observed in *E. coli* in the current decade. Due to its large genomic fragments, MGEs are involved in the transfer of resistance genes in the enterobacteriaceae family, particularly among *E. coli* strains. Plasmids are normally present in most of resistant strains, help in the transfer of genetic material among bacterial species. Transposons another MGEs, are being considered as one of the major sources of resistance transmission. In *E. coli* several antimicrobial resistance trends are associated with plasmid-mediated colistin resistance *Mcr-1* gene [6]. But horizontal gene transfer [HGT] are mainly involved in resistance dissemination [7]. It is estimated that almost 700,000 deaths are attributed yearly, and this could increase to 10 million deaths worldwide annually by 2050. Almost 2.8 million people are suffering, and approximately 35,000 peoples die each year in the USA alone due to antimicrobial resistance [8].

2. AMR trends

Capability of bacterial species to resist the action of a particular antimicrobial agent is referred to as antimicrobial resistance, and this phenomenon has been remarkably proliferated over the years. The availability and usage of antimicrobial have contributed in the increased incidence of resistant strains [9]. Though antimicrobial resistance is a natural phenomenon and was considered under control in the past but recently it is envisaged a high-level risk for world health [10]. Mainly three reasons responsible for antimicrobial resistance are; (a) increase usage of antibiotics, (b) due to unseriousness of the patients about treatments being suggested, (c) replacement of the existing class of antibiotics with a new one. Bacterial resistance to antimicrobial agents is classified into three types, namely intrinsic resistance, adopted resistance, and acquired resistance see in **Figure 1**.

The most common example of an intrinsic resistance system is the Acr AB/Tol C EPs in *E. coli*, which has a wide substrate specificity and can export antibiotics, detergents, dyes, and various disinfectants [11]. *E. coli*, Tol C has many efflux systems including the resistance-nodulation-division (RND) pumps as well as the main facilitator superfamily (MFS) systems [12]. RND pumps function as proton antiporters and confer resistance to tetracyclines, chloramphenicol, some β-lactams, vancomycin, and fluoroquinolones being supported by intrinsic resistance [13, 14]. While adopted

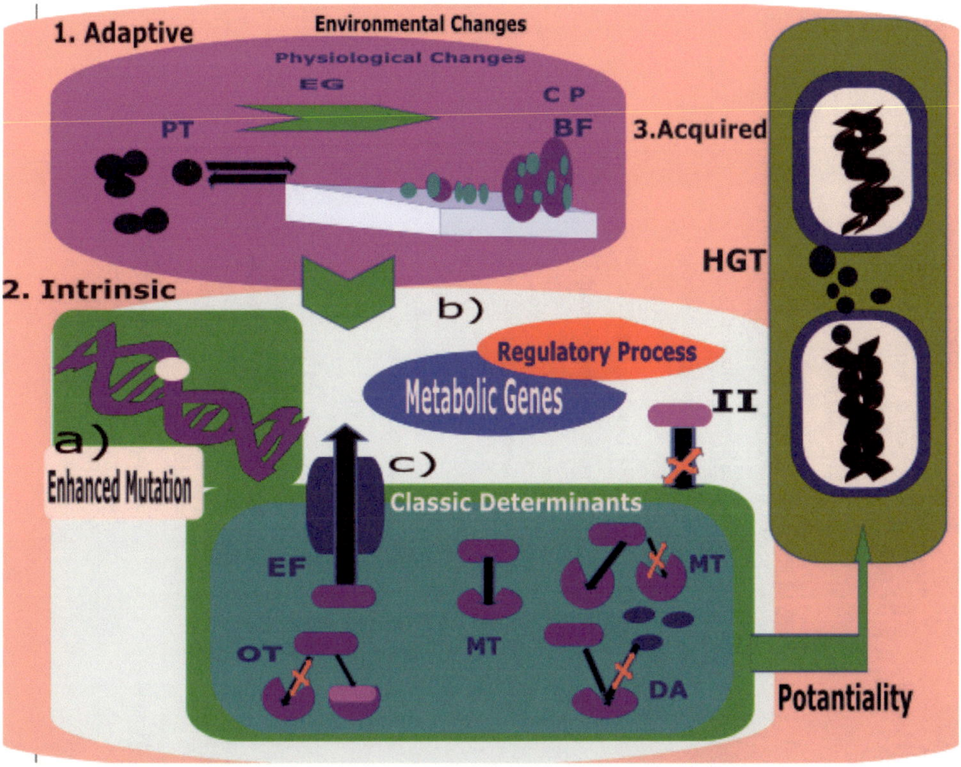

Figure 1.
Three types of antimicrobial resistance transmission and virulence factors can be classified into 1. adaptive resistance, 2. intrinsic resistance, and 3. acquired resistance. The adaptive resistance includes, environmentally induced EG (encoded genes) as two phases of bacteria 1) PT represents (planktonic), and BF (biofilms) can induce physiological changes at the cellular level (CP represents cellular process), and cause (a) enhanced mutation levels, (b) modification in metabolic genes and processes of the regulation, (c) classic determinants and a host antibiotic inactivation. Where EF shows efflux, OT (overprotection at the target site), MT (modification at the target site), MT (mutation at the target), DA represents the degradation of antibiotics and II represents impaired influx. This type of resistance increased infections which can potentially be transferred between E. coli strains leading to acquired resistance. Acquired resistance is transmitted through HGT among bacteria.

resistance contains environmentally induced genetic variations such as biofilm and persisted development, enzymatic driven inactivation of antibiotic see in **Figure 1** [15]. Due to adopted resistance, *E. coli* revealed resistance toward aminoglycoside encoded by *arm-A*, *npm-A*, *rmt-A*, *rmt-B*, *rmt-C*, and *rmt-D* resistant genes [16, 17]. The *rmt* gene provides resistance to gentamicin and amikacin, while *npm-A* provides resistance to gentamicin, neomycin, amikacin, and apramycin. While the most common ESBL gene in *E. coli* isolates of human origin is *blaCTX-M-15* and ST-131 clone and are mainly involved in dissemination AMR [18]. Similarly, the acquired resistance is usually influenced by HGT and may include plasmid-encoded specific EPs and enzymes that alter antibiotics [19, 20]. The increase in carbapenems (CPE) is mainly associated with the extensive dissemination of acquired CPE. CPE encoding genes are usually located in mobile genetic elements (MGEs), implying in the emergence of MDR and XDR strains [21]. Furthermore, colistin believes as a choice of drug for the treatment of resistant pathogens its resistance is facilitated through variations in lipopolysaccharides (LPS). *E. coli* the first pathogen in which plasmid-mediated colistin resistance was observed, through the acquisition of the *MCR-1* gene [6] The *MCR-1* gene could swiftly propagate and can impart resistance to other strains. *MCR-1* protein expression leads to the addition of a phosphor-ethanolamine group to lipid A. This produces a change in the charge of LPS, which in turn reduces the affinity of

Group of antibiotics	Resistance antibiotics	Enzyme produced	Gene involved	MGEs	Implication in virulence	References
β-Lactams		β-Lactamases	*Amp C*	Plasmid	Mutation and overexpression of genes	[23]
	Penicillin	Penicillinases				
	Cephalosporin	Cephalosporinases	*CTX-Ms*		Plasmid encoded, increased virulence in urinary tract	[24]
			OXA		Changes in peptidoglycan composition	
	Cephamycin	Amp C β-lactamases	*Amp C*		induced hyperproduction	
Fluoroquinolones	Norfloxacin	β-Lactamases ESBL	*Gyr-gene*	Plasmid and Transposons	Interfere with nucleic acid synthesis caused mutation	[25]
	Ciprofloxacin		*Par-C*		To inhibit topoisomerase IV	
Aminoglycosides	Tobramycin		*Arm-A*		Higher risk of illness and death	[26]
	Gentamicin		*Rmt-B*			
	Tetracyclines		*Tet-A*		Increased expression of virulence genes	
			Tet-B			
			Acr-AB			
			ANTs	Plasmid and Genomic Islands	Completely deactivate the enzyme	
			AACs			
			APHs			
Carbapenems	Imipenem-cilastatin	Carbapenemases metallo-β-lactamases	*NDM-1*	Plasmid	Degrade the β-lactam, increase the risk for development of acquired resistance	[27]
			OXA-48			
Colistin	Polymyxin E		*Mcr-1*	Plasmid	Cause mutation and reduce the affinity of antibiotics to LPS	[28]
Trimethoprim/ sulfamethoxazole	Co-trimoxazole	ESBL	*DHFR* and *DHPS*	Plasmid	Block formation of nucleic acid inhibit dihydropteroate synthetase (DHPS)	[29]

Table 1.
Antimicrobial resistance, MGEs, and their associated virulence factors.

colistin for LPS [22]. Resistance to colistin can be due to mutations in chromosomal genes or it may be acquired. Furthermore, quinolones and fluoroquinolones are important antimicrobial agents implied for treating pathogenic microbes associated with humans and animals. Resistance to these antimicrobial agents is generally due to mutations in the drug targets, namely, DNA gyrase and topoisomerase IV genes seen in **Table 1** [30]. All such changes will lead to the transfer of resistance genes from chromosomal DNA into a plasmid, which will have more chances of dissemination in the human population. Additionally, it will be prone more harmful to human health due to variation in their resistance determinant transfer like from chromosome into plasmid, will definitely bring variation in expression pattern and dispersal [28, 31]. Another well-documented example is a transfer of the chromosomal β-lactamase gene *Amp C* to a plasmid and their subsequent global dissemination see in **Table 1** [28].

3. Mobile genetic elements of *E. coli* associated with antibiotic resistance genes

Mobile genetic material (MGEs) has an important role in transferring resistance. Mutation has a key role in bringing changes in a particular DNA fragment. Similarly, HGT, transfer of plasmid or transposons have the major contribution in developing resistance to the reagent. Considering if point mutation brings changes in a promotor region, it will have an impact on the expression of genes [32]. Similarly, a point mutation in the gyrase gene has developed to fluoroquinolone-resistant phenotype [30]. Exogenic resistance genes encoded on plasmids, phage, integrons, and transposons can transfer horizontally through conjugation, transformation, or transduction and can encode all the 3-resistance mechanism (intrinsic, adopted, acquired) Details of genes, their mechanisms, and pathways are explained in the following section.

Resistant pathogens are a major source of infectious diseases worldwide. Infections due to MDR bacteria have considerably increased health care costs. Due to resistant pathogens, morbidity and mortality have been reported in different parts of the world. Molecular characterization showed that extensive multi drug-resistant has commonly been accomplished by the acquisition of pre-existing causes followed by amplification in response to selection. The accumulation, retention, and transfer of resistant genes are frequently due to the activities of MGEs of *E. coli*, MGEs are known as non-core genes, and have a significant contribution to the plasticity of bacterial genomes. Transposable elements, integrons (In), Plasmids, gene cassettes, insertion sequences (IS), bacteriophages, and genomic islands (GIs), all are considered as MGEs. Though, from 20 sequenced *E. coli* genomes, almost 2000 genes were detected to be noncore genes [33]. Transposons (Tn) and IS are discrete segments of DNA that can almost randomly transfer themselves within a DNA molecule. Other mobile elements, like integrons (In), use site-specific recombination to transfer resistance genes among distinct sites. Similarly, these types of MGEs are mostly present in different locations in the form of multiple copies in the genome, they can also facilitate homologous recombination (interchange of sequences between same or different segments). Genetic exchange of Intercellular mechanisms contain transduction (facilitated by bacteriophages), conjugation/mobilization (facilitated by plasmids and integrative conjugative elements [ICE]), and transformation (uptake of various superfluous types of MGEs) support the rapid development of various multi-resistant bacteria in the aspect of antibiotics chemotherapy [34]. Within genomic DNA the presence/absence of MGEs can lead to modification in pathotypes of *E. coli*. In fact, strains of *E. coli* have been identified as part of the normal microbiota of the human gastrointestinal tract. In addition, there are also pathogenic strains, and thereby the *E. coli* strains are characterized either as (i) non-pathogenic, which are commensal (ii) intestinal pathogenic strains (IPEC), or (iii) extraintestinal pathogenic (EXPEC) strains. Integration, excision, and rearrangements of the DNA fragments can be the mechanisms behind the rapid evolution of pathogenic *E. coli* strains [35].

3.1 Transposons

Transposons (Tn) can be defined as a DNA sequence that has potential to jump into different locations of the genome hence, they are called jumping genes. Transposons are divided into two-main groups: class I (Retrotransposons) and class II (DNA transposons). Retrotransposons are mostly found in eukaryotic organisms while DNA transposons can be found in both prokaryotes and eukaryotes. Prokaryotic DNA transposons harbor antibiotic resistance genes. It has the potential to move from plasmid to plasmids or from chromosomal DNA into a plasmid, as a result, it became the source of resistant genes dissemination [36, 37]. Transposon's elements have two major characteristics that differentiate them from other genetic elements. on basis of its mobile nature, it can move from one place to another and bring variation in the genetic makeup of the organism. During transpositions process, transposons can transmit resistance genes and can multiply intracellularly. Despite its large number, only few copies get access into an integral part of the genome. Transposons have stability and are maintained by their capability to replicate and maintain their existence [38]. Transposable elements have an important role in genome evolution and organization [39]. *E. coli* transposable elements are divided into three different types: (a) composite transposons, (b) non-composite transposons, (c) insertion sequence elements (ISE). Composite and non-composite transposons have extra genetic material not related to transposition, for example, antibiotic resistance genes. Composite transposons are lined by the IS. IS elements are the simplest type of transposable elements and do not carry extra genetic information apart from those needed for their mobility [40].

3.2 Plasmids

Plasmids are circular, self-replicating extra-chromosomal DNA elements. Besides the genetic information required for the autonomous multiplications, it has extra genetic information needed for suppression of antibiotic actions. It also encodes genes for virulency, involves in the removal of hazardous material, or is required for regulation of other metabolic functions [41]. Plasmids are commonly used cloning vectors and are categorized into different incompatibility (Inc) groups. Inc. groups are designated on basis of the incapability of two plasmids to co-exist together [42]. Same Inc. group of the Inc. plasmids have the same type of replication region and thus have incompatible replication, it cannot co-exist. Plasmids belonging to the IncX family encode different resistance genes, mostly circulated among *Enterobacteriaceae* [43]. For example, an IncX plasmid, which is responsible for encoding *bla-SHV-12* resistant gene was reported in *E. coli*. The *bla-IMP-2* gene, encoding an imipenem-hydrolyzing β-lactamase, is carried by pRJ-18, an IncFIB plasmid [44]. In Europe, ESBL-encoding plasmids belonging to the Inc. F, A/C, N, H12, 11, and K type have been reported. Another important ESBL genes, *CTX-M-1*, is reported in Inc1 or IncN plasmids. For example, *CTX-M-1* ß-lactamase was derived from an animal source disseminated through Inc-1 ST3 plasmid [45]. Similarly, F plasmid, has been reported in *Enterobacteriaceae* [34]. F-like plasmids are also reported in nonpathogenic as well as in pathogenic *E. coli* strains. The whole genome sequence of *E. coli* ST-131 showed the *CTX-M* resistance gene dissemination and mainly conjugative F plasmid was involved [46]. *Mcr-1* gene conferring resistance to colistin is also spread with help of F Plasmid. Furthermore, this *Mcr-1* gene was found to be carried by 13 various plasmid incompatibility groups and these are Incl-2, Inc-X4, and Inc-HI2 [47]. Some studies have reported transposons involvement in the dissemination of *Mcr-1*. Other *Mcr* genes comprising *Mcr-2*, *Mcr-3*, *Mcr-4*, *Mcr-5* have been seen in a plasmid [48]. Recently in Denmark, a strain identified as *E. coli* ST-410 has been reported harboring resistance toward fluoroquinolones, 3rd-generation, carbapenems, and cephalosporins. Other variants like Inc-X3 plasmid carrying *blaOXA-181* resistant gene and Inc-FII plasmid carrying *blaNDM-5* resistant gene [49]. Plasmids can transfer between bacteria through the conjugation process, that is transfer of genetic material between recipient and donor cell. Conjugative plasmids can transmit transposons or integrons, and such genetic information can be further disseminated horizontally by the conjugation process [50, 51]. For example, *E. coli* is isolated from pig have a conjugative plasmid with *cfr* gene, which conferred resistance to lincosamides, phenicol, pleuromutilin, oxazolidinones, streptogramin [52]. Another important plasmid, a ColV (pCERC3) from a commensal *E. coli* ST95 strains have been reported and revealed resistance against sulfonamide encoded by *sul-3* associated with a class 1 integrons [53] The pE80 plasmid from a foodborne *E. coli* strain encodes multiple resistance determinants *oqx-AB*, *fos-A3*, *blaCTX-M-55*, and *blaTEM-1* and therefore confers resistance to streptomycin, tetracycline, kanamycin and olaquindox/quinolone [54]. In addition to antibiotic resistance genes, plasmids are involved in the transfer of virulence-associated genes. In Germany, outbreaks of enteroaggregative *E. coli* (EAEC)-enterohemorrhagic *E. coli* (EHEC)-O104:H4 strain was reported harboring have three-different plasmids: p-AA (7.4 kb), p-ESBL (89 kb), and p-G (1.5 kb) [55, 56]. p-AA plasmid harboring information for different virulence factors like fimbriae for adherence, diffusion in surface protein, protease, and the virulence regulator A and R [57]. Moreover, an EHEC O104:H7 strain, being isolated from animal's waste, possessed Inc-B/O/K/Z and IncFIB plasmids. It encodes genes responsible for the expression of main virulence genes, including, entero hemolysin and auto transporter [58]. Another important *E. coli* serotype is the O103 serotype, the 2nd most common serogroup main causative agent of human foodborne disease. It has pO157 plasmid encoding different virulence factors including entero hemolysin and type II secretion protein [59].

3.3 Bacteriophages

Bacterial viruses that cause infections in bacterial cells are called as bacteriophages. it has an important role in the dissemination of virulence-associated and antibiotic resistance genes among foodborne pathogens, As we know viruses are found ubiquitously and are present in oceans, sewage, soils, and various microbial communities [60, 61]. Phages have an important role in protecting the bacterial colonization of mucosal surfaces [62]. In the case of lytic phages, there has an important role in bacterial DNA transfer, and the process is called transduction (generalized transduction), while temperate phages can transmit only some particular genes in the bacterial chromosome (specialized transduction). During this some segments of bacterial DNA are co-edited with the prophage DNA for example tetracycline resistance gene from the *E. coli* O157:H7 to the K-12 AB-1157 strain of *E. coli* [60, 63]. Bacteriophages are actively in the acquisition of β-lactamase genes such as *blaCTX-M, blaSHV, blaTEM, qnr A, qnr B*, and *qnr S*. like P1 bacteriophage with *SHV-2* gene has been reported [64]. Additionally, phages are also involved in the dissemination and transformation of staphylo-kinase, superantigens, and phosphor-lipase or DNase virulence factors. Bacteriophage λ, transmit not only adhesion genes of bacteria but also transfer the housekeeping genes of bacteria. Cytolethal distending toxins (Cdts) are inhibitory cyclomodulins, which prevent eukaryotic cell proliferation, *E. coli* strains are also associated with its production and it has been established that *Cdt-I* produced by EPEC strains the main source was lambdoid prophage [65]. Additionally, *E. coli* phage (lambdoid prophage) transfers the *Cdt* gene group encoding the *Cdt-A*, *Cdt-B*, and *Cdt-C* subunits of the *Cdt-I* holotoxin. One of the important toxins known as Shiga toxin 2, which is a virulence factor in *E. coli* O157:H7 strain being transferred by temperate phage. Furthermore, some other variants of Shiga toxin comprising of the infective *E. coli* O157 strain another variant Shiga toxin 2-c are also encoded by phages. For example, some phages such as phi-C119 can be used as biological control mediators, as they can lyse and infect their bacterial hosts [66].

3.4 Genomic islands (GIs)

Genomic islands (GIs) comprise of more than 10 kb DNA in length, exchanged frequently among bacterial isolates. GIs encode proteins for transfer, restriction/alteration, or other proprieties and recombination, for example, gene groups for metabolic adaptation, virulence, and or bacterial resistance [67]. GIs that are involved in the expression of virulence factors is called pathogenicity islands (PAIs) [68]. It encodes VFs comprising of adhesins, invasions, capsule formation, toxins, uptake system of iron, distinct secretion systems. Their GC contents vary in comparison to the genome. Their integration site is situated on the tRNA genes and repeated sequences, which is comprising at least one MGEs containing plasmids remnants, integrons, insertion sequences, and related gene cassettes. For the integration of foreign DNA, tRNA-encoding genes are considered as the hot spot. By site-specific recombination, some PAIs can be edited from bacterial chromosome [69]. Primarily, PAIs have been described in the uropathogenic *E. coli* genome and later cases were reported in other pathogenic bacteria [70]. Currently, PAIs are spread between plants and animals associated with bacterial pathogens, have a great influence on the rapid evolution of virulent and resistant strains. In *E. coli*, the locus of enterocyte effacement (LEE) is best example of PAIs, and its size is about 35 kb. It has a main role in bacterial adherence to the epithelial cells of the intestine [71]. High-pathogenicity Island (HPI) was found in enteroaggregative, enteropathogenic, entero-invasive, and enterotoxigenic *E. coli* [72].

4. Global antibiotics resistance in *E. coli*

As earlier described, *E. coli* is one of the important bacteria, causing infections in the gastrointestinal tract [73]. Worldwide, AMRs in *E. coli* have been reported which show significant geographic variation as well as differences in various populations and environments. The evolving of ESBL and fluoroquinolones resistance and lack of availability of effective treatment in infections in *E. coli* strains spread over the last few years. However, if *E. coli* resistance is not tackle will restrict out treatment strategies, and resistant clones spread in the general population [74].

4.1 Emergence of *E. coli* resistance in Europe

In European countries, particularly in *E. coli* AMRs are increasing [75, 76]. AMR is a worldwide threat, with an approximately 25,000 deaths occurring in Europe and 23,000 in the United States each year [77]. Due to MDR strains treatment becomes complicated and there are more chances of its spread. In addition, particularly *E. coli* is mostly involved in community and hospital-acquired infections [78, 79]. The severity of the disease differs considerably depending upon the *E. coli* strains [80]. Considering the case of Europe, faced two epidemics of the hemolytic uremic syndrome (HUS) and bloody diarrhea between May and July 2011. One major epidemic occurred in Germany (almost 4000 cases of bloody diarrhea, 850 of HUS, and 50 death cases were reported), while few cases were reported in southwest France (15 cases of bloody diarrhea and 9 cases of HUS) [81–83]. Commonly, these outbreaks were caused by a strain of Stx producing *E. coli* [84] which possesses a plasmid encoding ESBL [83]. The ratio of *E.coli* O104-H4-infected patients with complications such as HUS are more prevalent than in earlier epidemics [85]. AMRs *E. coli* strains are observed all around Europe. According to the European center for disease prevention and control (ECDC), the resistance in human sources varies significantly between countries [86]. Though, in each country mostly the prevalence of *E. coli* strains were observed resistant to all antibiotic classes such as 3rd generation cephalosporins, fluoroquinolones, and aminoglycosides. The ratio of isolates resistant to cephalosporins was observed highest in Cyprus (36.2%), Slovakia (31%), and Bulgaria (22.9%) and lowest in Sweden (3.0%) and Norway (3.6%) respectively. While less resistant were found against fluoroquinolones in Sweden (7.9%) and Estonia (9.9%) but fluoroquinolones resistance is more prevalent in Cyprus (47.4%) and Italy (40.5%) and furthermore high prevalence rate of isolates resistant to aminoglycosides were observed in the United States (23.9%), Romania (19.6%), Slovakia (17.9%) and Greece (16.8%), Sweden (3.7%). *E. coli* strains resistant to widespread Penicillin were found in 28 countries. Besides this, 0.04% of *E. coli* strains were observed to be resistant to carbapenems. In Europe, according to a current study resistance due to carbapenemases producing are still circulating [87].

4.2 Emergence of *E. coli* resistance in America

In America, increased resistance of fluoroquinolones and cephalosporin in *E. coli* has been reported [88]. In most patients, *E. coli* ST-131 strains have been reported [89]. ST-131 *E. coli* clone is thought mainly involve in AMRs spreading. The most common clinical manifestation associated with *E. coli* is intraabdominal infection (IAIs). Overall, 26% of *E. coli* infections associated IAI in the Latin American region produced ESBLs compared to with all over the world [90]. Region-wise prevalence of ESBL producing *E. coli* within America varies as in Latin America it was higher in 2008 than earlier according to data being shared by the Study for Monitoring AMR trends (SMART). Many surveillance studies have presented that ESBL-producing bacteria are common in Latin America. According to Tigecycline Evaluation and Surveillance Trial (TEST) in Latin America, during the years 2004–2006, where total of 13.5% of *E. coli* isolates with ESBL phenotypes were identified [18]. According to the Meropenem Yearly Susceptibility Test Information Collection (MYSTIC) surveillance study performed in 1997 and 2003, South America had a higher ratio of ESBL producing *E. coli* than North America [91]. Similarly in Colombia in 2002 higher cases of ESBLs producers'strains were documented. Based on available data [21–22%], the percentage of *E. coli* isolates in Latin American was higher as compared to other developed countries of the world [88].

4.3 Emergence of *E. coli* resistance e in Africa

Proper prescription of drugs is not strictly followed in the developing world. A similar case is Africa countries where no such policy is implemented. There are several challenges to implement sustainable and effective AMRs monitoring programs in the sub-Saharan Africa to encounter the rapid dissemination of AMRs [92]. Around 50–60% of *E. coli* infections reported in patients have a resistance nature to most of the available antibiotics i-e amoxicillin, ciprofloxacin, cefixime [93]. A current study reported the 48% prevalence of AMR *E. coli* in hospitalized patients in Brazzaville, Republic of Congo [94]. Similarly in another study where 65% to ceftazidime, 57% to amoxicillin, 51% to piperacillin, and 11% to ofloxacin resistant respectively in *E. coli* were documented [95]. First reported case of EHEC *E. coli* O157-H7 first case was reported in 1982 in the USA while in South Africa and parts of the world in 1990 cases were reported. Was found in 1982 in the United State while, in 1990 in throughout the world [96, 97]. Besides this, many infrequent cases of EHEC have been reported in different parts of South Africa. A total of 40,912 patients under the age of 5 years was hospitalized in 1992 due to the onset of diarrhea [98]. In South Africa, the most common strains detected are EPEC with detection rates ranging from 14.8% to 41.7%. Several pathotypes of *E. coli* are significant causes of diarrhea in children particularly in sub-Saharan Africa [94]. Most of the AMR genes are encoded in *E. coli* on MGEs that are transmissible among bacteria permitting the rapid spread and maintenance of resistance genes among species [99].

4.4 Emergence of *E. coli* resistance in Asia

E. coli is the most common bacterial pathogen associated with UTIs and IAIs, leading to bacteremia in severe cases, Infections caused by AMRs *E. coli* are becoming a serious threat over the last few years [100]. Strain ST-131 is reported worldwide and its infections rate is soaring. In addition, ST-131 strains have been associated with the increased rate of AMRs with *CTX-M* type ESBLs variant [101]. In Asian countries, *CTX-M* and ST-131producing *E. coli* have evolved as a foremost cause of hospital and community-acquired infections [102]. According to an earlier surveillance study, the occurrence of ESBL producing *E. coli* in Asian countries ranged from 2.3% to 40% [103]. *CTX-M* ESBLs are considered the most dominant phenotype. *CTX-M* producing *E. coli* pose a serious threat for densely populated cities and regions [41, 42]. Additionally *KPC* and *NDM* beta-lactamase-producing *E. coli* have been found to be on the rise in certain parts of Asian countries [104]. In some of the most Asian countries particularly [China, Malaysia, Macau, and Thailand], the prevalence rate in newborn sepsis due to AMR *E. coli* was found about 26.1% [105].

4.4.1 China

China was the 2nd largest consumer of antibiotics in 2010 around the world. According to the available data, the prescription of antibiotics for outpatient and inpatient was 52.9% and 77.5% respectively and only 39.4% and 24.6% were considered appropriate respectively. Among BRICS countries only in China usage of antibiotics has been escalated [57%] as compared to other nations [106]. According to European Antibiotic Resistance Monitoring Network (EARS-Net), in *E. coli* resistance to third-generation cephalosporins has surged from 1.7% to 8% in the period between 2002 and 2009 [107]. Similarly in other findings were conducted on bloodstreams infections where *E. coli* is the most common bacteria. Moreover, *CTX-M-14* was reported as the most persistent ESBL while ST-131 was the most prevalent sequence type [108]. China has the world's fast proliferation of antibiotics resistance, the ratio of *E. coli* resistant to 3rd generation cephalosporins was reported 54.2% in China in 2017 which was higher than Europe (54.2%) [109, 110] According to one study, antimicrobial resistance is potentially responsible for 214,000 of 690,000 annual neonatal deaths (31%) caused by sepsis. Carbapenems are β-lactam antibiotics that are used to cure severe infections caused by MDRs bacteria particularly *E. coli* [111].

4.4.2 Bangladesh

In 2004 a study conducted in Bangladesh, observed a high frequency of almost 43.2% of ESBL producing *E. coli* in an urban hospital in Dhaka [112]. In addition, the prevalence rate of *CTX-M* among ESBL was high (76%). In another study conducted on ESBL, 11% positive ESBL cases were reported, and all these belonged to *CTX-M-1* group [112]. Similarly, *NDM* producing *E. coli* has been reported in diarrhea patients [113]. The environment being contaminated human feces which might be the source to affect the bird's fecal flora [114]. The *E. coli* isolates that were detected in water samples were found resistant to almost one antibiotic of the tested antibiotics, Similarly, there are other reports where *E. coli* isolates were found resistant to cefuroxime, nalidixic acid, ciprofloxacin, and tetracycline see in **Table 1** [115].

4.4.3 India

India is one of the world most populated country with weak health care systems are exposed to resistant pathogens. In comparison to Pakistan, China, and Iran it has a similar prevalence rate of resistance, In India, the prevalence of ESBLs producing *E. coli* has been observed in a range between 45 and 79% [116]. *CTX-M-15* produces *E. coli* colonization was most common especially among children [55%] who were admitted in intensive care unit (ICU) [117]. Another report from South India where *CTX-M* positive cases were reported in more than 60% of *E. coli* isolates [118]. In all reported the most frequent isolated group was *CTX-M-1* group *E. coli*. In India, the frequency of ESBL producing *E. coli* was 23% among UTIs patients, with *CTX-M-15* and ST-131 *E. coli* strain was the highest [119].

4.4.4 Pakistan

Pakistan is the 6th largest most populous country in the world. Resistance has increased in *E. coli* around the world and sensitivity patterns significantly vary across geographic settings and within the populations [120]. Early in 2000, *CTX-M* producing *E. coli* has been found the most widespread uropathogenic in Pakistan [121]. ESBL and Amp C were observed in 35 and 64% of the *E. coli* isolates [122]. Furthermore, pandemic *CTX-M* producing *E. coli* ST-131 were also reported. *NDM*-producing *E. coli* was predominantly found in hospitalized patients with resistance to ceftriaxone. From 2013 to 2017, a comprehensive report was released on the susceptibility pattern of *E. coli* isolates in hospitalized and non-hospitalized patients. Where *E. coli* isolates of hospitalized patients were more resistant to all antibiotics [123]. The variation in the ratio of resistance between hospitalized and non-hospitalized patients forces us for prior antibiotic susceptibility screening [124]. In Enterobacteriaceae *E. coli* has a high resistance to β-lactam antibiotics of having ESBL phenotype [125]. ESBL positive cases have been reported in different parts of the country [126].

5. Conclusions

Antimicrobial-resistant in *E. coli* has become a serious and complex problem worldwide in clinical treatment as well as in veterinary medicine. *E. coli* is intrinsically vulnerable to all clinically important antimicrobial agents, but it has great potential to accumulate resistance genes, through acquired resistance (HGT).

Acquired resistance plays an essential role in the acquisition of new properties, such as antimicrobial resistance, emphasizing the remarkable adaptive potential of *E. coli*. In addition, among all the MGEs, Transposons and plasmid have a significant role in the spread of antimicrobial resistance with high potential in resistance gene transmission. In *E. coli* plasmid and transposons mediated genes are involved in the spread of quinolone and *Mcr* resistance genes. The epidemiological study of AMR in *E. coli* revealed that *CTX-M* beta-lactamase and ST-131 clone have emerged as the main cause of hospital and community-acquired infections across the globe noteworthy in developing countries. This is being linked with lack of proper prescription of antibiotics and no such strict policy is in place. There are several challenges to implement sustainable and effective AMRs monitoring programs in Africa as well as in Asian countries to encounter the rapid dissemination of AMR. There is a dire need to support and develop antimicrobial policy, standard therapy guidelines for control of AMR in hospitals as well as in the community. To promote and regulate the balanced use of medicines and ensure proper patient care at all stages, antibiotics without doctor's prescription should be discouraged and ensure continuous access to essential medicines of guaranteed quality at the hospital and community.

Author details

Waheed Ullah[1*] and Shandana Ali[2]

1 Department of Microbiology, Kohat University of Science and Technology, Kohat, Pakistan

2 Department of Zoology, Kohat University of Science and Technology, Kohat, Pakistan

*Address all correspondence to: waheedwazir@gmail.com

References

[1] Boolchandani M, D'Souza AW, Dantas G. Sequencing-based methods and resources to study antimicrobial resistance. Nature Reviews Genetics. 2019;**20**(6):356-370

[2] Bryce A, Hay AD, Lane IF, Thornton HV, Wootton M, Costelloe C. Global prevalence of antibiotic resistance in paediatric urinary tract infections caused by *Escherichia coli* and association with routine use of antibiotics in primary care: Systematic review and meta-analysis. BMJ. 2016;**352**

[3] Goossens H. Antibiotic consumption and link to resistance. Clinical Microbiology and Infection. 2009;**15**: 12-15

[4] Rasheed MU, Thajuddin N, Ahamed P, Teklemariam Z, Jamil K. Antimicrobial drug resistance in strains of *Escherichia coli* isolated from food sources. Revista do Instituto de Medicina Tropical de São Paulo. 2014;**56**:341-346

[5] Meyer E, Schwab F, Schroeren-Boersch B, Gastmeier P. Dramatic increase of third-generation cephalosporin-resistant *E. coli* in German intensive care units: Secular trends in antibiotic drug use and bacterial resistance, 2001 to 2008. Critical Care. 2010;**14**(3):1-9

[6] Liu Y-Y, Wang Y, Walsh TR, Yi L-X, Zhang R, Spencer J, et al. Emergence of plasmid-mediated colistin resistance mechanism MCR-1 in animals and human beings in China: A microbiological and molecular biological study. The Lancet Infectious Diseases. 2016;**16**(2):161-168

[7] Poirel L, Madec J-Y, Lupo A, Schink A-K, Kieffer N, Nordmann P, et al. Antimicrobial resistance in *Escherichia coli*. Microbiology Spectrum. 2018;**6**(4):14

[8] Nji E, Kazibwe J, Hambridge T, Joko CA, Larbi AA, Damptey LAO, et al. High prevalence of antibiotic resistance in commensal *Escherichia coli* from healthy human sources in community settings. Scientific Reports. 2021;**11**(1):1-11

[9] Fair RJ, Tor Y. Antibiotics and bacterial resistance in the 21st century. In: Perspectives in Medicinal Chemistry. New York: Journals.sagepub.com; Vol. 6. 2014. p. PMC.S14459

[10] Aslam B, Wang W, Arshad MI, Khurshid M, Muzammil S, Rasool MH, et al. Antibiotic resistance: A rundown of a global crisis. Infection and Drug Resistance. 2018;**11**:1645

[11] Nikaido H, Takatsuka Y. Mechanisms of RND multidrug efflux pumps. Biochimica et Biophysica Acta (BBA)-Proteins and Proteomics. 2009;**1794**(5):769-781

[12] Kobayashi N, Nishino K, Yamaguchi A. Novel macrolide-specific ABC-type efflux transporter in *Escherichia coli*. Journal of Bacteriology. 2001;**183**(19):5639-5644

[13] Fajardo A, Martinez-Martin N, Mercadillo M, Galán JC, Ghysels B, Matthijs S, et al. The neglected intrinsic resistome of bacterial pathogens. PLoS One. 2008;**3**(2):e1619

[14] Cox G, Wright GD. Intrinsic antibiotic resistance: Mechanisms, origins, challenges and solutions. International Journal of Medical Microbiology. 2013;**303**(6-7):287-292

[15] Schroeder M, Brooks BD, Brooks AE. The complex relationship between virulence and antibiotic resistance. Genes. 2017;**8**(1):39

[16] Salverda ML, Koomen J, Koopmanschap B, Zwart MP, de Visser JAG. Adaptive benefits from small mutation supplies in an antibiotic

resistance enzyme. Proceedings of the National Academy of Sciences. 2017;**114**(48):12773-12778

[17] Galimand M, Courvalin P, Lambert T. Plasmid-mediated high-level resistance to aminoglycosides in Enterobacteriaceae due to 16S rRNA methylation. Antimicrobial Agents and Chemotherapy. 2003;**47**(8):2565-2571

[18] Xia J, Sun J, Li L, Fang L-X, Deng H, Yang R-S, et al. First report of the IncI1/ST898 conjugative plasmid carrying rmtE2 16S rRNA methyltransferase gene in *Escherichia coli*. Antimicrobial Agents and Chemotherapy. 2015;**59**(12):7921-7922

[19] Bismuth R, Zilhao R, Sakamoto H, Guesdon J, Courvalin P. Gene heterogeneity for tetracycline resistance in Staphylococcus spp. Antimicrobial Agents and Chemotherapy. 1990;**34**(8):1611-1614

[20] Van Hoek AH, Mevius D, Guerra B, Mullany P, Roberts AP, Aarts HJ. Acquired antibiotic resistance genes: An overview. Frontiers in Microbiology. 2011;**2**:203

[21] Paitan Y. Current trends in antimicrobial resistance of *Escherichia coli*. In: *Escherichia coli*, a Versatile Pathogen. United states: Springer; 2018. pp. 181-211

[22] Aghapour Z, Gholizadeh P, Ganbarov K, Bialvaei AZ, Mahmood SS, Tanomand A, et al. Molecular mechanisms related to colistin resistance in Enterobacteriaceae. Infection and Drug Resistance. 2019;**12**:965

[23] Andersson DI, Hughes D. Persistence of antibiotic resistance in bacterial populations. FEMS Microbiology Reviews. 2011;**35**(5):901-911

[24] Bonnet R. Growing group of extended-spectrum β-lactamases: The CTX-M enzymes. Antimicrobial Agents and Chemotherapy. 2004;**48**(1):1-14

[25] Bagattini M, Crivaro V, Di Popolo A, Gentile F, Scarcella A, Triassi M, et al. Molecular epidemiology of extended-spectrum β-lactamase-producing *Klebsiella pneumoniae* in a neonatal intensive care unit. Journal of Antimicrobial Chemotherapy. 2006;**57**(5):979-982

[26] Lindemann PC, Risberg K, Wiker HG, Mylvaganam H. Aminoglycoside resistance in clinical *Escherichia coli* and *Klebsiella pneumoniae* isolates from Western Norway. APMIS. 2012;**120**(6):495-502

[27] Gauthier L, Dortet L, Cotellon G, Creton E, Cuzon G, Ponties V, et al. Diversity of carbapenemase-producing *Escherichia coli* isolates in France in 2012-2013. Antimicrobial Agents and Chemotherapy. 2018;**62**(8):e00266-e00218

[28] Dantas G, Sommer MO. Context matters—The complex interplay between resistome genotypes and resistance phenotypes. Current Opinion in Microbiology. 2012;**15**(5):577-582

[29] Talan DA, Krishnadasan A, Abrahamian FM, Stamm WE, Moran GJ, Group EINS. Prevalence and risk factor analysis of trimethoprim-sulfamethoxazole—and fluoroquinolone-resistant *Escherichia coli* infection among emergency department patients with pyelonephritis. Clinical Infectious Diseases. 2008;**47**(9):1150-1158

[30] Hopkins KL, Davies RH, Threlfall EJ. Mechanisms of quinolone resistance in *Escherichia coli* and Salmonella: Recent developments. International Journal of Antimicrobial Agents. 2005;**25**(5):358-373

[31] Martínez JL. Ecology and evolution of chromosomal gene transfer between environmental microorganisms and pathogens. Microbiology Spectrum. 2018;**6**(1):06

[32] Tracz DM, Boyd DA, Bryden L, Hizon R, Giercke S, Van Caeseele P, et al. Increase in ampC promoter strength due to mutations and deletion of the attenuator in a clinical isolate of cefoxitin-resistant *Escherichia coli* as determined by RT-PCR. Journal of Antimicrobial Chemotherapy. 2005;**55**(5):768-772

[33] Rankin DJ, Rocha EP, Brown SP. What traits are carried on mobile genetic elements, and why? Heredity. 2011;**106**(1):1-10

[34] Partridge SR, Kwong SM, Firth N, Jensen SO. Mobile genetic elements associated with antimicrobial resistance. Clinical Microbiology Reviews. 2018;**31**(4):e00088-e00017

[35] Javadi M, Bouzari S, Oloomi M. Horizontal gene transfer and the diversity of *Escherichia coli*. In: *Escherichia coli* Recent Advances on Physiology, Pathogenesis and Biotechnological Applications. Australia: Book.google.com; 2017. pp. 317-331

[36] Babakhani S, Oloomi M. Transposons: The agents of antibiotic resistance in bacteria. Journal of Basic Microbiology. 2018;**58**(11):905-917

[37] Bao W, Jurka MG, Kapitonov VV, Jurka J. New superfamilies of eukaryotic DNA transposons and their internal divisions. Molecular Biology and Evolution. 2009;**26**(5):983-993

[38] Martin SL, Garfinkel DJ. Survival Strategies for Transposons and Genomes. United states: Springer; 2003

[39] Sousa A, Bourgard C, Wahl LM, Gordo I. Rates of transposition in *Escherichia coli*. Biology Letters. 2013; **9**(6):20130838

[40] Snyder L, Champness W, Champness W. Molecular Genetics of Bacteria. Washington, DC: ASM Press; 1997

[41] Su L-H, Chen H-L, Chia J-H, Liu S-Y, Chu C, Wu T-L, et al. Distribution of a transposon-like element carrying bla CMY-2 among Salmonella and other Enterobacteriaceae. Journal of Antimicrobial Chemotherapy. 2006; **57**(3):424-429

[42] Thomas CM, Smith CA. Incompatibility group P plasmids: Genetics, evolution, and use in genetic manipulation. Annual Reviews in Microbiology. 1987;**41**(1):77-101

[43] Liakopoulos A, Van Der Goot J, Bossers A, Betts J, Brouwer MS, Kant A, et al. Genomic and functional characterisation of IncX3 plasmids encoding bla SHV-12 in *Escherichia coli* from human and animal origin. Scientific Reports. 2018;**8**(1):1-13

[44] Zhang F, Wang X, Xie L, Zheng Q, Guo X, Han L, et al. A novel transposon, Tn6306, mediates the spread of blaIMI in Enterobacteriaceae in hospitals. International Journal of Infectious Diseases. 2017;**65**:22-26

[45] Irrgang A, Hammerl JA, Falgenhauer L, Guiral E, Schmoger S, Imirzalioglu C, et al. Diversity of CTX-M-1-producing *E. coli* from German food samples and genetic diversity of the blaCTX-M-1 region on IncI1 ST3 plasmids. Veterinary Microbiology. 2018;**221**:98-104

[46] Koraimann G. Spread and persistence of virulence and antibiotic resistance genes: A ride on the F plasmid conjugation module. EcoSal Plus. 2018;**8**(1)

[47] Matamoros S, Van Hattem JM, Arcilla MS, Willemse N, Melles DC, Penders J, et al. Global phylogenetic analysis of *Escherichia coli* and plasmids carrying the mcr-1 gene indicates bacterial diversity but plasmid restriction. Scientific Reports. 2017;**7**(1):1-9

[48] Xavier BB, Lammens C, Ruhal R, Kumar-Singh S, Butaye P, Goossens H,

et al. Identification of a novel plasmid-mediated colistin-resistance gene, mcr-2, in *Escherichia coli*, Belgium, June 2016. Eurosurveillance. 2016; **21**(27):30280

[49] Phornphisutthimas S, Thamchaipenet A, Panijpan B. Conjugation in *Escherichia coli*: A laboratory exercise. Biochemistry and Molecular Biology Education. 2007;**35**(6):440-445

[50] Norman A, Hansen LH, Sørensen SJ. Conjugative plasmids: Vessels of the communal gene pool. Philosophical Transactions of the Royal Society B: Biological Sciences. 2009;**364**(1527): 2275-2289

[51] Grohmann E, GN M, Espinosa M. Conjugative plasmid transfer in gram-positive bacteria. Microbiology and Molecular Biology Reviews. 2003;**67**(2): 277-301

[52] Zhang W-J, Wang X-M, Dai L, Hua X, Dong Z, Schwarz S, et al. Novel conjugative plasmid from *Escherichia coli* of swine origin that coharbors the multiresistance gene cfr and the extended-spectrum-β-lactamase gene bla CTX-M-14b. Antimicrobial Agents and Chemotherapy. 2015;**59**(2): 1337-1340

[53] Moran RA, Holt KE, Hall RM. pCERC3 from a commensal ST95 *Escherichia coli*: A ColV virulence-multiresistance plasmid carrying a sul3-associated class 1 integron. Plasmid. Netherlands: Elsevier; 2016;**84**:11-19

[54] Wong MH, Xie M, Xie L, Lin D, Li R, Zhou Y, et al. Complete sequence of a F33: A-: B-conjugative plasmid carrying the oqxAB, fosA3, and blaCTX-M-55 elements from a foodborne *Escherichia coli* strain. Frontiers in Microbiology. 2016;**7**:1729

[55] Brzuszkiewicz E, Thürmer A, Schuldes J, Leimbach A, Liesegang H, Meyer F-D, et al. Genome sequence analyses of two isolates from the recent *Escherichia coli* outbreak in Germany reveal the emergence of a new pathotype: Entero-aggregative-haemorrhagic *Escherichia coli* (EAHEC). Archives of Microbiology. 2011;**193**(12): 883-891

[56] Mellmann A, Harmsen D, Cummings CA, Zentz EB, Leopold SR, Rico A, et al. Prospective genomic characterization of the German enterohemorrhagic *Escherichia coli* O104:H4 outbreak by rapid next generation sequencing technology. PLoS One. 2011;**6**(7):e22751

[57] Kampmeier S, Berger M, Mellmann A, Karch H, Berger P. The 2011 German enterohemorrhagic *Escherichia coli* O104:H4 outbreak—The danger is still out there. In: *Escherichia coli*, a Versatile Pathogen. 2018. pp. 117-148

[58] Shridhar PB, Patel IR, Gangiredla J, Noll LW, Shi X, Bai J, et al. Genetic analysis of virulence potential of *Escherichia coli* O104 serotypes isolated from cattle feces using whole genome sequencing. Frontiers in Microbiology. 2018;**9**:341

[59] Noll LW, Worley JN, Yang X, Shridhar PB, Ludwig JB, Shi X, et al. Comparative genomics reveals differences in mobile virulence genes of *Escherichia coli* O103 pathotypes of bovine fecal origin. PLoS One. 2018;**13**(2):e0191362

[60] Colavecchio A, Cadieux B, Lo A, Goodridge LD. Bacteriophages contribute to the spread of antibiotic resistance genes among foodborne pathogens of the Enterobacteriaceae family—A review. Frontiers in Microbiology. 2017;**8**:1108

[61] Penadés JR, Chen J, Quiles-Puchalt N, Carpena N, Novick RP. Bacteriophage-mediated

spread of bacterial virulence genes. Current Opinion in Microbiology. 2015;**23**:171-178

[62] Barr JJ, Auro R, Furlan M, Whiteson KL, Erb ML, Pogliano J, et al. Bacteriophage adhering to mucus provide a non–host-derived immunity. Proceedings of the National Academy of Sciences. 2013;**110**(26):10771-10776

[63] Marinus MG, Poteete AR. High efficiency generalized transduction in *Escherichia coli* O157:H7. F1000 Research. 2013;**2**

[64] Billard-Pomares T, Fouteau S, Jacquet ME, Roche D, Barbe V, Castellanos M, et al. Characterization of a P1-like bacteriophage carrying an SHV-2 extended-spectrum β-lactamase from an *Escherichia coli* strain. Antimicrobial Agents and Chemotherapy. 2014;**58**(11):6550-6557

[65] Asakura M, Hinenoya A, Alam MS, Shima K, Zahid SH, Shi L, et al. An inducible lambdoid prophage encoding cytolethal distending toxin (Cdt-I) and a type III effector protein in enteropathogenic *Escherichia coli*. Proceedings of the National Academy of Sciences. 2007;**104**(36): 14483-14488

[66] Amarillas L, Chaidez C, González-Robles A, Lugo-Melchor Y, León-Félix J. Characterization of novel bacteriophage phiC119 capable of lysing multidrug-resistant Shiga toxin-producing *Escherichia coli* O157:H7. PeerJ. 2016;**4**:e2423

[67] Hacker J, Carniel E. Ecological fitness, genomic islands and bacterial pathogenicity. EMBO Reports. 2001;**2**(5):376-381

[68] Lloyd AL, Rasko DA, Mobley HL. Defining genomic islands and uropathogen-specific genes in uropathogenic *Escherichia coli*. Journal of Bacteriology. 2007;**189**(9):3532-3546

[69] Nieto PA, Pardo-Roa C, Salazar-Echegarai FJ, Tobar HE, Coronado-Arrázola I, Riedel CA, et al. New insights about excisable pathogenicity islands in Salmonella and their contribution to virulence. Microbes and Infection. 2016;**18**(5): 302-309

[70] Blum G, Falbo V, Caprioli A, Hacker J. Gene clusters encoding the cytotoxic necrotizing factor type 1, Prs-fimbriae and α-hemolysin form the pathogenicity island II of the uropathogenic *Escherichia coli* strain J96. FEMS Microbiology Letters. 1995; **126**(2):189-195

[71] Perna NT, Mayhew GF. Pósfai Gr, Elliott S, Donnenberg MS, Kaper JB, et al. Molecular evolution of a pathogenicity island from enterohemorrhagic *Escherichia coli* O157:H7. Infection and Immunity. 1998;**66**(8):3810-3817

[72] Schubert S, Rakin A, Karch H, Carniel E, Heesemann J. Prevalence of the "high-pathogenicity island" of Yersinia species among *Escherichia coli* strains that are pathogenic to humans. Infection and Immunity. 1998;**66**(2): 480-485

[73] Kaper JB, Nataro JP, Mobley HL. Pathogenic *escherichia coli*. Nature Reviews Microbiology. 2004;**2**(2):123-140

[74] Sáenz Y, Zarazaga M, Briñas L, Lantero M, Ruiz-Larrea F, Torres C. Antibiotic resistance in *Escherichia coli* isolates obtained from animals, foods and humans in Spain. International Journal of Antimicrobial Agents. 2001;**18**(4):353-358

[75] Van Duijn PJ, Dautzenberg MJ, Oostdijk EA. Recent trends in antibiotic resistance in European ICUs. Current Opinion in Critical Care. 2011;**17**(6): 658-665

[76] Balode A, Punda-Polić V, Dowzicky MJ. Antimicrobial

susceptibility of gram-negative and gram-positive bacteria collected from countries in Eastern Europe: Results from the tigecycline evaluation and surveillance trial (TEST) 2004-2010. International Journal of Antimicrobial Agents. 2013;**41**(6):527-535

[77] Li B, Webster TJ. Bacteria antibiotic resistance: New challenges and opportunities for implant-associated orthopedic infections. Journal of Orthopaedic Research. 2018;**36**(1):22-32

[78] Pitout JD. Multiresistant Enterobacteriaceae: New threat of an old problem. Expert Review of Anti-Infective Therapy. 2008;**6**(5):657-669

[79] Paterson DL. Resistance in gram-negative bacteria: Enterobacteriaceae. American Journal of Infection Control. 2006;**34**(5):S20-SS8

[80] Bilinski P, Kapka-Skrzypczak L, Posobkiewicz M, Bondaryk M, Holownia P, Wojtyla A. Public health hazards in Poland posed by foodstuffs contaminated with *E. coli* O104:H4 bacterium from the recent European out break. Annals of Agricultural and Environmental Medicine. 2012;**19**(1)

[81] Frank C, Werber D, Cramer JP, Askar M, Faber M. an der Heiden M, et al. Epidemic profile of Shiga-toxin–producing *Escherichia coli* O104:H4 outbreak in Germany. New England Journal of Medicine. 2011;**365**(19): 1771-1780

[82] Gault G, Weill F-X, Mariani-Kurkdjian P, Jourdan-da Silva N, King L, Aldabe B, et al. Outbreak of haemolytic uraemic syndrome and bloody diarrhoea due to *Escherichia coli* O104:H4, south-west France, June 2011. Eurosurveillance. 2011;**16**(26):19905

[83] Bielaszewska M, Mellmann A, Zhang W, Köck R, Fruth A, Bauwens A, et al. Characterisation of the *Escherichia coli* strain associated with an outbreak of haemolytic uraemic syndrome in Germany, 2011: A microbiological study. The Lancet Infectious Diseases. 2011;**11**(9):671-676

[84] Scheutz F, Nielsen EM, Frimodt-Møller J, Boisen N, Morabito S, Tozzoli R, et al. Characteristics of the enteroaggregative Shiga toxin/ verotoxin-producing *Escherichia coli* O104:H4 strain causing the outbreak of haemolytic uraemic syndrome in Germany, May to June 2011. Eurosurveillance. 2011;**16**(24):19889

[85] Jansen A, Kielstein J. The new face of enterohaemorrhagic *Escherichia coli* infections. Eurosurveillance. 2011; **16**(25):19898

[86] European Centre for Disease Prevention and Control. Antimicrobial Resistance Surveillance in Europe 2015. Stockholm: ECDC; 2017.

[87] Albiger B, Glasner C, Struelens MJ, Grundmann H, Monnet DL. Carbapenemase-producing Enterobacteriaceae in Europe: Assessment by national experts from 38 countries, May 2015. Eurosurveillance. 2015;**20**(45):30062

[88] Turner P, Greenhalgh J, Edwards J, McKellar J. The MYSTIC (meropenem yearly susceptibility test information collection) programme. International Journal of Antimicrobial Agents. 1999;**13**(2):117-125

[89] Vigil KJ, Johnson JR, Johnston BD, Kontoyiannis DP, Mulanovich VE, Raad II, et al. *Escherichia coli* pyomyositis: An emerging infectious disease among patients with hematologic malignancies. Clinical Infectious Diseases. 2010;**50**(3):374-380

[90] Cha R, Michienzi S, Hsaisky L. Antimicrobial pharmacokinetics and pharmacodynamics in the treatment of nosocomial gram negative infections. Advances in Pharmacoepidemiological Drug Safety S. 2012;**1**:005

[91] Sathyavathy K, Madhusudhan BK. Isolation, identification, speciation and antibiotic susceptibility pattern of Klebsiella species among various clinical samples at tertiary care hospital. Journal of Pharmaceutical Research International. 2021:78-87

[92] Elton L, Thomason MJ, Tembo J, Velavan TP, Pallerla SR, Arruda LB, et al. Antimicrobial resistance preparedness in sub-Saharan African countries. Antimicrobial Resistance & Infection Control. 2020;**9**(1):1-11

[93] Laxminarayan R, Duse A, Wattal C, Zaidi AK, Wertheim HF, Sumpradit N, et al. Antibiotic resistance—the need for global solutions. The Lancet Infectious Diseases. 2013;**13**(12):1057-1098

[94] Mpelle FL, Ngoyi ENO, Kayath CA, Nguimbi E, Moyen R, Kobawila SC. First report of the types TEM, CTX-M, SHV and OXA-48 of beta-lactamases in *Escherichia coli*, from Brazzaville, Congo. African Journal of Microbiology Research. 2019;**13**(8):158-167

[95] Mapanguy CCM, Adedoja A, Kecka LGV, Vouvoungui JC, Nguimbi E, Velavan TP, et al. High prevalence of antibiotic-resistant *Escherichia coli* in Congolese students. International Journal of Infectious Diseases. 2021;**103**:119-123

[96] Szabo RA, Todd EC, Jean A. Method to isolate *Escherichia coli* O157:H7 from food. Journal of Food Protection. 1986;**49**(10):768-772

[97] Browning N, Botha J, Sacho H, Moore P. *Escherichia coli* O157:H7 haemorrhagic colitis. Report of the first South African case. South African Journal of Surgery. Suid-Afrikaanse tydskrif vir chirurgie. 1990;**28**(1):28-29

[98] Raji MA, Minga U, Machangu R. Current epidemiological status of enterohaemorrhagic *Escherichia coli* O157:H7 in Africa. Chinese Medical Journal. 2006;**119**(3):217-222

[99] Anderson RM. Antimicrobial resistance: Addressing the threat to global health. Preface. Philosophical Transactions of the Royal Society of London Series B, Biological Sciences. 2015;**370**(1670):20140305

[100] Barber AE, Norton JP, Wiles TJ, Mulvey MA. Strengths and limitations of model systems for the study of urinary tract infections and related pathologies. Microbiology and Molecular Biology Reviews. 2016;**80**(2):351-367

[101] Nicolas-Chanoine M-H, Bertrand X, Madec J-Y. *Escherichia coli* ST131, an intriguing clonal group. Clinical Microbiology Reviews. 2014; **27**(3):543-574

[102] Kang C-I, Song J-H. Antimicrobial resistance in Asia: Current epidemiology and clinical implications. Infection & Chemotherapy. 2013;**45**(1):22-31

[103] Lagamayo EN. Antimicrobial resistance in major pathogens of hospital-acquired pneumonia in Asian countries. American Journal of Infection Control. 2008;**36**(4):S101-S1S8

[104] Ma L, Siu LK, Lin J-C, Wu T-L, Fung C-P, Wang J-T, et al. Updated molecular epidemiology of carbapenem-non-susceptible *Escherichia coli* in Taiwan: First identification of KPC-2 or NDM-1-producing *E. coli* in Taiwan. BMC Infectious Diseases. 2013;**13**(1):1-8

[105] Al-Taiar A, Hammoud MS, Cuiqing L, Lee JK, Lui K-M, Nakwan N, et al. Neonatal infections in China, Malaysia, Hong Kong and Thailand. Archives of Disease in Childhood-Fetal and Neonatal Edition. 2013;**98**(3): F249-FF55

[106] Van Boeckel TP, Gandra S, Ashok A, Caudron Q, Grenfell BT, Levin SA, et al. Global antibiotic consumption 2000 to 2010: An analysis of national pharmaceutical sales data.

The Lancet Infectious Diseases. 2014;**14**(8):742-750

[107] Gagliotti C, Balode A, Baquero F, Degener J, Grundmann H, Gür D, et al. *Escherichia coli* and *Staphylococcus aureus*: Bad news and good news from the European antimicrobial resistance surveillance network (EARS-Net, formerly EARSS), 2002 to 2009. Eurosurveillance. 2011;**16**(11):19819

[108] Zhao S-Y, Wang Y-C, Xiao S-Z, Jiang X-F, Guo X-K, Ni Y-X, et al. Drug susceptibility and molecular epidemiology of *Escherichia coli* in bloodstream infections in Shanghai, China, 2011-2013. Infectious Diseases. 2015;**47**(5):310-318

[109] Wang C, Hao W, Yu R, Wang X, Zhang J, Wang B. Analysis of pathogen distribution and its antimicrobial resistance in bloodstream infections in hospitalized children in East China, 2015-2018. Journal of Tropical Pediatrics. 2021;**67**(1):fmaa077

[110] European Centre for Disease Prevention and Control. Surveillance of Antimicrobial Resistance in Europe—Annual Report of the European Antimicrobial Resistance Surveillance Network (EARS-Net) 2017. Stockholm: ECDC; 2018

[111] Lee C-R, Lee JH, Park KS, Kim YB, Jeong BC, Lee SH. Global dissemination of carbapenemase-producing *Klebsiella pneumoniae*: Epidemiology, genetic context, treatment options, and detection methods. Frontiers in Microbiology. 2016;7:895

[112] Lina TT, Khajanchi BK, Azmi IJ, Islam MA, Mahmood B, Akter M, et al. Phenotypic and molecular characterization of extended-spectrum beta-lactamase-producing *Escherichia coli* in Bangladesh. PLoS One. 2014;**9**(10):e108735

[113] Islam MA, Nabi A, Rahman M, Islam M, Ahmed D, Faruque ASG, et al. Prevalence of faecal carriage of NDM-1-producing bacteria among patients with diarrhoea in Bangladesh. Journal of Medical Microbiology. 2014;**63**(4): 620-622

[114] Feare CJ, Sanders M, Blasco R, Bishop J. Canada goose (*Branta canadensis*) droppings as a potential source of pathogenic bacteria. The Journal of the Royal Society for the Promotion of Health. 1999;**119**(3): 146-155

[115] Rashid M, Rakib MM, Hasan B. Antimicrobial-resistant and ESBL-producing *Escherichia coli* in different ecological niches in Bangladesh. Infection Ecology & Epidemiology. 2015;**5**(1):26712

[116] Kaur N, Sharma S, Malhotra S, Madan P, Hans C. Urinary tract infection: Aetiology and antimicrobial resistance pattern in infants from a tertiary care hospital in northern India. Journal of Clinical and Diagnostic Research: JCDR. 2014;**8**(10):DC01

[117] Roy S, Krishnan R, Mukherjee S, Schneiders T, Niyogi SK, Basu S. Prevalence of ST131 virulence-associated strains among CTX-M-producing *Escherichia coli* in the gut of hospitalized neonates in India. Diagnostic Microbiology and Infectious Disease. 2013;**77**(2):158-159

[118] Parveen RM, Manivannan S, Harish B, Parija S. Study of CTX-M type of extended spectrum β-lactamase among nosocomial isolates of *Escherichia coli* and *Klebsiella pneumoniae* in South India. Indian Journal of Microbiology. 2012;**52**(1):35-40

[119] Hussain A, Ewers C, Nandanwar N, Guenther S, Jadhav S, Wieler LH, et al. Multiresistant uropathogenic *Escherichia coli* from a region in India where urinary tract infections are endemic: Genotypic and phenotypic characteristics of sequence type 131 isolates of the

CTX-M-15 extended-spectrum-β-lactamase-producing lineage. Antimicrobial Agents and Chemotherapy. 2012;**56**(12):6358-6365

[120] Von Baum H, Marre R. Antimicrobial resistance of *Escherichia coli* and therapeutic implications. International Journal of Medical Microbiology. 2005;**295**(6-7):503-511

[121] Mirza SH, Salman M, Khurshid U, Wiqar MA. CTX-M ESBL enzyme in *Escherichia coli* from urology patients in Rawalpindi, Pakistan. Journal of Pakistan Medical Association. 2006;**56**(12):576

[122] Hussain M, Hasan F, Shah AA, Hameed A, Jung M, Rayamajhi N, et al. Prevalence of class A and AmpC b-lactamases in clinical *Escherichia coli* isolates from Pakistan Institute of Medical Science, Islamabad, Pakistan. Japanese Journal of Infectious Diseases. 2011;**64**(3):249252

[123] Al-Tawfiq JA. Occurrence and antimicrobial resistance pattern of inpatient and outpatient isolates of *Pseudomonas aeruginosa* in a Saudi Arabian hospital: 1998-2003. International Journal of Infectious Diseases. 2007;**11**(2):109-114

[124] Al Yousef SA. Surveillance of antibiotic-resistant bacteria in King Khalid Hospital, Hafr Al-Batin, Saudi Arabia, during 2013. Jundishapur Journal of Microbiology. 2016;**9**(9)

[125] Tanvir R, Hafeez R, Hasnain S. Prevalence of multiple drug resistant *Escherichia coli* in patients of urinary tract infection registering at a diagnostic laboratory in Lahore Pakistan. Pakistan Journal of Zoology. 2012;**44**(3):707-712

[126] Zaman G, Karamat KA, Abbasi SA, Rafi S, Ikram A. Prevalence of extended-spectrum beta-lactamase [ESBL] producing enterobacteriaceae in nosocomial isolates. Pakistan Armed Forces Medical Journal. 1999;**49**(2): 91-96

Chapter 10

Mechanisms of Antimicrobial Resistance of *E. coli*

Rodney C. Jariremombe

Abstract

Escherichia coli has become a major significant pathogen behind infections, many researches have been conducted on possible drugs that can successfully eradicate the pathogenic isolates. To ensure survival, *E. coli* strains improvised resistant mechanisms to allow them to maneuver through with life among bactericidal agents. The chapter gives an overview of the antimicrobial resistance mechanisms found in major groups of antimicrobial drugs. *E. coli* uses enzymes in defying drug susceptibility for example aminoglycoside modifying enzymes in modifying drug recognition sites, in cephalosporin, penicillin the pathogen indulged in the use of β-lactamases to break down the β-lactam ring on the structure of the drugs. In fluoroquinolones, the pathogen uses efflux pumps, DNA gyrase mutation as a mechanism of resistance. The continuous use of drugs induces resistance mechanisms to increase, there is a need for continuous researches on drugs effectivity and the discovery of new and better medication to fight against *E. coli* pathogens.

Keywords: mutation, ESBL, efflux-pumps, genes, enzyme

1. Introduction

Escherichia coli is one of the most primitive microorganisms that are affecting the normal body functionality, bringing sickness, attributed by infections that are becoming difficult to cure since the microorganisms are evolving with time they tend to mutate and produce different species which are resistant to drugs that were previously effective in fighting and eradicating the bacterial species. The aspect of antimicrobial resistance has become a non-healing wound in as much as health is concerned, with time the wound continues to deepen and expand bringing in more confusion, sickness as well as problems in medicinal drug references [1].

E. coli Resistant germs are emerging at an alarming rate, posing a growing threat to human society. Antibiotic misuse and overuse, as well as antibiotic buildup in the environment, have been blamed for the growth of antimicrobial resistance (AMR). With the pharmaceutical industry's lack of new medication development, it is becoming very difficult to tackle diseases behind the infection [2].

It is believed the long-term use of drugs on *E. coli* has brought problems in curing the infections it causes because of many adaptive mechanisms the pathogen has developed to discard drug susceptibility over time thereby allowing its survival and perpetuation [3]. According to the World Health Organization (WHO) [4], the long-term use of drugs, misuse and abuse of drugs are the foundation of creating resistant mechanisms that may lead to difficulties in prevention as well as treatment.

 IntechOpen

2. Mechanisms of resistance

2.1 Mechanisms of resistance to cephalosporin drugs

Third-generation Cephalosporin has been used as a successor of penicillin which has been resisted by many drugs due to long-term use on the pathogens. Third-generation cephalosporin drugs are a much-improved version that has been useful in eradicating bacterial species such as *E. coli* [5]. However, with time, the bacteria tend to become resistant to the drugs due to the improvising of mechanisms to create barriers and different structures that are not recognized by the drugs for disruption of the bacterial cell. Since some drugs recognize specific polypeptide sequences where the chemical drugs cleave for destruction. Such actions involve acquired mechanisms of resistance which involve the passing down of resistant plasmids from cell to cell, another way includes the intrinsic mechanism of resistance whereby the cell creates ways of denouncing the drug susceptibility by adjusting or improving structures within the cell. They can change the polypeptide sequences also creating structures that limit the uptake of the drug from the environment, by modifying specific sites targeted by the drugs the cell automatically deprives recognition thus inducing resistance [6].

2.1.1 Penicillin and cephalosporin β-lactam mode of action

The gram-negative bacterial cell wall is made up of a complex structure which is made of a thinner peptidoglycan layer with a structure of crosslinking peptidoglycan precursors made by adjoining N-acetyl glucosamine and the N-acetyl muramic acid proteins which are then cross-linked to form several layers of peptidoglycan catalyzed by Transpeptidase and de-alanyl carboxypeptidase. The penicillin-binding proteins form the D-ala D-ala cross-linkages of the peptidoglycan wall in cell wall synthesis. The β-lactam ring in penicillin and cephalosporin will bind to the enzymes (Trans-peptidase and D-ala- carboxyl peptidase) thereby preventing bacterial cell synthesis leading to bacterial cell wall damage that will cause bursting after being subjected to the low osmotic pressure of the surrounding environment. The antibiotic penicillin-binding complex will stimulate the release of autolytic compounds that are capable of digesting the cell wall [7].

2.1.2 Resistance mechanisms

Quite several researches have outlined that multidrug-resistant species which include *E. coli* have been a long-term migraine problem with the drastic increase in resistance as of date, with special attention on the development of the extended-spectrum β-lactamase (ESBL). The genes which are encoded by the ESBLs are located in the plasmid of the bacterium cell and most cases, they are transferred through horizontal transfer to other cells. *E. coli* has acquired resistance to β-lactam antibiotics through the production of the β-lactamase enzyme which is used to break down the β-lactam ring of most penicillin derivatives [8]. β-lactamase enzymes are the biggest and greatest reason why penicillin drugs are failing to eradicate infections behind *E. coli* bacteria.

With this problem being pointed out by scientists, new-generation drugs of the cephalosporin class were invented which were believed to defy the stability of many bacterial β-lactamases on the drug, thereby allowing the drug to temper with the bacterial structure and eliminate them. With persistent use and exposure to the third-generation drugs which include; cedox, cefixime, cefotsxime and avycaz which have been successfully superior to older penicillin drugs in terms of effectiveness on treatment assays. In the early 1980s, in response to the increasing prevalence and spread of β-lactamase, third-generation cephalosporins or oxyimino groups were introduced into clinical practice. Resistance to these broad-spectrum cephalosporins quickly emerged. As early as 1983, Germany published the first report on the SHV2 enzyme that can hydrolyze these antibiotics [8]. The continuous use of these third-generation cephalosporins has brought along dynamic inducement on the production of many mutated lactamases in many bacteria, allowing survival and denying drug effects. The β-lactamase in ESBLs contains serine chemicals at their active site which hydrolyzes the spectrum of cephalosporins using an oxyminoside chain [9].

The TEM1, TEM2 are genes that aid in coding for the ESBLs through mutation to alter the amino acid configuration of the β-lactamases, thereby extending the degree of affinity and complementarity for the spectrum of the β-lactam antibiotics to be susceptible for hydrolysis. There are several groups of ESBLs with similar behaviors but different evolutionary histories. The largest population is TEM and sulfhydryl reagent variable (SHV) β-lactamase mutants, with members exceeding 150 [10]. Mutations affecting a small number of key amino acids expand the active site of the enzyme, allowing it to bypass the oxyimino substitution that normally protects the β-lactam ring. Therefore, although the classic TEM and SHV enzymes cannot significantly hydrolyze the oxyiminocephalosporin, the mutant can do so, thereby conferring resistance to its host strain [10].

The CTX-M enzyme is another type of ESBLs. Based on sequence homology, they are divided into five subgroups. Most of these subgroups have evolved due to the leak of the chromosomal β-lactamase gene of Kluvera spp., which is a less clinically significant Enterobacter spp.). After migrating to mobile DNA, CTX-M β-lactamase can further evolve. *E. coli* isolates that produces CTX-M enzyme have been identified as the cause of urinary tract infections. Some reports indicate that the CTX-M ESBL may now be the most common ESBL type in the world [10, 11].

Figure 1; Showing the mechanisms in which gram-negative bacteria can be resistant to penicillin and third-generation cephalosporin drugs. The penicillin-binding protein is being modified in such a way to prevent complementary pairing

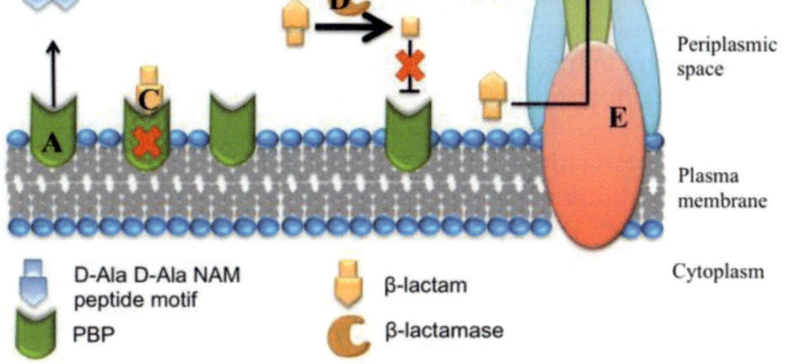

Figure 1.
B1-Metallo-β-Lactamases: Where do we stand? Adapted from Mojica et al. [12].

with the drug (C) that is the modification of the drug target. The B-lactamase enzyme (D) cleaving the B lactam structure of the drug defying its susceptibility and action. At (E) showing efflux pumping of the drugs from the cell [12].

3. Resistance on fluoroquinolones

Quinolones are the most frequently used drugs against *E. coli* infection because they are highly bioavailable meaning they have a good tissue distribution once administrated in the body orally [13]. This fact has caused several doctors to prefer referencing quinolones once an *E. coli* infection is detected. However, the major factor behind *E. coli* resistance in Fluoroquinolones is through mutations in the genome of the bacteria that is DNA gyrase [14].

3.1 Fluoroquinolone mode of action

The mode of action of fluoroquinolones is by making complexes with DNA Gyrase and topoisomerase IV on the DNA chromosome thereby allowing for the disruption of the DNA sequence of *E.coli*. Fluoroquinolone antibiotics have a chemical structure that allows them to interrupt *E. coli* activity through the alteration of the DNA Gyrase and the Topoisomerase IV protein structure, thereby preventing any form of replication and translation processes for protein synthesis bringing for the destruction of *E. coli* microorganisms. The interruption with DNA Gyrase affects the conversion of the relaxed double-stranded DNA into a negatively super twisted form that allows the replication to commence, this diminishes relegation through entrapping of the enzymes changing their protein arrangement in their active site preventing complementarity with the DNA strand. The replication fork is held steady by Topoisomerase IV and the interruption of its structure affects the replication fork formation, therefore, prohibiting replication to proceed [14].

3.2 Mechanisms of resistance

3.2.1 DNA Gyrase and Topoisomerase Base substitution

The mutation in the genome results in amino acid-base substitution in the Gyrase A Gene (GyrA) and topoisomerase IV proteins [14]. Changes on those two genomic structures have been termed the quinolone resistance determining regions. The research conducted by Friedman [15], outlined that the amino acid substitution happens between 67 and 106 bases specifically at bases 83 and 87, therefore altering the drug targets. Further researches proved that there are other sites found on the nalidixic acid-resistant mutant that was not thermo-tolerant had a 5′ base change of guanine to thiamine, in the codon 87 which is expected to reduce the susceptibility of quinolone to nalidixic acid due to the substitution of tyrosine for aspartic acid [15].

However, some investigations are taking place with the intention to defend and uplift the bactericidal status and this includes recent studies done on nybomycin, where investigations on the susceptibility of fluoroquinolone-sensitive and fluoroquinolone-resistant strains were conducted and discovered that nybomycin was successively efficient in destroying the bacterial species [16].

It is important to determine whether the *E. coli* mutants are thermotolerant or non-tolerant because this aids in determining how they can be susceptible to drugs and how temperatures can affect the genomic structures [17]. *E. coli* strains have adaptation characteristics such as physiological, metabolic and proton consuming acid-resistant mechanisms that allow their survival and perpetuation in acid environments below pH 2. They reduce the effects of acid damage by modifying the membrane, altering membrane porins to reduce proton influx and periplasmic chaperons [18].

3.2.2 Efflux pumping

Active efflux pumping is a mechanism by which a substance that is not needed in the cell is pumped out to prevent the damages that the substance or chemical may bring to the cell, they are used in moving a variety of different toxic compounds out of the cell and in bacteria they use it to pump out antibiotics. It is a major fundamental characteristic in antimicrobial resistance of gram-negative bacteria including *E. coli* [19]. The efflux pump family in enterobacteriaceae called the resistance nodulation division (RND), is the most significant factor behind multidrug resistance and one of the most characterized RND systems in enterobacteriaceae is the AcrAB-TolC efflux systems. Expression of the AcrAB and TolC genes are regulated by the MarA protein in *E. coli* [20].

In the *E. coli* operon, the expression of AcrAB is controlled or mediated by AcrR which is a repressor located at the upstream part of the acrAB operon where the expression can be transcribed or repressed [20, 21]. Studies have shown that mutations are taking place in the acrR which means there is no repression of expression for AcrAB which means the more the expression the higher the rate of pumping out of the drugs [22]. They also have a specific modification of the porin membrane channel proteins which have a specific mediated width that allows the in and outflow of the substance of the cell. The porins for example against vancomycin are modified in such a way that the vancomycin molecules cannot pass through into the cell [23]. The major porins in *E. coli* such as OmpF and OmpC protein were believed to be the drug binding sites, however recent studies show that there have been changes in their structural arrangements making the drugs unable to bind to the proteins. The presence of mutant porins can even cause resistance to carbapenems which are believed to be the most efficient and reliable drugs against *E. coli* and other bacterial species [23, 24].

Most enterobacteriaceae gram-negative bacteria have developed specialized genes which aid in resisting carbapernemdrugs which are called the ndm genes which are often found branch host range conjugative plasmids which work in conjunction with other resistance genes. The ndm genes have a transposon mechanism which means they are found on plasmids as well as the host chromosomes and can move between the two at a much higher frequency thereby enabling the resistance build-up in many cells in a much-minimized time range via transformation mechanisms [25].

4. Resistance on aminoglycosides

Aminoglycoside drugs have been part of the fight against *E. coli* pathogenic species, however with the changes in the phenotype of the *E. coli* isolates for example the ever-evolving changes in the ESBL structure as well as the genes that the aminoglycoside drugs encode for susceptibility [26].

4.1 Aminoglycoside mode of action

Aminoglycosides are bactericidal agents that inhibit the synthesis of bacterial proteins through the interruption of the ribosomes. They interfere by binding to the 30S and 50S ribosomal subunits, they inhibit the translocation process of moving the peptidyl-tRNA from the A site to the P site thereby causing mRNA misreading in that way denouncing the translation process forwarding zero protein synthesis, which entails no budding, multiplication and denouncing perpetuation and survival [26].

4.2 Mechanism of resistance on aminoglycoside drugs

There are more than 50 types of aminoglycosides modifying enzymes which include, acetyltransferases (aac), phosphotransferases (aph) and nucleotidyl-transferases (ant), these enzymes are capable of modifying aminoglycosides at their respective drug recognition sites [27]. The main cause of resistance to the aminoglycoside drugs is that more than one aminoglycoside modifying enzymes may be found in a single isolate bringing in a high probability rate of resistance to the drugs [28].

Mancin [28], conducted a population analysis of genetic and enzymatic resistance of *E. coli* to aminoglycosides, they concluded that the genes aac(6′) and aac(3) can cause significant resistance to amikacin and kanamycin. Another study on genes was conducted by Bodendoerfer [29], where they concluded that in Switzerland the most prevalent resistance genes included the aph(3′)-la, aac(3)-lld and aac(6′)-lb-cr, they also alluded that the genes tend to change in their structural arrangement which is a mediated by the action of the transposon mechanisms of the resistant genes. The mechanisms of the jumping gene continues to be a threat to human health, because the transposon genes can be transferred to the next cell through transformation and conjugation processes which will cause the development of more resistant genes prior to transcription and translation, this information has caused migraine headaches to researchers and scientists who are working tirelessly in the attempt of denouncing resistance [29].

5. Conclusion

E. coli bacteria continues to be a nuisance in the medical field bringing endless prospects in resistance against the drugs that are used in an attempt of eradicating the bacteria. The bacterial species has so many different forms of isolates that differ from one another both structurally and genetically, hence the drug that is susceptible to one *E. coli* may be found ineffective to another *E. coli* variant species. The review showed that there are many different mechanisms in which *E. coli* strains are becoming more and more difficult to treat, some evolving to possess enzymes that work in a conjunctive manner to denounce the effect of the drug.

The evolution brings forth the production of many different isolates with different protein structures for drug resistance and allows perpetuation and survival. This alarms for continuous assessments to provide information on the drugs and the interference with the pathogenic microorganism, how the bacteria respond in susceptibility, which will act as a fortress in tracing the reason behind resistance tomorrow. Studies are of importance to help and provide practical proof on how to tackle pathogens which will aid in improving health, denouncing long hospital stays and even patients from succumbing to infections caused by the microorganism.

Conflict of interest

The authors declare no conflict of interest.

Author details

Rodney C. Jariremombe[1,2]

1 Bachelor of Science Honors Degree in Biological Sciences (Genetics and Microbiology), Harare, Zimbabwe

2 Safety, Health and Environmental Officer, Harare, Zimbabwe

*Address all correspondence to: rodneyjari@gmail.com

References

[1] Erjavec MS. Introductory Chapter: The Versatile *Escherichia coli*. The Universe of Escherichia coli. Rijeka: IntechOpen; 2019. p. 3. DOI: 10.5772. intechopen/88882

[2] Lauxen A et al. Mechanisms of resistance development in TCAT a trimethoprim- based photo switchable antibiotic. Pharmaceuticals. 2021;**14**:392

[3] Sondi I, Salopek-Sondi B. Silver nanoparticles as antimicrobial agent: A case study on *E. coli as* a model for Gram-negative bacteria. Journal of Colloid and Interface Science. 2004; **275**(1):177-182

[4] Li D, Reid CJ, Kudinha T, Jarocki VM, Djordjevic SP. Genomic analysis of trimethoprim-resistant extraintestinal pathogenic *Escherichia coli* and recurrent urinary tract infections. Microbial Genomics. 2020;**6**(12)

[5] Beneduce L, Spano G, Massa S. *Escherichia coli 0157: H7* general characteristics, isolation and identification techniques. Annals of Microbiology. 2003;**53**(4):511-528

[6] World Health Organization. Antimicrobial Resistance Global Report on Surveillance: 2014 Summary. Geneva, Switzerland: World Health Organization; 2014

[7] Moellering RC Jr, Swartz MN. The newer cephalosporins. New England Journal of Medicine. 1976;**294**(1):24-28

[8] Reygaert WC. Antimicrobial Mechanisms of *Escherichia coli*. London, UK: IntechOpen; 2017. pp. 81-97. DOI: 10.5772/67363

[9] Sauvage E, Kerff F, Terrak M, Ayala JA, Charlier P. The penicillin-binding proteins: Structure and role in peptidoglycan biosynthesis. FEMS Microbiology Reviews. 2008;**32**(2):234-258. DOI: 10.1111/j.1574-6976.2008. 00105.x

[10] Rawat D, Nair D. Extended-spectrum β-lactamases in gram negative bacteria. Journal of Global Infectious Diseases. 2010;**2**(3):263-274. DOI: 10.4103/0974-777X.68531

[11] Paterson DL. Recommendation for treatment of severe infections caused by Enterobacteriaceae producing extended-spectrum β-lactamases (ESBLs). Clinical Microbiology and Infection. 2000;**6**:460-463

[12] Mojica MF, Bonomo RA, Fast W. B1-Metallo-β-Lactamases: Where do we stand? Current Drug Targets. 2016;**17**(9):1029-1050. DOI: 10.2174/138 9450116666151001105622

[13] Oliphant CM, Green GM. Quinolones: A comprehensive review. American Family Physician. 2002;**65**(3):455-464

[14] GrugerT NLJ, Maxwell A, et al. A mutation in *Escherichia coli* DNA gyrase conferring quinolone resistance results in sensitivity to drugs targeting eukaryotic topoisomerase II. Antimicrob Agents Chemother. 2004;**48**(12):4495-4504. DOI: 10.1128/ AAC.48.12.4495-4504

[15] Friedman SM, Lu T, Drlica K. Mutation in the DNA gyrase A Gene of *Escherichia coli* that expands the quinolone resistance-determining region. Antimicrobial Agents and Chemotherapy. 2001;**45**(8):2378-2380. DOI: 10.1128/AAC.45.8.2378-2380

[16] Shiriaev DI, Sofronova AA, Berdnikovich EA, Lukianov DA, Komarova ES, Marina VI, et al. Nybomycin inhibits both types of *E. coli* DNA gyrase – fluoroquinolone-sensitive and fluoroquinolone-resistant. Antimicrobial Agents and Chemotherapy. 2021;**65**(5):e00777-e00720. DOI: 10.1128/AAC.00777-20

[17] Xu Y, Zhao Z, Tong W, et al. An acid-tolerance response system protecting exponentially growing *Escherichia coli*. Nature Communications. 2020;**11**:1496. DOI: 10.1038/s41467-020-15350-5

[18] Kanjee U, Houry WA. Mechanisms of acid resistance in *Escherichia coli*. Annual Review of Microbiology. 2013;**67**:65-81. DOI: 10.1146/annurev-micro-092412-155708

[19] Sharma A, Gupta VK, Pathania R. Efflux pump inhibitors for bacterial pathogens: From bench to bedside. The Indian Journal of Medical Research. 2019;**149**(2):129-145. DOI: 10.4103/ijmr.IJMR_2079_17

[20] Weston N, Sharma P, Ricci V, Piddock LJV. Regulation of the AcrAB-TolC efflux pump in Enterobacteriaceae. Research in Microbiology. 2018;**169**(7-8):425-431. DOI: 10.1016/j.resmic.2017.10.005

[21] Li XZ, Plésiat P, Nikaido H. The challenge of efflux-mediated antibiotic resistance in Gram-negative bacteria. Clinical Microbiology Reviews. 2015;**28**(2):337-418. DOI: 10.1128/CMR.00117-14

[22] Lawler AJ, Ricci V, Busby SJ, Piddock LJ. Genetic inactivation of acrAB or inhibition of efflux induces expression of ramA. Journal of Antimicrobial Chemotherapy. 2013;**68**(7):1551-1557. DOI: 10.1093/jac/dkt069

[23] Blair JM, Webber MA, Baylay AJ, Ogbolu DO, Piddock LJ. Molecular mechanisms of antibiotic resistance. Nature Reviews. Microbiology. 2015;**13**(1):42-51. DOI: 10.1038/nrmicro3380

[24] Baroud M, Dandache I, Araj GF, Wakim R, Kanj S, Kanafani Z, et al. Underlying mechanisms of carbapenem resistance in extended-spectrum β-lactamase-producing Klebsiella pneumoniae and *Escherichia coli* isolates at a tertiary care centre in Lebanon: Role of OXA-48 and NDM-1 carbapenemases. International Journal of Antimicrobial Agents. 2013;**41**(1):75-79. DOI: 10.1016/j.ijantimicag.2012.08.010

[25] Nordmann P, Poirel L, Carrër A, Toleman MA, Walsh TR. How to detect NDM-1 producers. Journal of Clinical Microbiology. 2011;**49**(2):718-721. DOI: 10.1128/JCM.01773-10

[26] Soleimani N, Aganj M, Ali L, et al. Frequency distribution of genes encoding aminoglycoside modifying enzymes in uropathogenic *E. coli* isolated from Iranian hospital. BMC Research Notes. 2014;7:842. DOI: 10.1186/1756-0500-7-842

[27] Ojdana D, Sieńko A, Sacha P, Majewski P, Wieczorek P, Wieczorek A, et al. Genetic basis of enzymatic resistance of *E. coli* to aminoglycosides. Advances in Medical Sciences. 2018;**63**(1):9-13. DOI: 10.1016/j.advms.2017.05.004

[28] Mancini S, Marchesi M, Imkamp F, Wagner K, Keller PM, Quiblier C, et al. Population-based inference of aminoglycoside resistance mechanisms in *Escherichia coli*. eBioMedicine. 2019;**46**:184-192. DOI: 10.1016/j.ebiom.2019.07.020

[29] Bodendoerfer E, Marchesi M, Imkamp F, Courvalin P, Böttger EC, Mancini S. Co-occurrence of aminoglycoside and β-lactam resistance mechanisms in aminoglycoside- non-susceptible *Escherichia coli* isolated in the Zurich area, Switzerland. International Journal of Antimicrobial Agents. 2020;**56**(1):106019. DOI: 10.1016/j.ijantimicag.2020.106019

Chapter 11

Antibiotic Resistance among *Escherichia coli* Isolates, Antimicrobial Peptides and Cell Membrane Disruption to the Control of *E. coli* Infections

Sara Kadkhodaei and Gelareh Poostizadeh

Abstract

The treatment of *Escherichia coli* infections has been seriously complicated due to the appearance of multidrug-resistant isolates and the rapid distribution of extended-spectrum β-lactamase-producing species. In recent years there has been considerable effort to develop alternative therapies to traditional antibiotics for infection diseases caused by antimicrobial agents. The mechanisms by which antimicrobial compounds induce bacterial damage have been suggested to be interaction with membranes, formation of pores lined by both lipids and peptides, or by a more general "Anionic lipid clustering," and other specific mechanisms. The major constituents of the lipid bilayer on the outer membrane of *E. coli* as a Gram-negative bacteria are lipopolysaccharide, zwitterionic core oligosaccharides, saturated fatty acid chains with zwitterionic phospholipid head groups, and lipid A functionalized with anionic phosphate groups. Research findings emphasize the importance of the membrane composition of *E. coli* in determining the susceptibility to certain antimicrobial agents, such as antimicrobial peptides (AMPs) and successful treatment.

Keywords: *E. coli*, antibiotic resistance, membrane, antimicrobial peptides, novel therapy

1. Introduction

By the discovery of penicillin in 1928, the twentieth century was the golden age of antibiotics based on small molecule natural products, for instance, tetracyclines, β-lactams, and aminoglycosides [1]. These products were successful in the treatment of infectious diseases, and they saved the lives of many human beings from different types of bacterial infections. However, common antibiotics have become ineffective due to the constant evolution of most bacterial strains against them [2]. These bacterial strains can spread all around the world and lead to fatal infectious diseases because of antimicrobial resistance (AMR), which is currently one of the most crucial global health concerns [3]. Not only the antibiotics misuse and overuse have an important role in increasing AMR, but also the relatively slow pace of the development of novel antibiotics has aggravated this problem [4]. The latter reason, the so-called "discovery void," occurred because no major class of antibiotics has been introduced since the introduction of lipopeptide antibiotics (e.g., daptomycin) in the mid-1980s. While in over 50 years no new class of antibiotics has been approved, the antibacterial treatments for Gram-negative bacteria become more difficult [2–5]. In 2016, multidrug resistance (MDR) was announced as one of the major health challenges of that time by the World Economic Forum (WEF). The foundation believed without urgent action, the estimated global death because of MDR could reach 10 million by 2050 [6]. Hence, the design and synthesis of new antibiotics with new antimicrobial mechanisms is evident [7]. According to the literature, bacterial cell membranes have a critical role in modulating antibiotic resistance, during recent years, studies on bacterial cell membranes perturbed by new compounds to overcome antibiotic resistance have been developed [8].

In comparison to Gram-positive bacteria, all Gram-negative bacteria have an extra membrane that surrounds them and is called the outer membrane (OM) (**Figure 1**) [8]. Unlike the cytoplasmic membrane (CM), the OM is very asymmetric, containing phospholipids on the inner leaflet, and lipopolysaccharides (LPSs) on the outer leaflet [9]. LPSs are the major constituents on the OM of Gram-negative bacteria which have zwitterionic oligosaccharides as core [10], zwitterionic phospholipid head groups on saturated fatty acid chains [11], and lipid A with anionic phosphate groups [12].

The OM is essential for cell viability and prevents the entry of harmful toxic substances by blocking permeability. LPSs play a central role in the selective permeability and integrity of OM. While many hydrophobic molecules are able to limit diffusion [13], LPSs play an important factor in providing selectivity to them. Because of the anionic phosphate groups, LPS molecules are able to form intermolecular electrostatic bonds with neighbors. The cross-bridging of neighboring LPS molecules significantly contributes to the resistance against hydrophobic antimicrobial agents. The anionic nature of lipid A seems to be the Achilles heel for OM integrity. The OM of *E. coli* is composed not only of LPS but also outer membrane proteins (OMP), lipoproteins (LPP), and porins [14–16]. Porins are charged proteins that allow the penetration of drugs, nutrients, and small molecules inside bacterial cells [17].

So far, intracellular processes are the target of many antibiotics to create holes in the bacterial cell envelope. In particular, there is a formidable barrier on the outer membrane (OM) of Gram-negative bacteria that must be overcome by antibiotics. There are two different pathways that help antibiotics to take through the OM, which are general diffusion porins for hydrophilic antibiotics and a lipid-mediated pathway for hydrophobic antibiotics. Some outer membrane structures such as

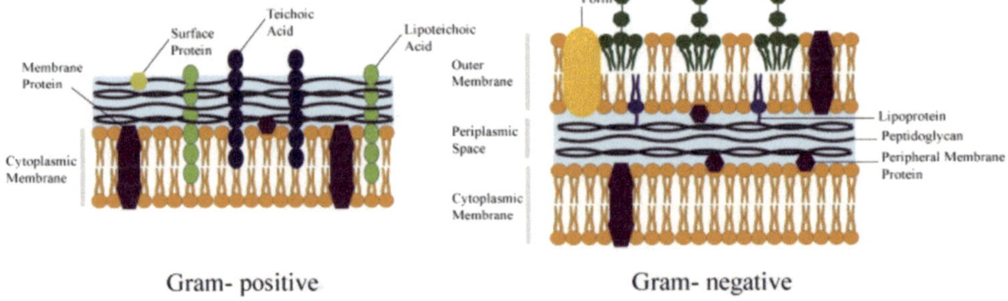

Figure 1.
Comparing gram-positive and gram-negative bacterial cell membranes.

protein and lipid, and their modifications have a striking influence on the bacterial antibiotics sensitivity and resistance [18]. The ability of OM disruption to change the rules of Gram-negative entry, overcome pre-existing and spontaneous resistance. Disruption of the OM expands the threshold of hydrophobicity compatible with Gram-negative activity to include hydrophobic molecules. Together, OM disruption overcomes many of the traditional hurdles encountered during antibiotic treatment and is a high-priority approach for further development [19].

2. Factors associated with antibiotic resistance in *Escherichia coli*

It is absolutely clear to us today that the antibiotic resistance of *E. coli* and some other bacteria involves a combination of different factors [20]. Drug-resistant *E. coli* can be transmitted to human beings from the environment through direct or indirect contact (e.g., consumption of contaminated food and water) [21]. The uncontrolled use of antibiotics in domestic animals, as well as dietary supplements, could be one of the main reasons for high antimicrobial resistance [22]. In addition, colonization of healthy adult workers with extended-spectrum β-lactamases (ESBLs) producing *E. coli* may be related to consumption of food and water contaminated with ESBL-producing bacteria [23].

The main causes of antibiotic resistance may involve aimless antibiotic use, deficiencies in health centers and infection-control programs in hospitals, insufficient staff training, poor hygiene and other preventive measures in veterinary medicine, and lack of right management steps in animal farms, that may cause a high frequency of ESBL producing *E. coli* isolates in human (42%) and animal sample (63%) [24].

According to the worldwide antibiotic sales database, comparing antibiotic use for the years 2000 to 2015, an evident rise from about 11 doses per 1000 inhabitants per day to almost 16 is noticed [25]. Analyzing research findings with the statistics demonstrates that the mean value of antibiotic consumption was largely impacted by mid-income and low-income countries [26]. The highest number of MDR bacterial infections were observed in these countries [20]. In the past 10 years, a growing number of resistance genes have been identified in *E. coli* isolates, and many of these resistance genes were received by horizontal gene transfer. *E. coli* acts as a donor and a recipient of resistance genes in the enterobacterial gene pool, and as a result can acquire resistance genes from other bacteria but can also pass on its resistance genes to other bacteria. AMR in *E. coli* is considered one of the most important disputes in both animals and humans on a global scale as a real public health concern [20].

According to research studies, proper monitoring of disposal processes in hospitals, systematic surveillance of hospital-associated infections, monitoring the consumption of antibiotics in animals, evaluation and monitoring of antibiotic-sensitivity patterns, and preparation of safe antibiotic strategies may ease more corrective steps for the control and inhibition of *E. coli* infections in all around the world [24].

The three most common manners by which Gram-negative bacteria develop antibiotic resistance are: (i) cleaving the antibiotic drugs such as β-lactam in the periplasmic space by secretion of enzymes like β-lactamase [27]; (ii) decreasing the size and number of the porins that facilitate drug transport [12]; and, (iii) changing the selective permeation and electrostatic field within the constriction zones of porins where the antibiotic docks first and then translocate inside the cells [28]. In attention to these facts, it seems the nature charge of the OM plays definitive roles in the interactions of electrostatic binding, charged molecules transportation, and drugs killing/inhibitory actions of wild-type and antibiotic-resistant species [29].

3. Antimicrobial peptides and alternative antimicrobial agents

The two major classes of alternate antimicrobial candidates are cationic, gene-encoded antimicrobial peptides (AMPs) [30], and, π-conjugated oligo/polyelectrolytes [29]. Initial studies suggested that bacteria still find it difficult to show resistance against the second class of antimicrobial molecules [31].

The bounded ability to cross the bacterial cell membrane is the major limitation in the subsequent development of peptides as antimicrobial agents. To overcome this problem and achieve a more efficient cellular uptake, peptides and a delivery vector were combined in a single molecule. For this purpose, proline-rich antimicrobial peptides (PrAMPs), as a part of the innate immune response [32] via the inner membrane transporters SbmA and MdtM are transported into a large panel of Gram-negative bacterial cells. Results showed that PrAMPs could be suitable carriers to transfer the other non-penetrating AMPs into the bacterial cells [33, 34]. According to this, PrAMPS are considered as a novel class of antibiotics [32–35].

Antimicrobial peptides (AMPs) are one of the most promising candidates for a novel class of antibiotics [36]. Antimicrobial peptides (also called host-defense peptides) occur in nature as an ancient class of polypeptides [37]. AMPs are part of the innate immune system and exhibit antibacterial activity against Gram-negative and Gram-positive bacteria. According to this, AMPs serve as templates for the design of new antibacterial agents against multidrug resistance. Gramicidin and defense are natural AMPs that were discovered at the beginning of the twentieth century [38]. Today, lots of cationic AMPs are known to permeabilize real bacterial membranes [39]. After the emergence of antimicrobial-resistant bacteria, AMPs were considered as potential antibiotic drugs. The advantages of AMPs over conventional antibiotics and exigent need for the development of novel antibiotics lead to the upsurge of AMP research and their clinical trials activity in recent years [36]. In addition, synthetic AMPs or a variety of peptidomimetic antimicrobials have been very investigated to overcome the inherent drawbacks (e.g., stability) of peptides in physiological conditions [7].

Antimicrobial peptides have terrific chemical diversity and are based on some common structural characteristics set apart from traditional antibiotics. AMPs generally contain less than 100 amino acids, most of them including positively charged residues, such as lysine, arginine, and histidine, and more than 50% of them have a large portion of hydrophobic. In addition to the structural differences, AMPs directly target the bacterial cell membrane in most cases. Antimicrobial peptides based on their structure are classified into four different groups—α-helical, β-sheet, extended, and cyclic. For example, while some AMPs consist of a single helix or sheet entirely, others have a more complicated structure. The extended peptides are characterized by non-recognizable structural motifs and consist of specific amino acids, such as arginine, tryptophan, glycine, and histidine [40].

Natural antimicrobial peptides isolated are effective against Gram-positive and Gram-negative bacteria, enveloped and non-enveloped viruses, yeasts, fungi, molds, and parasites [41]. A single AMP may not be effective against all pathogens, however, may exhibit the same antimicrobial activity between different germs with anionic membranes. In addition, due to their mechanism of action, some isolated AMPs from natural sources can display species-specific antimicrobial activity [42]. This may be an outcome of a highly specialized environmental niche and evolutionary advantage that specific antimicrobial peptides present for survival [41]. As many antimicrobial peptides act on lipid components of the bacterial cell membrane, they often demonstrate broad-spectrum antimicrobial activity [43].

4. Molecular mechanisms of antimicrobial peptides action

Antimicrobial peptides can alter bacterial membrane properties by different mechanisms. Alteration of the bulk physical properties of the membrane is one way (**Figure 2**) [8]. AMPs can modify bulk properties while not having a specific target on the membrane. Changes in the spatial distribution of cell membrane molecules within or modification of a bulk physical property as intrinsic curvature or fluidity are examples of these alterations. Contrary to this, altering the bulk biophysical properties by AMPs can occur by targeting a class of particular lipids. Specific phospholipids are the potential targets which AMPs being effective against both Gram-negative and Gram-positive bacteria. These mechanisms are not completely independent of each other. For example, membrane clustering will lead to packing defects at the boundary between domains and physical curvature can drive clustering. In addition, directly targeting lipids can lead to any of the three phenomena that are mentioned.

Antimicrobial agents also can target membrane phospholipids. Cardiolipin; CL, phosphatidylglycerol; PG, and phosphatidylethanolamine; PE, are the three main phospholipids in most bacteria. PG is the most abundant of them. It is an anionic lipid, and therefore, attracts cationic antimicrobial peptides. Modifying the PG head group by adding lysine and or alanine and reducing the negative charge on the membrane is one of the ways that bacteria use to protect themselves from these peptides and thus will be more resistant to the cationic antimicrobial peptides. In Gram-negative bacteria, PE is generally abundant. Several cyclic peptides are able to specifically bind to PE, and therefore, can be used to target these bacteria. In addition, PE and anionic lipid mixtures can create segregated clusters when the anionic

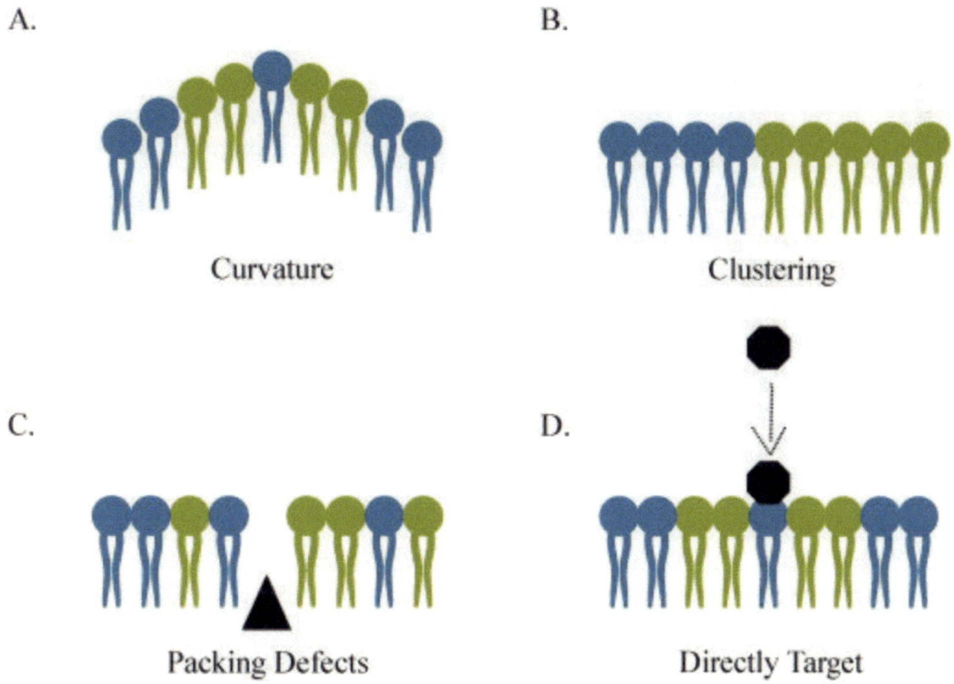

Figure 2.
Different outcomes of AMPs on properties of the bacterial cell membrane. AMPs can affect the physical properties of the cellular membrane, such as (A) induction of membrane physical curvature [44], (B) lipid clustering, (C) prompting packing defects resulting in complete or partial loss of the permeability barrier, and (D) directly targeting components of a membrane such as lipids leading to a variety of consequences.

lipid is bound to an AMP. The activity of antimicrobial agents that bind to cardiolipin is also based on a clustering mechanism [9]. Some AMPs can bind to either CL or PE and target specific bacteria. So, the cell membrane is a multipurpose target for AMPs that serves as targets or wards them off in resistance and provides a crucial site for toxic activities [8].

However, the head group structure of phospholipids is the same in bacteria and eukaryotes, the acyl chains in bacteria are shorter and more saturated [45]. In addition, while anionic lipids and PE are sequestered to the cytoplasmic surface of eukaryotic membranes, they are exposed to the external surface of bacterial membranes. These differences provide the feasibility of designing antibacterial agents that target specific bacterial lipids [8].

The antimicrobial activities of AMPs against various types of pathogens, including Gram-positive and Gram-negative bacteria, viruses, and fungi occur through a wide range of mechanisms, for example, membrane disruption, intracellular penetration, and immunomodulation [46]. Although AMPs may have different mechanisms of action, it is thought that their ability to act against such diverse cellular organisms is related to membrane activity [7]. The positive charge of cationic amino acid AMPs enables electrostatic interaction to the negatively charged microbial membranes [47, 48] and that the hydrophobic region is involved in the penetration of the cells [49]. The nature of the cell surface, in particular, the composition of the OM of Gram-negative bacteria has a major impact on the antimicrobial activity and efficacy of antimicrobial agents including cationic AMPs [50]. Antimicrobial peptides penetrate the bacterial membranes through several different mechanisms [51]. Briefly, AMPs binding to cell membrane break down the membrane potential, lead to alteration membrane permeability and metabolite leakage, and finally cause bacterial cell death [41].

Structural antimicrobial peptide studies have strongly suggested that the physicochemical properties of AMPs are responsible for their microbiological activities, rather than any specific amino acid residues. Since the amphiphilic topology is fundamental for insertion into the cytoplasmic membrane and disruption of cells, AMPs and their mimics have been considered as attractive targets for drug development. In particular, they are able to kill bacteria quickly and development of the bacterial antibiotic resistance is relatively difficult [7].

There is a clear phenomenological link between anionic lipid clustering and the bacterial species specificity of a number of antimicrobial agents. Direct activity on the bacterial cell membrane is the most prevalent mechanism of antimicrobial peptides [52]. Antimicrobial peptides can interact with the bacterial membranes due to their amphipathic nature. Most AMPs have a net positive charge, and therefore, are named cationic antimicrobial peptides. The binding of cationic antimicrobial peptides to the bacterial membranes is stabilized through electrostatic interactions between the cationic parts of AMPs and anionic compounds on bacterial membranes. Consequently, the bacterial membrane integrity is disrupted, causing antimicrobial peptides penetration into the membranes, and in most cases, finally forming the pores [53].

The clustering of anionic lipids to a region of the bacterial membrane would concentrate negative charge in a domain to which cationic peptides would congregate, possibly leading to the formation of a pore. After increasing the concentration of cationic antimicrobial agents on the anionic surface of the membrane, the rest of the membrane will surround the domain of anionic lipids and lead to less membrane stability under line tension. It seems that there are always domains with phase boundary defects in bacterial membranes [54] and those that would form in the presence of lipid clustering AMPs would appear suddenly [55]. Under these conditions, bacteria would not have enough time to repair this rearrangement and would be damaged as a result of the redistribution of membrane's lipids. Consequently, disruption of functional natural domains or decreasing the availability of anionic lipids that may be necessary for the specific protein function in the cytoplasmic membrane would happen [8].

5. Antimicrobial peptides in *Escherichia coli* clinical trials

Naturally occurring cationic antimicrobial peptides (AMPs) and their mimics form a diverse class of antibacterial agents currently validated in preclinical and clinical settings for the treatment of infections caused by antimicrobial-resistant bacteria [56]. Clinical trials have been cautious of toxicity at these doses, as other peptide-based antibiotics (such as colistin) are toxic in high concentrations.

AMPs can be classified into three distinct approaches based on their clinical development: (i) direct antimicrobial effect on the cell membrane, (ii) indirect antimicrobial activity through immune regulation, and (iii) blocking the intracellular functions. Among forty-four peptides that have been undergoing clinical and preclinical trials, 35 target the bacterial cell membrane directly, eight affect the immune system to regulate the response of the body to infection, and three act on intracellular targets. Sixteen of these that show broad-spectrum activity, have been considered for treatment of Gram-negative infections [50].

Until 2020, FDA approves seven AMPs for clinical usages. Vancomycin and dalbavancin (vancomycin derivative) block bacterial wall synthesis, while oritavancin and telavancin (other derivatives) have both membranolytic and cell wall synthesis inhibition actions. Gramicidin D is a linear peptide that forms in the membrane. Daptomycin, colistin (polymyxin E), and cyclic lipopeptide lysis the membrane [7]. Among the AMPs that FDA approves, Buforin II that binds to nucleic acid, Colicin E1 and Bac8c that disrupt the bacterial membrane, specifically are used for *E. coli* treatment [41]. LPS in the outer membrane of Gram-negative bacteria, act as a protective shield and prevent from transporting the large glycopeptide antibiotics, such as vancomycin to intracellular targets. Recent studies have shown that vancomycin, when given together with other AMPs acts against vancomycin-resistant Gram-positive bacteria [57]. Corbett et al. reported SPR741 (polymyxin B derivative) that potentiates the efficacy of conventional antibiotics on Gram-negative bacteria whose spectrum of activity is limited because of bacterial outer membrane permeability obstacles [58]. Studies show the MICs in eight out of 35 antibiotics while combined with SPR741 were reduced 32 to 8000-fold against *E. coli* and *Klebsiella pneumoniae*. Interestingly, based on research *E. coli* becomes susceptible to vancomycin under cold stress. Moreover, the mechanism of vancomycin action to eradicate *E. coli* is similar to the Gram-positive bacteria, which is through inhibition of peptidoglycan biosynthesis [59]. It was also shown that silver ions can increase membrane permeability of Gram-negative bacteria and can potentiate the Gram-positive-specific antibiotic vancomycin against Gram-negative bacteria [60].

The permeability barrier of the outer membrane of Gram-negative bacteria limits the efficacy of vancomycin. So, a synergistic mechanism of action for P-113 derivatives (e.g., Bip-P-113, Dip-P-113, and Nal-P-113) and vancomycin was proposed. Study results showed, however, P-113 derivatives could perturb the outer membrane of Gram-negative bacteria and increase vancomycin entry into the resistant species. In addition, P-113 derivatives bind to the extra hydrophobic motif of lipid A and neutralize LPS protective actions [61].

Employing long-chain amino acid sequences increases the output cost of peptides and thereby the cost of research; hence, synthetic short-chain cationic peptides with potential antimicrobial activity have been attempted [62].

In particular, Indolicidin, a tridecapeptide isolated from the cytoplasmic granules of bovine neutrophils, was reported to exhibit membrane permeabilization effects and antimicrobial activity against Gram-negative and Gram-positive bacteria, fungi, HIV-1 virus, and protozoa [63, 64].

Enteroaggregative *Escherichia coli* (EAEC), an emerging foodborne pathogen, is implicated in endemic and epidemic diarrheal episodes. Multidrug resistance toward the antibiotics of first-line empirical therapy (fluoroquinolones and b-lactams) has been evident globally among the EAEC isolates [65, 66]. Indolicidin as an antimicrobial peptide exhibited a complete elimination of multidrug-resistant EAEC isolates in the time-kill kinetic assay by 2 h pi, while meropenem represented a similar effect after 60 min. These results indicate a unique advantage of AMPs over conventional antibiotics for better treatment of resistant antibacterial species. Studies about the antimicrobial effect of Indolicidin against MDR-EAEC strains in the *G. mellonella* larval model reported that Indolicidin is stable at high temperatures, in the presence of proteinase K and at physiological concentration of cationic salts. In addition, results demonstrated that while Indolicidin could eliminate MDR-EAEC completely, to be safe for commensal gut flora and eukaryotic cells [67].

Peptide 35,409 contains 20 amino acid residues and has been exhibited antibacterial activity against *Escherichia coli* ML35 at 22 ìM minimum inhibitory concentration (MIC). In spite, this peptide did not have cytotoxic activity against human cell lines such as HeLa and HepG2, showing hemolytic effects on human red blood cells at 1.5 μM minimum concentration. According to the low selectivity of peptide 35,409 at the therapeutic index for *E. coli* ML35 (calculated equal 0.045), its therapeutic use is restricted [7]. However, considering the essential need for developing new compounds with activity against microorganisms, 17-residue-long peptide 35,409-1 was obtained from peptide 35,409. This shorter peptide synthesized chemically with less charge but had greater hydrophobicity and amphipathicity properties than the original sequence. Peptide 35,409-1 sequence could inhibit *E. coli* multiresistant isolates and seemed to be highly selective for Gram-negative *E. coli* bacteria because it does not act against Gram-positive bacteria or human red blood cells. Peptide 35,409-1 permeabilizes into the bacterial membrane and leads to *E. coli* cytoplasmatic content leakage [7]. The interactions of AMPs with membranes have been very considered due to serious implications regarding AMPs therapeutic advantages [68, 69]. Five of the seven AMPs that are approved by the FDA are active on the membrane [70], so this mechanism must be surveyed for 35,409-1 profoundly. In comparison to conventional antibiotics, peptide 35,409-1 exhibited a lower potential for inducing resistance significantly. Therefore, it seems that peptide 35,409-1 could be a potential candidate for clinical therapy usages or developing highly selective new AMPs against Gram-negative *E. coli*. The stability in the presence of sera, efficacy against MRD- *E. coli*, and low inducing resistance of peptide 35,409-1 propose its significant clinical advantages for overcoming recent antibacterial *E. coli* resistance [71].

Some substitutes of histidine-rich antimicrobial peptide P-113 were developed recently [72]. Among them, Bip-P-113 showed serum proteolytic stability, enhanced salt resistance, peptide-induced permeabilization, zeta potential measurements, LPS condensed, and *in vitro* and *in vivo* neutralizing activities against LPS [70].

Polymyxin B and its derivatives are able to interact with anionic LPS in the outer membrane (OM) of Gram-negative bacteria. The derivatives of polymyxin B act as "permeabilizers" or "potentiators" and sensitize bacteria to antibiotics, Moreover, reinforce the action of other antibiotics [58]. Studies showed synergistic effects between colistin and bacteriocins that led to inhibit Gram-negative bacteria and reduction of antibiotic toxicity [73]. Ionic silver (Ag+) in silver nitrate salt (AgNO3) was found to increase the permeability of the bacterial outer membrane and sensitize Gram-negative bacteria to vancomycin [60]. Synergistic effects also have been proved between highly membrane-active AMPs and intracellular targeting antibiotics [61].

Stationary phase bacteria are much more resistant than exponentially growing cells to killing by conventional antibiotics, such as ampicillin, tetracycline, ciprofloxacin, and streptomycin [74]. The susceptibility of *E. coli* to human α-defensin 5 (HD5ox) was shown to be lower in the stationary phase compared to mid-log phase cells [75]. The authors suspected a correlation between bacterial susceptibility and altered cellular morphology [39]. Treated β-lactam resistant *E. coli* with ampicillin displayed changes in cell elasticity, membrane permeability, nanoscale morphology, and hydrophilic/hydrophobic interactions. Moreover, different ampicillin-resistant *E. coli* strains exhibited different traits phenotypically [76]. Therefore, exploring the interactions of conjugated molecules with wild-type and ampicillin-resistant bacterial strains is crucial since the cell drug interaction is highly dependent on the type of strains and the drug molecules applied [29].

6. Future perspectives

The development of novel antimicrobial compounds is critical for averting multidrug-resistant bacterial strains. Clinical trials showed that a large number of antimicrobial peptides have clinical potential. While AMPs have antimicrobial activity, in many cases, their clinical use has not been yet fully confirmed. Because of improper trial study design or lack of enough efficacy, many of the AMPs in clinical trials failed to progress to market. Thus, more research into the interaction between antimicrobial peptides and the human host would help to assess the true potential of these compounds.

Indeed, many of the antimicrobial compounds in clinical trials have some sort of chemical modification to improve their drug ability. The sophisticated digital libraries and modeling software would be useful for further optimization of the development of these compounds. In the future, we must try seriously to reduce the resistance to novel antimicrobial compounds. While AMPs have shown a lower tendency for resistance, this is an inevitable phenomenon due to evolutionary consequences. In fact, following the development of diverse antimicrobial agents and their mechanisms of antimicrobial action will impact antimicrobial resistance. In conclusion, it seems that through detailed monitoring and analysis of new antimicrobial drugs, limiting the use of antimicrobials in nonessential cases, and coadministration with antibiotics, the risk of appearance resistant bacterial strains will decrease in the future.

7. Conclusion

Outer membrane targeting is a revolutionary strategy for antibiotic discovery. Gram-negative pathogens will become sensitive to the range of clinically approved Gram-positive active antibiotics when their OM is perturbed. In this way, chemical space compatible with novel antimicrobial peptides would be expanded. However, there are many obstacles before performing this approach in the clinic successfully. The success or failure of this approach depends on the correct selection or development of the outer membrane perturbant and antibiotic adjutant combination. In comparison to monotherapy approaches as other combination therapies, dosage optimizing for adequate overlap in bioavailability approves difficulty and needs more complicated clinical trials [77]. Spontaneous resistance development, horizontally acquired resistance genes, and biofilm formations are all significant barriers to successful antibiotic treatment. The capacity for OM disruption to overcome many of these challenges, uniquely positioning this approach among discovery efforts in the Gram-negative resistance crisis [19].

Author details

Sara Kadkhodaei[1*] and Gelareh Poostizadeh[2]

1 H. pylori Department, CoLife Integrated Medical Laboratories, Tehran, Iran

2 Department of Microbiology, School of Biology, University College of Sciences, Tehran University, Tehran, Iran

*Address all correspondence to: saramicro1998@gmail.com

References

[1] Penesyan A, Gillings M, Paulsen IT. Antibiotic discovery: Combatting bacterial resistance in cells and in biofilm communities. Molecules. 2015;**20**(4):5286-5298

[2] Koo HB, Seo J. Antimicrobial peptides under clinical investigation. Peptide Science. 2019;**111**(5):e24122

[3] Prestinaci F, Pezzotti P, Pantosti A. Antimicrobial resistance: A global multifaceted phenomenon. Pathogens and Global Health. 2015;**109**(7):309-318

[4] Butler MS, Blaskovich MA, Cooper MA. Antibiotics in the clinical pipeline in 2013. The Journal of Antibiotics. 2013;**66**(10):571-591

[5] Silver LL. Challenges of antibacterial discovery. Clinical Microbiology Reviews. 2011;**24**(1):71-109

[6] Chou H-T, Wen H-W, Kuo T-Y, Lin C-C, Chen W-J. Interaction of cationic antimicrobial peptides with phospholipid vesicles and their antibacterial activity. Peptides. 2010;**31**(10):1811-1820

[7] Barreto-Santamaría A, Rivera ZJ, García JE, Curtidor H, Patarroyo ME, Patarroyo MA, et al. Shorter antibacterial peptide having high selectivity for E. Coli membranes and low potential for inducing resistance. Microorganisms. 2020;**8**(6):867

[8] Epand RM, Walker C, Epand RF, Magarvey NA. Molecular mechanisms of membrane targeting antibiotics. Biochimica et Biophysica Acta (BBA)-Biomembranes. 2016;**1858**(5):980-987

[9] Scherer KM, Spille J-H, Sahl H-G, Grein F, Kubitscheck U. The lantibiotic nisin induces lipid II aggregation, causing membrane instability and vesicle budding. Biophysical Journal. 2015;**108**(5):1114-1124

[10] Clifton LA, Skoda MW, Daulton EL, Hughes AV, Le Brun AP, Lakey JH, et al. Asymmetric phospholipid: Lipopolysaccharide bilayers; a Gram-negative bacterial outer membrane mimic. Journal of the Royal Society Interface. 2013;**10**(89):20130810

[11] Malanovic N, Lohner K. Gram-positive bacterial cell envelopes: The impact on the activity of antimicrobial peptides. Biochimica et Biophysica Acta (BBA)-Biomembranes. 2016;**1858**(5):936-946

[12] Fernández L, Hancock RE. Adaptive and mutational resistance: Role of porins and efflux pumps in drug resistance. Clinical Microbiology Reviews. 2012;**25**(4):661-681

[13] Nikaido H. Molecular basis of bacterial outer membrane permeability revisited. Microbiology and Molecular Biology Reviews. 2003;**67**(4):593-656

[14] Vaara M. Antibiotic-supersusceptible mutants of Escherichia coli and Salmonella typhimurium. Antimicrobial Agents and Chemotherapy. 1993;**37**(11):2255-2260

[15] Schneck E, Schubert T, Konovalov OV, Quinn BE, Gutsmann T, Brandenburg K, et al. Quantitative determination of ion distributions in bacterial lipopolysaccharide membranes by grazing-incidence X-ray fluorescence. Proceedings of the National Academy of Sciences. 2010;**107**(20):9147-9151

[16] Neidhardt F. Chemical Composition of Escherichia coli. Vol. 2. Washington, DC: ASM Press; 1996

[17] Mahendran KR, Kreir M, Weingart H, Fertig N, Winterhalter M. Permeation of antibiotics through Escherichia coli OmpF and OmpC porins: Screening for influx on a

single-molecule level. Journal of Biomolecular Screening. 2010; **15**(3):302-307

[18] Delcour AH. Outer membrane permeability and antibiotic resistance. Biochimica et Biophysica Acta (BBA)-Proteins and Proteomics. 2009; **1794**(5):808-816

[19] MacNair CR, Brown ED. Outer membrane disruption overcomes intrinsic, acquired, and spontaneous antibiotic resistance. MBio. 2020; **11**(5):e01615-e01620

[20] Puvača N, de Llanos FR. Antimicrobial resistance in Escherichia coli strains isolated from humans and pet animals. Antibiotics. 2021;**10**(1):69

[21] Reinthaler FF, Galler H, Feierl G, Haas D, Leitner E, Mascher F, et al. Resistance patterns of Escherichia coli isolated from sewage sludge in comparison with those isolated from human patients in 2000 and 2009. Journal of Water and Health. 2013;**11**(1):13-20

[22] Alali WQ, Scott H, Harvey R, Norby B, Lawhorn D, Pillai S. Longitudinal study of antimicrobial resistance among Escherichia coli isolates from integrated multisite cohorts of humans and swine. Applied and Environmental Microbiology. 2008;**74**(12):3672-3681

[23] Boonyasiri A, Tangkoskul T, Seenama C, Saiyarin J, Tiengrim S, Thamlikitkul V. Prevalence of antibiotic resistant bacteria in healthy adults, foods, food animals, and the environment in selected areas in Thailand. Pathogens and Global Health. 2014;**108**(5):235-245

[24] Pormohammad A, Nasiri MJ, Azimi T. Prevalence of antibiotic resistance in Escherichia coli strains simultaneously isolated from humans, animals, food, and the environment: A systematic review and meta-analysis. Infection and Drug Resistance. 2019;**12**:1181

[25] Chui CS, Cowling BJ, Lim WW, Hui CK, Chan EW, Wong IC, et al. Patterns of inpatient antibiotic use among public hospitals in Hong Kong from 2000 to 2015. Drug Safety. 2020 Mar 12:1

[26] Osoro AA, Atitwa EB, Moturi JK. Universal Health Coverage. World Journal of Social Science Research. 2020

[27] Munita JM, Arias CA. Mechanisms of antibiotic resistance. Microbiology Spectrum. 2016;**4**(2):15

[28] Lou H, Chen M, Black SS, Bushell SR, Ceccarelli M, Mach T, et al. Altered antibiotic transport in OmpC mutants isolated from a series of clinical strains of multi-drug resistant E. coli. PLoS One. 2011;**6**(10):e25825

[29] Zamani E, Chatterjee S, Changa T, Immethun C, Sarella A, Saha R, et al. Mechanistic understanding of the interactions of cationic conjugated oligo-and polyelectrolytes with Wild-type and Ampicillin-resistant Escherichia coli. Scientific Reports. 2019;**9**(1):1-12

[30] Moravej H, Moravej Z, Yazdanparast M, Heiat M, Mirhosseini A, Moosazadeh Moghaddam M, et al. Antimicrobial peptides: Features, action, and their resistance mechanisms in bacteria. Microbial Drug Resistance. 2018;**24**(6): 747-767

[31] Chilambi GS, Gao IH, Yoon BK, Park S, Kawakami LM, Ravikumar V, et al. Membrane adaptation limitations in Enterococcus faecalis underlie sensitivity and the inability to develop significant resistance to conjugated oligoelectrolytes. RSC Advances. 2018;**8**(19):10284-10293

[32] Graf M, Mardirossian M, Nguyen F, Seefeldt AC, Guichard G, Scocchi M, et al.

Proline-rich antimicrobial peptides targeting protein synthesis. Natural Product Reports. 2017;**34**(7):702-711

[33] Mattiuzzo M, Bandiera A, Gennaro R, Benincasa M, Pacor S, Antcheva N, et al. Role of the Escherichia coli SbmA in the antimicrobial activity of proline-rich peptides. Molecular Microbiology. 2007;**66**(1):151-163

[34] Krizsan A, Knappe D, Hoffmann R. Influence of the yjiL-mdtM gene cluster on the antibacterial activity of proline-rich antimicrobial peptides overcoming Escherichia coli resistance induced by the missing SbmA transporter system. Antimicrobial Agents and Chemotherapy. 2015;**59**(10):5992-5998

[35] Mardirossian M, Sola R, Beckert B, Collis DW, Di Stasi A, Armas F, et al. Proline-rich peptides with improved antimicrobial activity against E. coli, K. pneumoniae, and A. baumannii. ChemMedChem. 2019;**14**(24):2025

[36] Sochacki KA, Barns KJ, Bucki R, Weisshaar JC. Real-time attack on single Escherichia coli cells by the human antimicrobial peptide LL-37. Proceedings of the National Academy of Sciences. 2011;**108**(16):E77-E81

[37] Brogden KA. Antimicrobial peptides: Pore formers or metabolic inhibitors in bacteria? Nature Reviews Microbiology. 2005;**3**(3):238-250

[38] Choi H, Rangarajan N, Weisshaar JC. Lights, camera, action! Antimicrobial peptide mechanisms imaged in space and time. Trends in Microbiology. 2016;**24**(2):111-122

[39] Agrawal A, Rangarajan N, Weisshaar JC. Resistance of early stationary phase E. coli to membrane permeabilization by the antimicrobial peptide Cecropin A. Biochimica et Biophysica Acta (BBA)-Biomembranes. 2019;**1861**(10):182990

[40] Lei J, Sun L, Huang S, Zhu C, Li P, He J, et al. The antimicrobial peptides and their potential clinical applications. American journal of Translational Research. 2019;**11**(7):3919

[41] Browne K, Chakraborty S, Chen R, Willcox MD, Black DS, Walsh WR, et al. A new era of antibiotics: The clinical potential of antimicrobial peptides. International Journal of Molecular Sciences. 2020;**21**(19):7047

[42] Franzenburg S, Walter J, Künzel S, Wang J, Baines JF, Bosch TC, et al. Distinct antimicrobial peptide expression determines host species-specific bacterial associations. Proceedings of the National Academy of Sciences. 2013;**110**(39):E3730-E37E8

[43] Hancock RE. Cationic peptides: Effectors in innate immunity and novel antimicrobials. The Lancet infectious Diseases. 2001;**1**(3):156-164

[44] Epand RM, D'Souza K, Berno B, Schlame M. Membrane curvature modulation of protein activity determined by NMR. Biochimica et Biophysica Acta (BBA)-Biomembranes. 2015;**1848**(1):220-228

[45] Ratledge C. Microbial lipids. In: Biotechnology Set. 2001. pp. 133-197

[46] Henzler-Wildman KA, Martinez GV, Brown MF, Ramamoorthy A. Perturbation of the hydrophobic core of lipid bilayers by the human antimicrobial peptide LL-37. Biochemistry. 2004;**43**(26):8459-8469

[47] Yeaman MR, Yount NY. Mechanisms of antimicrobial peptide action and resistance. Pharmacological Reviews. 2003;**55**(1):27-55

[48] Jenssen H, Hamill P, Hancock RE. Peptide antimicrobial agents. Clinical Microbiology Reviews. 2006;**19**(3): 491-511

[49] Aoki W, Ueda M. Characterization of antimicrobial peptides toward the

development of novel antibiotics. Pharmaceuticals. 2013;**6**(8):1055-1081

[50] Ebbensgaard A, Mordhorst H, Aarestrup FM, Hansen EB. The role of outer membrane proteins and lipopolysaccharides for the sensitivity of Escherichia coli to antimicrobial peptides. Frontiers in Microbiology. 2018;**9**:2153

[51] Huan Y, Kong Q, Mou H, Yi H. Antimicrobial peptides: Classification, design, application and research progress in multiple fields. Frontiers in Microbiology. 2020;**11**:2559

[52] Epand RM, Epand RF. Bacterial membrane lipids in the action of antimicrobial agents. Journal of Peptide Science. 2011;**17**(5):298-305

[53] Zhang L-j, Gallo RL. Antimicrobial peptides. Current Biology. 2016; **26**(1):R14-RR9

[54] Matsumoto K, Kusaka J, Nishibori A, Hara H. Lipid domains in bacterial membranes. Molecular Microbiology. 2006;**61**(5):1110-1117

[55] Oreopoulos J, Epand RF, Epand RM, Yip CM. Peptide-induced domain formation in supported lipid bilayers: direct evidence by combined atomic force and polarized total internal reflection fluorescence microscopy. Biophysical Journal. 2010;**98**(5):815-823

[56] Findlay B, Zhanel GG, Schweizer F. Cationic amphiphiles, a new generation of antimicrobials inspired by the natural antimicrobial peptide scaffold. Antimicrobial Agents and Chemotherapy. 2010;**54**(10): 4049-4058

[57] Wu EL, Fleming PJ, Yeom MS, Widmalm G, Klauda JB, Fleming KG, et al. E. coli outer membrane and interactions with OmpLA. Biophysical Journal. 2014;**106**(11):2493-2502

[58] Corbett D, Wise A, Langley T, Skinner K, Trimby E, Birchall S, et al. Potentiation of antibiotic activity by a novel cationic peptide: Potency and spectrum of activity of SPR741. Antimicrobial Agents and Chemotherapy. 2017;**61**(8):e00200-e00217

[59] Stokes JM, French S, Ovchinnikova OG, Bouwman C, Whitfield C, Brown ED. Cold stress makes Escherichia coli susceptible to glycopeptide antibiotics by altering outer membrane integrity. Cell Chemical Biology. 2016;**23**(2):267-277

[60] Morones-Ramirez JR, Winkler JA, Spina CS, Collins JJ. Silver enhances antibiotic activity against gram-negative bacteria. Science Translational Medicine. 2013;**5**(190):190ra81

[61] Iwu CD, Korsten L, Okoh AI. The incidence of antibiotic resistance within and beyond the agricultural ecosystem: A concern for public health. Microbiology. 2020;**9**(9):e1035

[62] Anunthawan T, De La Fuente-Núñez C, Hancock RE, Klaynongsruang S. Cationic amphipathic peptides KT2 and RT2 are taken up into bacterial cells and kill planktonic and biofilm bacteria. Biochimica et Biophysica Acta (BBA)-Biomembranes. 2015;**1848**(6):1352-1358

[63] Falla TJ, Karunaratne DN, Hancock RE. Mode of action of the antimicrobial peptide indolicidin. Journal of Biological Chemistry. 1996;**271**(32):19298-19303

[64] Rokitskaya TI, Kolodkin NI, Kotova EA, Antonenko YN. Indolicidin action on membrane permeability: Carrier mechanism versus pore formation. Biochimica et Biophysica Acta (BBA)-Biomembranes. 2011;**1808**(1):91-97

[65] Lima AA, Medeiros PH, Havt A. Enteroaggregative Escherichia coli

subclinical and clinical infections. Current Opinion in Infectious Diseases. 2018;**31**(5):433-439

[66] Guiral E, Gonçalves Quiles M, Muñoz L, Moreno-Morales J, Alejo-Cancho I, Salvador P, et al. Emergence of resistance to quinolones and β-lactam antibiotics in enteroaggregative and enterotoxigenic Escherichia coli causing traveler's diarrhea. Antimicrobial Agents and Chemotherapy. 2019;**63**(2): e01745-e01718

[67] Vergis J, Malik SS, Pathak R, Kumar M, Ramanjaneya S, Kurkure NV, et al. Antimicrobial efficacy of indolicidin against multi-drug resistant enteroaggregative Escherichia coli in a Galleria mellonella model. Frontiers in Microbiology. 2019;**10**:2723

[68] Epand RF, Maloy L, Ramamoorthy A, Epand RM. Amphipathic helical cationic antimicrobial peptides promote rapid formation of crystalline states in the presence of phosphatidylglycerol: Lipid clustering in anionic membranes. Biophysical Journal. 2010;**98**(11): 2564-2573

[69] Epand RF, Maloy WL, Ramamoorthy A, Epand RM. Probing the "charge cluster mechanism" in amphipathic helical cationic antimicrobial peptides. Biochemistry. 2010;**49**(19):4076-4084

[70] Chen CH, Lu TK. Development and challenges of antimicrobial peptides for therapeutic applications. Antibiotics. 2020;**9**(1):24

[71] Barreto-Santamaría A, Patarroyo ME, Curtidor H. Designing and optimizing new antimicrobial peptides: All targets are not the same. Critical Reviews in Clinical Laboratory Sciences. 2019;**56**(6):351-373

[72] Yu H-Y, Chen Y-A, Yip B-S, Wang S-Y, Wei H-J, Chih Y-H, et al. Role of β-naphthylalanine end-tags in the enhancement of antiendotoxin activities: Solution structure of the antimicrobial peptide S1-Nal-Nal in complex with lipopolysaccharide. Biochimica et Biophysica Acta (BBA)-Biomembranes. 2017;**1859**(6):1114-1123

[73] Naghmouchi K, Baah J, Hober D, Jouy E, Rubrecht C, Sané F, et al. Synergistic effect between colistin and bacteriocins in controlling Gram-negative pathogens and their potential to reduce antibiotic toxicity in mammalian epithelial cells. Antimicrobial Agents and Chemotherapy. 2013;**57**(6):2719-2725

[74] Levin BR, Rozen DE. Non-inherited antibiotic resistance. Nature Reviews Microbiology. 2006;**4**(7):556-562

[75] Chileveru HR, Lim SA, Chairatana P, Wommack AJ, Chiang I-L, Nolan EM. Visualizing attack of Escherichia coli by the antimicrobial peptide human defensin 5. Biochemistry. 2015;**54**(9):1767-1777

[76] Uzoechi SC, Abu-Lail NI. The effects of β-Lactam antibiotics on surface modifications of multidrug-resistant Escherichia coli: A multiscale approach. Microscopy and Microanalysis. 2019;**25**(1):135-150

[77] Foucquier J, Guedj M. Analysis of drug combinations: Current methodological landscape. Pharmacology Research & Perspectives. 2015;**3**(3):e00149

Chapter 12

Escherichia coli (*E. coli*) Resistance against Last Resort Antibiotics and Novel Approaches to Combat Antibiotic Resistance

Rana Elshimy

Abstract

An important feature complicating the treatment of infections caused by *E. coli* is the increase in resistance to different antibiotics, even to last resort antibiotics. When resistant bacteria spread to the community, resistance creates comprehensive infection control issues, increasing morbidity for non-hospitalized patients of all ages and sexes. New resistance mechanisms are constantly being described, and new genes and vectors of transmission are identified on a regular basis. This chapter reviews different mechanisms of *E. coli* resistance against different classes of last resort antibiotics such as fosfomycin, nitrofurantoin, and polymixins. In addition, *E. coli* vaccines, epidemiology, and novel approaches to combat antibiotic resistance will be discussed throughout the chapter. In the age of antibiotic resistance and precise microbial genome engineering, many new strategies are now being used to combat multidrug-resistant bacteria, hoping to be our end game weapon. These strategies include CRISPR-Cas antimicrobials, nanobiotics, phage therapy, and probiotics, which promise to have a substantial impact on the way we treat diseases in the future, as we will discuss in the chapter.

Keywords: multidrug-resistant *E. coli*, colistin resistance, *mcr-1*, hospitals, fosfomycin, plasmid, ARGs, probiotics, CRISPR-Cas

1. Introduction

Escherichia coli is the best-known member of the normal microbiota of the human intestine and a versatile gastrointestinal pathogen. *E. coli* infection is a major global problem in the clinical and community setting. The prevalence of *E. coli* among clinical specimens varies from country to country and even among two different institutions in the same country and continuously changes over time [1–4]. More and more people die each year from hospital infections caused by multidrug-resistant *E. coli* [5].

According to the WHO, *E. coli* is considered a global critical pathogen that possesses the highest priority for research, discovery, and development of new antibiotics [6]. When antibiotics are consumed during bacterial infection treatment, the resistance of the commensal *E. coli* is developed after exposure [7]. Undeniably, commensal *E. coli* is one of the main reservoirs for antibiotic resistance transmission to other pathogenic bacteria through plasmid exchange, for example (**Figure 1**) [8–10].

Via contact with livestock or a contaminated natural environment, humans can be exposed to viable commensal antibiotic-resistant *E. coli* [11].

Indiscriminate use of antimicrobials and antibiotic overuse has led to the treacherous resistance rates in recent years, creating a very complicated therapeutic challenge that threatens to return clinicians and patients to a "pre-antibiotic era". Furthermore, mobile genetic elements (plasmids, bacteriophages) carrying antibiotic resistance genes (ARGs) play a major role in transferring resistance to both human and nonhuman, contribute to the spread of antimicrobial-resistant organisms, and increases the risk factor of infections and diseases in both animals and humans [12].

This chapter will discuss various mechanisms, epidemiology, vaccines, and novel approaches to combat *E. coli* antibiotic resistance.

2. Antimicrobial resistance against last resort antibiotics in *E. coli*

Generally, there are five main mechanisms by which gram-negative organisms develop resistance: First, bacteria can carry genes coding for enzymes, such as beta-lactamases, hydrolyzing, and inactivating beta-lactam antibiotics. Second, mutations can occur in the genes for binding sites for antibiotics changing the specific target or its function. Third, alterations of the membrane porins result in reduced permeability. Fourth, bacteria can express efflux pumps to actively transport antibiotics out of the cell, and finally, fifth, alternate metabolic pathways can bypass paths inhibited by antibiotics [13, 14].

Resistance in gram-negative bacteria can be intrinsic, arise, or be acquired and is often composed of a combination of resistance mechanisms like beta-lactamases, porin deletions, and efflux pumps [15].

Acquired bacterial resistance may be due to mutations in chromosomal genes and by horizontal gene transfer. The intrinsic resistance appears due to inherent structural or functional characteristics (**Figure 2**).

The intrinsic resistance of some gram-negative bacteria to many compounds is due to an inability of these agents to cross the outer membrane: For example, the glycopeptide antibiotic vancomycin inhibits peptidoglycan cross-linking by binding

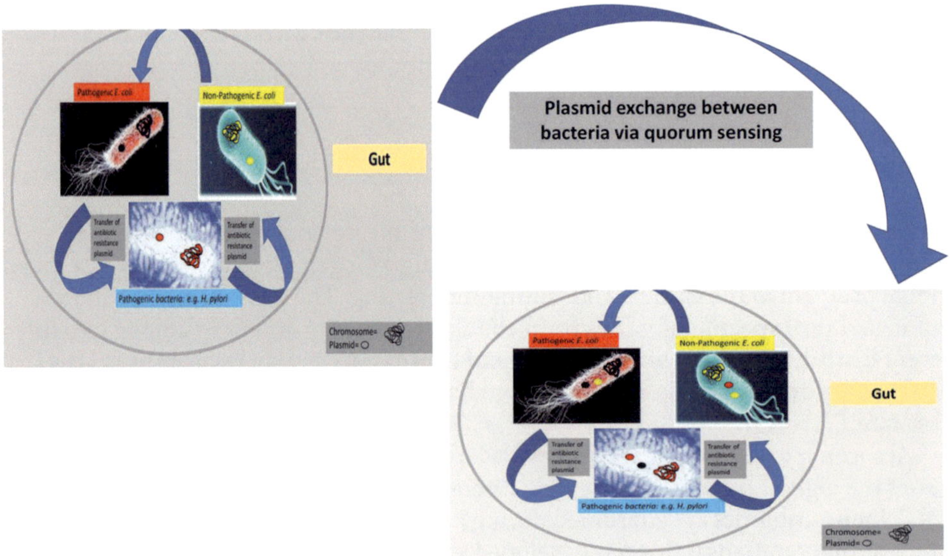

Figure 1.
Transfer of resistance between bacteria through plasmid exchange.

to target d-Ala-d-Ala peptides but is only normally effective in gram-positive bacteria as, in gram-negative organisms, it cannot cross the outer membrane and access these peptides in the periplasm [17].

2.1 Resistance to fosfomycin

Fosfomycin is receiving renewed worldwide attention as one of the most active agents for sparing carbapenems in extended-spectrum β-lactamase (ESBL)–producing isolates and for treatment of carbapeneme-resistant *Enterobacteriaceae* (CRE) in combination with colistin [18].

The mechanism of *E. coli* resistance to fosfomycin is through the production of *fos*A, a glutathione S-transferase that inactivates fosfomycin by the addition of a glutathione residue [19]. The mechanism of action of Fosfomycin is inhibition of the initial step in peptidoglycan synthesis by blocking *Mur*A irreversibly in both gram-positive and -negative bacteria. It is imported through the inner membrane through the glycerol-3-phosphate (G3P) transporter GlpT and the glucose-6-phosphate (G6P) transporter UhpT. Reduced expression or mutations in *glpT* or *uhpT* genes are the most common causes leading to lowered susceptibility [20]. Another mechanism is the production of *fos*A, a glutathione *S*-transferase that inactivates fosfomycin by addition of a glutathione residue. This mechanism is particularly relevant because it is disseminative and frequently associated with ESBL-producing *Escherichia coli*. Plasmid-mediated *fosA3* and, less frequently, *fosA5* (formerly *fosKp96*), are mostly associated with *CTX*-M and co-harbored on a conjugative plasmid. The possible dissemination of this gene is worrisome because *fos*A3 is generally surrounded by the IS26 insertion sequence on a composite transposon borne by the IncFII conjugative plasmid, which is known to be a dissemination vector of resistance genes worldwide [21].

2.2 Resistance to nitrofurantoin

Nitrofurans are a group of compounds characterized by the presence of one or more nitro-groups on a nitroaromatic or nitroheterocyclic backbone. Examples of compounds belonging to this group include furazolidone, nitrofurazone, and

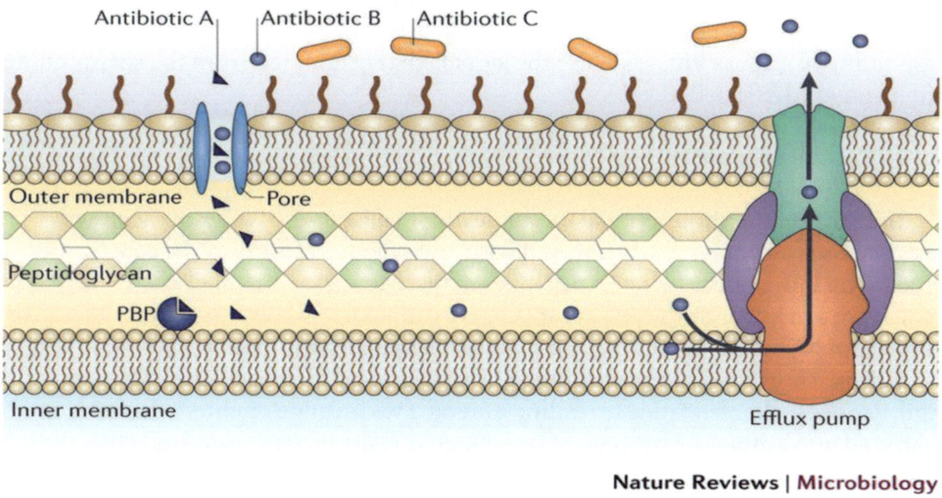

Figure 2.
Proportion of anionic phospholipids in the cytoplasmic membrane is lower in gram-negative bacteria than gram-positive bacteria so the efficiency of the Ca^{2+}-mediated insertion of daptomycin into the cytoplasmic membrane that is required for its antibacterial activity is reduced [16].

nitrofurantoin: drugs that all display antimicrobial activity and are used clinically to treat different types of infections [22].

Nitrofurans need to be activated by *E. coli* nitroreductase reducing activity to show their antibiotic effect. *E. coli* nitroreductase activities may be insensitive to oxygen (type I) or inhibited by oxygen (type II). In type 1, reduction occurs via a sequence of toxic intermediates, including a nitroso and hydroxylamine state, to a biologically inactive end product where one of the intermediates is thought to be responsible for toxicity as it binds and disrupts bacterial DNA and protein. Increasing resistance is accompanied by a decrease in the activity of their reductive capacity [23].

Sequential increase in resistance was genetically shown to result from sequential inactivation of the diverse nitro-reducing activities present in *E. coli*. The mutations were genetically mapped and named *nfs*A and *nfs*B. The direct link between these genes, and the sequential loss of nitro-reducing activity, was established by mutant isolation and sequencing of *nsf*A and *nsf*B. Nitrofuran resistance has been mapped only to type I nitroreductase genes [23, 24].

2.3 Resistance to polymyxin (last defense line)

Colistin and the other polymyxins are cationic antimicrobial peptides. These agents interfere with the negatively charged outer membrane of gram-negative bacteria. When polymyxins bind to the outer bacterial membrane it will disrupt the membrane. Thereby promotes the killing of the bacteria [25].

The clinicians have reconsidered the value of colistin due to the rising number of hospital outbreaks with carbapenem-resistant gram-negative bacteria along with the deficiency of the development of new antimicrobial agents directed toward such MDR strains. Upon this, colistin systemic administration has been reintroduced as a final treatment option. In the light of this the WHO reclassified colistin in 2012 as a critically essential antibacterial agent for mankind's remedy [26].

Worldwide, the increased use of colistin led to the appearance of colistin resistance. Colistin resistance rates have been noticed to increase more often [27].

2.3.1 Intrinsic resistance

Colistin resistance occurs normally via alterations n the two-component regulatory system phoPQ-PmrAB, both contain a sensor kinase (PhoQ and PmrB, correspondingly), these kinases sense the signals that originated from the surrounding environment [28].

2.3.2 Acquired resistance

Similarly, to the chromosomal mechanisms of colistin resistance, the acquired resistance to colistin is mainly involved with lowering the affinity of the colistin to bind to the LPS by decreasing its negative charge. MCR gens (mobile colistin resistance genes) are a member of the phosphor-ethanol-amino transferase enzyme family, with expression of this gene resulting in ethanolamine moiety addition to the lipid A [29].

This plasmid-mediated mechanism of resistance is of special due to the possibility of colistin resistance spreading among a wide range of enteric bacteria in mankind and animals. This type of resistance is resistance is associated with the low level of MIC (4–8 mg/L) [30].

2.4 Plasmid-mediated colistin resistance

On the 18th of November 2015, **Liu *et al.*** reported the first description of plasmid-mediated colistin resistance (*mcr*-1 gene) among samples from food-producing animals, food, and humans in China, the detection of the mcr-1 gene the horizontally transferred plasmid-mediated colistin resistance gene altered the previous idea about the colistin resistance in gram-negative bacteria which stats that Enterobacteriaceae only develop colistin resistance through chromosomal mutations or other adaptive mechanisms. *In vitro* studies on *mcr-1* gene showed self-transfer of the gene from conjugative plasmids [31]. After the first detection of the *mcr*-1 gene in *Enterobacteriaceae* in China, within the 6 months after its first detection, the plasmids which carry the *mcr*-1 gene were found in isolates from animals, food, the environment, and humans worldwide [32]. On 3 March 2016, a literature review published in Euro surveillance showed that during 3 months of its discovery the *mcr*-1 gene had been spread to many parts of the world and found in isolates from different sources of food and environment and also from infected patients as well as asymptomatic human carriers [33]. It is worthy to mention that, the *mcr*-1 gene detected in a human was in the U.S on the 26th of May 2016 in an *E. coli* isolate [32].

Although the first detection of *mcr* gene was in 2015 it is believed that this gene was existed among Enterobacterial isolates for many years before, but it wasn't identified and it was transmitted silently for years [29].

Surprisingly, a retrospective study by [34], on isolates from Chicken origin indicated that the emergence of *mcr*-1 gene among enterobacterial isolates was when the colistin was first used in food-producing animals in 1980, but it did not appear again in the isolates from the next 20 years. However, the *mcr*-1 gene was noticed again in random isolates belonged to the period from 2004 to 2006. In isolates from 2009, the outbreak of the presence of *mcr*-1 gene among isolates recovered from the chicken was noted [35].

Until now, there are 8 variants of the *mcr* gene (*mcr*-1, *mcr*-2, *mcr*-3, *mcr*-4, *mcr*-5, *mcr*-6, *mcr*-7 and *mcr*-8). *mcr*-2 gene is a variant of *mcr* gene which share about 76.7% nucleotide (81% amino acid) with *mcr*-1, this gen was first detected in Belgium [25]. The *mcr*-3 gene has nucleotide sequence 45.0 and 47.0% identity to *mcr*-1 and *mcr*-2 respectively [29].

3. Classification of *E. coli* according to their antimicrobial resistance pattern

The continuous emergence of resistance to antimicrobial agents among the prevalent pathogens is the most dangerous obstacle facing the treatment of infectious diseases. Many different definitions for multidrug-resistant (MDR), extensively drug-resistant (XDR), and pandrug-resistant (PDR) bacteria are being used in the medical literature to characterize the different patterns of resistance found in healthcare-associated, antimicrobial-resistant bacteria [36].

Lists of antimicrobial categories proposed for antimicrobial susceptibility testing were created using documents and breakpoints from the Clinical Laboratory Standards Institute (CLSI), the European Committee on Antimicrobial Susceptibility Testing (EUCAST) and the United States Food and Drug Administration (FDA) [37].

I. MDR E. coli:

MDR was defined as acquired non-susceptibility to at least one agent in three or more antimicrobial categories.

II. XDR E. coli:

XDR was defined as non-susceptibility to at least one agent in all but two or fewer antimicrobial categories (i.e., bacterial isolates remain susceptible to only one or two categories).

III. PDR E. coli:

PDR was defined as non-susceptibility to all agents in all antimicrobial categories. To ensure the correct application of these definitions, bacterial isolates should be tested against nearly all of the antimicrobial agents within the antimicrobial categories and selective reporting and suppression of results should be avoided [37].

4. Epidemiology of resistance in *E. coli*

The World Health Organization (WHO), through its Global Antimicrobial Surveillance System (GLASS) report, bared that there are treacherous levels of antibiotic resistance in both low- and high-income countries (LMIC) [38].

The European Centre for Disease Prevention and Control (ECDC) conveyed that 25,000 people died due to antibiotic-resistant bacterial infections in 2007, which is over half the number caused by road traffic accidents in the same countries [39]. In 2015, this number increased to about 33,000 deaths resulting from an estimated 671,689 infections of selected antibiotic-resistant bacteria leading to 874,541 total disability-adjusted life-years (DALYs) [40]. This indicates that the burden on the European Union and European Economic Area is on the rise. By 2050, the World Health Organization (WHO) predicted that, death because of antibiotic resistance would upsurge from 700,000 to 10 million per year globally [39]. As a result of antibiotic resistance, more than 2.8 million people are infected, and more than 35,000 die each year in the USA [41].

The estimated number of cases of uncomplicated cystitis per year, caused by *E. coli* alone, is 130–175 million globally and 2–300.000 in Denmark alone [42]. Consequently, infections caused by *E. coli*, susceptible and resistant, collectively result in considerable morbidity as well as direct and indirect financial costs seen as increased healthcare expenses, antibiotic treatment, and loss of productivity [43].

Furthermore, UTI patients experience morbidity and impaired quality of life with an estimated 20–40% of women having at least one UTI during their lifetime [43].

It is difficult to determine the precise incidence of UTI, but by using self-reported medical history the annual incidence in the USA was 13% among women and 3% among men [44]. Resistance in *E. coli*, besides β-lactam resistance, includes sulphonamides, trimethoprim, and ciprofloxacin [45].

In 2008, UPEC isolates from five countries, were commonly resistant to ampicillin 28%), sulfonamides (25%), trimethoprim (17%), and nalidixic acid (10%), with a significant increase in resistance to nalidixic acid and trimethoprim from 2000 to 2009. A total of 60%, only, of the UPEC isolates, were found to be fully susceptible [42].

Antibiotic resistance continued to increase throughout Europe, with 41% being fully susceptible in 2012, only. Especially, the current increase in resistance to extended-spectrum cephalosporins mean = 12% and aminoglycosides (mean = 10%) in combination with increased resistance to at least three antibiotic classes, is worrisome. The increased resistance is likewise worrying in Denmark. In 2012, the resistance in *E. coli* isolated from urine (primary health care) was 40% for ampicillin with 33% for sulphonamide and 10% were resistant to ciprofloxacin and 6% to mecillinam [42, 46, 47].

The continual increase in resistant *E. coli* has added to the enormous economic and human costs of infections with 400.000 infections caused by MDR bacteria in Europe in 2007 [46]. The economic costs associated with these infections, counted as extra hospital costs and productivity losses exceeds €1.5 billion in Europe and $20 billion per year in the United States [48, 49].

5. Novel technique for detection of antibiotic resistance

For appropriate treatment of antibiotic-resistant *E. coli* infected patients, it is crucial to recognize the pathogen species and drug-resistant gene accurately in a timely manner [50]. Traditionally, the conventional culture-based plating assay was commonly used for antibiotic-resistant bacteria diagnosis. However, this method is very time-consuming as it takes several days to confirm the growth of the targeted bacterial colony [51]. On the other hand, a molecular characterization via polymerase chain reaction (PCR) requires relatively less time than the culture-based plating assay, but still cannot fully avoid separation and bacterial pre-enrichment [52]. Therefore, using a novel rapid and accurate technique to detect resistance was an urgent goal. Matrix-assisted laser desorption ionization time-of-flight spectrometry has captured the attention for the rapid identification of resistant pathogens by profiling bacterial proteins from the whole cells [53]. Moreover, endogenous H2S evolution was recently developed for drug-resistant bacteria via in situ hybridization [54].

Furthermore, fluorescence in situ hybridization (FISH) is a technique for the identification and analysis of diverse organisms such as bacteria and animal cells, based on the hybridization of a fluorescently labeled oligonucleotide probe to complementary target sequences from organisms using epifluorescence or confocal laser scanning microscopy [55]. Unfortunately, weak and unstable fluorescent signals due to quenching caused by natural and artificial light remain the limitation for the detection of a single microbe using fluorescence microscopy.

In 2020, Lee et al., could develop a novel fluorescent nanoparticle-based probe (nanoprobe) for FISH technique and successfully applied the nanoprobe for the detection of antibiotic-resistant bacteria [56]. The stable nanoprobe was prepared by the modified sol–gel chemistry and consisted of fluorescent dye-loaded poly (d,l-lactide-co-glycolide) (PLGA) and silica nanoparticles (NPs) [57, 58]. For the identification of ampicillin-resistant *E. coli*, the nanoprobe was functionalized with two kinds of biotinylated single-stranded DNAs (ssDNAs) which can conjugate to *E. coli*-specific gene and ampicillin-resistance *bla* gene that encodes beta-lactamase conferring beta-lactams (e.g., ampicillin) degrading enzyme, respectively. Finally, ampicillin-resistant *E. coli* was successfully detected using a nanoprobe-ssDNA.

6. Development of MDR *E. coli* vaccines

Since 1969, many strategies were applied to develop an effective vaccine against *E. coli* infections but they all have failed [59, 60]. In the 1990s, traditional vaccine strategies were based on single-purified virulence factors like Hemolysin [61] or on the O-specific polysaccharide (OPS) chain of the lipopolysaccharide (named O-antigen), conjugated to r Pseudomonas aeruginosa endotoxin A (TA) or cholera toxin (CT) as carrier proteins [62].

Although the prevalence of K-antigen and O-antigen is different among the different pathotypes, there is an association between K (K1, K5, 30, and 92) and O (O1, 2, 4, 6, 7, 8, 16, 16/72, 18, 25, 50, and 75) antigenic groups and uropathogenic strains [62].

However, because of the high antigenic heterogeneity of the surface polysaccharides, the design of a polysaccharide vaccine able to prevent ExPEC infections has been extremely difficult [62]. An O18-polysaccharide conjugated to either cholera toxin or to P. aeruginosa exoprotein A (EPA) was shown to be safe and able to induce antibodies with opsonophagocytic killing activity (OPK) in human volunteers. IgG purified from immunized individuals was protective in mice in an *E. coli* 018 challenge sepsis model [2].

Vaccines based on whole or lysed fractions of inactivated *E. coli* were evaluated in human clinical trials and were so far the most effective in inducing a high degree of protection in subjects suffering from recurrent urinary tract infections [62].

Extraintestinal *E. coli* vaccines are either in the preclinical or clinical stage as follows:

New antigens in preclinical studies

1. Antigens involved in iron acquisition: FyuA, IutA, ChuA, Iha, IreA, Hma, IroN.

2. Highly conserved antigens: SsIE(YghJ) and FdeC (EaeH.

3. Fimbrial-based antigens: MrpH-FimH.

Vaccines in clinical studies

1. Uromune: Inactivated *E. coli*, *Klebsiella pneumonae*, *Proteus vulgaris, and Enterococcus faecalis.* Uromune showed clinical benefit in reduction of UTI recurrence in females [11].

2. Solco-Urovac: Inactivated six *E. coli* serotype, *Proteus mirabilis*, *Morganella morganii*, *Klebsiella pneumonae* and *E. faecalis*. Solco-Urovac showed minimal efficacy in Phase 1 and two Phase 2 trials in recurrent UTIs in females [62, 63].

3. OM89/Uro-vaxom: Lyophilized lysate of 18 *E. coli* strains and ExPEC-4 V: 4-valent O antigens conjugated to exotoxin A from *P. aeruginosa*.

7. Synthetic microbiomes and engineered vaccine probiotics

The World Health Organization (WHO) has defined probiotics as "live microorganisms that, when administered in adequate amounts, confer a health benefit on the host". Dysbiosis of healthy gut microbiota plays a critical role in the dysregulation of microbial ecology that favors colonization of pathogenic bacterial strains and diseases [64, 65].

Interestingly, from healthy specific-pathogen-free chickens, competitive exclusion (CE) products could be designed and administered by crop gavage as feed supplements for broiler chickens [66, 67]. This type of bacteriotherapy effectively protects chickens from bacterial infections, notably salmonellosis [67]. Freeze-dried CE preparations are manufactured and commercialized in many countries [68].

Extensive basic and clinical studies with lactic acid bacteria strains such as *Bifidobacteria* spp., *Bacteroides,* and *Akkermansia* spp. have provided strong evidence of their health benefits. This is achieved through multiple mechanisms and effector molecules [69, 70]. In-feed supplementation with *Bacillus*, *Bifidobacterium,*

Enterococcus, Lactobacillus, Streptococcus, Lactococcus spp. And Yeast Saccharomyces have been given as probiotics to inhibit infection of enteric pathogens such as *E. coli* and mitigate antibiotic-associated diarrhea [71, 72].

8. Novel engineered probiotics/postbiotics/synbiotics

Novel engineered probiotics based on yeast strains, mainly saccharomyces boulardii, have been constructed to secrete multispecific and single-domain antibodies directly targeting bacterial virulence factors, in particular, enterotoxins [73, 74].

In the last century, the uncontrolled use of antimicrobials has led to a massive increase in *E. coli* resistance representing a threat to public health [75] so the use of an alternative including probiotics and acidifiers was a must [76]. One of the most common substitutes is probiotics *Lactobacillus* spp. and *Bifidobacterium* spp. which are commonly used to combat *E. coli* infections like gastroenteritis [77], antibiotic-associated diarrhea [77], necrotizing enterocolitis [78], inflammatory bowel diseases, [79]allergic disorders and others [80].

Furthermore, bioactive molecules secreted by probiotics can effectively down-regulate virulence gene expression in enterohemorrhagic *E. coli* O157: H7 [81]. They can also reduce.

E. coli O157: H7 and *E. coli* O127: H6 adhesion to epithelial cells monolayers [82, 83]. Unfortunately, probiotics need to be further studied to evaluate their efficacy as anti-biofilm against pathogenic *E. coli*. One of the most common examples of probiotics is Protexin which is a commercially available multistrain potential probiotic [84]. The Protexin contained bacteria that showed antimicrobial activity against *Salmonella Typhimurium* LT2, *E. coli* NCFB 1989, *Staphylococcus aureus* NCTC 8532, *E. faecalis* NCTC 775, and *Clostridium difficile* ATCC 43,594 [85].

Probiotics are capable of controlling MDR bacterial agents by enhancing immune response and competitive exclusion [86]. They augment the activities of macrophages and natural killer cells, modulate cytokine and immunoglobulin secretion, promote intestinal epithelial barrier integrity [87–90] and activate B lymphocytes [89, 91].

9. Novel approaches to control *E. coli* infections

Several innovative approaches have been developed to combat antibiotic resistance in MDR *E. coli*, including the use of peptide nucleic acid (PNA) as an ultra-narrow-spectrum antibiotic, phage therapy, zinc finger nucleases (ZFNs) [92], and clustered regularly interspaced short palindromic repeat – CRISPR-associated (CRISPR-Cas) systems, [93, 94] which are genomic engineering tools for gene knock-out and knock-in of sequence-specific DNA antibiotic targets. Fast development CRISPR-based synthetic biology may transfigure the way we treat disease in the upcoming years.

9.1 The clustered regularly interspaced short palindromic repeats – CRISPR-associated (CRISPR-Cas) system

The clustered regularly interspaced short palindromic repeats – CRISPR-associated (CRISPR-Cas) system is a bacterial adaptive immune system, which is used for controlling antibiotic-resistant strains. Moreover, the programmable Cas nuclease of this system can totally diminish or reduce the resistance of bacteria to antibiotics [93].

The Cas (CRISPR-associated) nucleases identify a specific sequence of DNA by establishing a complex with a CRISPR-RNA (crRNA) that has sequence homology to the target4,5. The crRNA-Cas complex binds to the target and leads to DNA damage [95].

Interestingly, the CRISPR-Cas system is precise and easily programmable, so CRISPR-based tools for genome editing are magnificently applied nowadays in eukaryotes and prokaryotes [96].

Eukaryotic cells can repair DNA breaks using the error-prone non-homologous end joining (NHEJ) mechanisms but most prokaryotes lack NHEJ mechanisms, wherefore continuous DNA damage leads to cell death if not repaired through homologous recombination. This phenomenon has been exploited for the development of CRISPR-Cas based antimicrobials [95]. The most important advantage for CRISPR-Cas antimicrobials is the discrimination and elimination of specific bacteria at the strain level such as *E. coli* [95, 96].

CRISPR-Cas provides acquired immunity against viruses and plasmids. In the treatment of *E. coli*, CRISPR-encoded immunity is provided by transcription of the repeat-spacer array, followed by transcript processing into small crRNAs (CRISPR RNAs), which are then used in combination with Cas proteins as guides to interfere with invasive DNA or RNA. In *E. coli*, few model systems have been established to study of CRISPR/Cas functionality [97].

The CRISPR2 and CRISPR4 systems present in the *S. thermophilus* DGCC7710 genome belong to the TypeIII (Mtube) and Type I (*E. coli*), respectively. Differences between types can be observed in terms of repeat, spacer, and Cas gene content and sequence. The multiplicity of CRISPR/Cas systems in *S. thermophilus* is explained by their susceptibility to horizontal gene transfer, and phage selective pressure [98].

9.2 Phage therapy

Bacteriophages are bacteria-specific viruses, which can specifically infect and lyse bacteria. Phage therapy has been used to treat MDR *E. coli* that are resistant to last resort antibiotics. It is considered one of the most effective weapons for combating MDR *E. coli* [99, 100].

An example of phages used to treat *E. coli* is VB_EcoS-Golestan which is a virulent phage that belongs to *Kagunavirus* genus of *Guernseyvirinae* subfamily, *Siphoviridae* family. VB_EcoS-Golestan has many advantages in the treatment of UPEC specifically such as broad host range specificity, a rapid adsorption time, large burst size, and high stability at a wide range of pH and temperatures, which makes it a promising agent against *E. coli* infections [101].

Since phage therapy is still an under-study therapeutic approach, further development of this method requires biological characterization of bacteriophages such as their host specificity, genome diversity, and adaption to their bacterial hosts.

9.3 Nanoparticles

Nowadays, nanoparticles are one of the safest, cost-effective, and most effective bactericidal materials, which can be efficiently used as carriers of therapeutic agents [102].

Unfortunately, one of the major obstacles facing us when using silver nanoparticles is their high toxicity toward mammalian cells but to a lesser extent than pathogenic bacterial cells [103]. On the other hand, silver is less dangerous to mammalian cells than other metals [104]. Silver nanoparticles (Ag NPs or nanosilver), a kind of nanosized silver particle, are widely used NPs and show strong broad-spectrum biocidal effects on pathogenic bacteria, including MDR *E. coli* [104, 105].

Furthermore, Gold NPs may become useful in the development of antibacterial strategies due to their polyvalent effects, versatility in surface modification, and nontoxicity [106, 107].

Cui et al. could develop a strategy to fight against MDR bacteria via presenting inactive small organic molecules, such as 4, 6-diamino-2-pyrimidinethiol on gold NPs (Au_DAPT NPs), which act on *E. coli* such as disorganizing cell membranes, binding to nucleic acids, and inhibiting protein synthesis [108].

The Gold NPs antibacterial mechanism of action is to change membrane potential and inhibit ATP synthase activities to decrease the ATP level, indicating a general decline in metabolism; and inhibition of the ribosome subunit for tRNA binding, resulting in a collapse of biological process [108, 109].

10. Conclusions

The treacherous *E. coli* resistance rates in recent years created a very complicated therapeutic challenge that threatens to return clinicians and patients to a "pre-antibiotic era". Clinicians must be alert to the possibility of nitrofurantoin, fosfomycin, and colistin resistance among MDR and XDR bacteria. Finally, vaccines and probiotics, CRISPR-Cas systems, and phage therapy may be the means to combat AMR of *E. coli*.

Author details

Rana Elshimy[1,2]

1 Department of Microbiology and Immunology, Egyptian Drug Authority (EDA), Giza, Egypt

2 Faculty of Pharmacy, Department of Microbiology and Immunology, Ahram Canadian University, Egypt

*Address all correspondence to: rana.elshimy@acu.edu.eg

References

[1] Osman KM, Kappell AD, Elhofy F, Orabi A, Mubarak AS, Dawoud TM, et al. Urinary tract infection attributed to *Escherichia coli* isolated from participants attending an unorganized gathering. Future Microbiology. 2018;**13**(7):757-769

[2] Poolman JT, Wacker M. Extraintestinal pathogenic *Escherichia coli*, a common human pathogen: Challenges for vaccine development and progress in the field. Journal of Infectious Diseases. 2016;**213**(1):6-13

[3] Vally H, Hall G, Dyda A, Raupach J, Knope K, Combs B, et al. Epidemiology of Shiga toxin producing *Escherichia coli* in Australia, 2000-2010. BMC Public Health. 2012;12(1), 63.

[4] Ahmed MYA. Characterization and molecular epidemiology of Extended-Spectrum-ß-Lactamase-Producing *Escherichia coli* derived from University Hospitals of Egypt and Germany [Doctoral dissertation]. Vol. 2013. Gießen: Justus-Liebig-Universität; 2013

[5] Andersen BM. Prevention and Control of Infections in Hospitals: Practice and Theory. Cham: Springer; 2019

[6] Tacconelli E, Carrara E, Savoldi A, Harbarth S, Mendelson M, Monnet DL, et al. Discovery, research, and development of new antibiotics: the WHO priority list of antibiotic-resistant bacteria and tuberculosis. The Lancet Infectious Diseases. 2018;**18**(3):318-327

[7] Nji E, Kazibwe J, Hambridge T, Joko CA, Larbi AA, Damptey LAO, et al. High prevalence of antibiotic resistance in commensal *Escherichia coli* from healthy human sources in community settings. Scientific Reports. 2021;**11**(1):1-11

[8] Li B, Zhao ZC, Wang MH, Huang XH, Pan YH, Cao YP. Antimicrobial resistance and integrons of commensal *Escherichia coli* strains from healthy humans in China. Journal of Chemotherapy. 2014;**26**:190-192

[9] Lamikanra A, Ako-Nai AK, Ogunniyi DA. Transferable antibiotic resistance in *Escherichia coli* isolated from healthy Nigerian school children. International Journal of Antimicrobial Agents. 1996;**7**(1):59-64

[10] Li B, Qiu Y, Song Y, Lin H, Yin H. Dissecting horizontal and vertical gene transfer of antibiotic resistance plasmid in bacterial community using microfluidics. Environment International. 2019;**131**:105007

[11] Gomes TAT, Elias WP, Scaletsky ICA, Guth BEC, Rodrigues JF, Piazza RMF, et al. Diarrheagenic *Escherichia coli*. Brazilian Journal of Microbiology. Elsevier Editora Ltda. 2016;**47**:3-30

[12] Dafale NA, Srivastava S, Purohit HJ. Zoonosis: An emerging link to antibiotic resistance under "one health approach.". Indian Journal of Microbiology. 2020;**60**:139-152

[13] Breijyeh Z, Jubeh B, Karaman R. Resistance of gram-negative bacteria to current antibacterial agents and approaches to resolve it. Molecules. 2020;**25**:1340

[14] Marchaim D, Gottesman T, Schwartz O, Korem M, Maor Y, Rahav G, et al. National multicenter study of predictors and outcomes of bacteremia upon hospital admission caused by Enterobacteriaceae producing extended-spectrum β-lactamases. Antimicrobial Agents and Chemotherapy. 2010;**54**(12):5099-5104

[15] Zango UU, Ibrahim M, Abdurrahman S, Shawai A, Shamsuddin IM. A review on β -lactam

antibiotic drug resistance. MOJ Drug Design Development & Therapy. 2019;**3**(2):52-58

[16] Blair JMA, Webber MA, Baylay AJ, Ogbolu DO, Piddock LJV. Molecular mechanisms of antibiotic resistance. Nature Reviews Microbiology. 2015;**13**(1):42-51

[17] Schaenzer AJ, Wright GD. Antibiotic resistance by enzymatic modification of antibiotic targets. Trends in Molecular Medicine. 2020;**26**:768-782

[18] Elshimy R, Zedan H, Elmorsy TH, Khattab RA. A study on multidrug-resistant *Escherichia coli* clinical isolates from different hospitals in greater Cairo. Microbial Drug Resistance. 2021;**27**(10):1420-1432

[19] Benzerara Y, Gallah S, Hommeril B, Genel N, Decré D, Rottman M, et al. Emergence of plasmid-mediated fosfomycin-resistance genes among *Escherichia coli* isolates, France. Emerging Infectious Diseases. 2017;**23**(9):1564

[20] Garallah ET, Al-Jubori SS. Surveillance of murA and the plasmid-mediated fosfomycin resistance fosA gene in uropathogenic *E. coli* isolates from UTI patients. Gene Reports. 2020;**21**:100872

[21] Zurfluh K, Treier A, Schmitt K, Stephan R. Mobile fosfomycin resistance genes in Enterobacteriaceae—An increasing threat. MicrobiologyOpen. 2020;**9**:e1135

[22] Issakhanian L, Behzadi P. Antimicrobial agents and urinary tract infections. Current Pharmaceutical Design. 2019;**25**(12):1409-1423

[23] van Hung LV, Davies IG, Moon CD, Wheeler D, Biggs PJ, Rakonjac J. Novel 5-nitrofuran-activating reductase in *Escherichia coli*. Antimicrobial Agents and Chemotherapy. 2019;**63**(11): e00869-19

[24] Sandegren L, Lindqvist A, Kahlmeter G, Andersson DI. Nitrofurantoin resistance mechanism and fitness cost in *Escherichia coli*. Journal of Antimicrobial Chemotherapy. 2008;**62**(3):495-503

[25] Moubareck CA. Polymyxins and bacterial membranes: A review of antibacterial activity and mechanisms of resistance. Membranes. 2020;**10**:181

[26] Andrade FF, Silva D, Rodrigues A, Pina-Vaz C. Colistin update on its mechanism of action and resistance, present and future challenges. Microorganisms. 2020;**8**:1716

[27] Wang Y, Xu C, Zhang R, Chen Y, Shen Y, Hu F, et al. Changes in colistin resistance and mcr-1 abundance in *Escherichia coli* of animal and human origins following the ban of colistin-positive additives in China: An epidemiological comparative study. The Lancet Infectious Diseases. 2020;**20**(10):1161-1171

[28] Hamel M, Rolain JM, Baron SA. The history of colistin resistance mechanisms in bacteria: Progress and challenges. Microorganisms. 2021;**9**(2):442

[29] Dalmolin TV, Martins AF, Zavascki AP, de Lima-Morales D, Barth AL. Acquisition of the mcr-1 gene by a high-risk clone of KPC-2-producing Klebsiella pneumoniae ST437/CC258, Brazil. Diagnostic Microbiology and Infectious Disease. 2018;**90**(2):132-133

[30] Gharaibeh MH, Shatnawi SQ. An overview of colistin resistance, mobilized colistin resistance genes dissemination, global responses, and the alternatives to colistin: A review. Veterinary World. 2019;**12**:1735-1746

[31] Wang R, van Dorp L, Shaw LP, Bradley P, Wang Q, Wang X, et al. The global distribution and spread of the

mobilized colistin resistance gene mcr-1. Nature Communications. 2018;**9**(1):1-9

[32] Doumith M, Godbole G, Ashton P, Larkin L, Dallman T, Day M, et al. Detection of the plasmid-mediated mcr-1 gene conferring colistin resistance in human and food isolates of *Salmonella enterica* and *Escherichia coli* in England and Wales. Journal of Antimicrobial Chemotherapy. 2016;**71**(8):2300-2305

[33] Skov RL, Monnet DL. Plasmid-mediated colistin resistance (mcr-1 gene): Three months later, the story unfolds. Eurosurveillance. 2016;**21**:30155

[34] Shen J, Rump L, Ju W, Shao J, Zhao S, Brown E, et al. Virulence characterization of non-O157 Shiga toxin-producing *Escherichia coli* isolates from food, humans and animals. Food Microbiology. 2015;**50**:20-27

[35] Liu YY, Wang Y, Walsh TR, Yi LX, Zhang R, Spencer J, et al. Emergence of plasmid-mediated colistin resistance mechanism MCR-1 in animals and human beings in China: A microbiological and molecular biological study. The Lancet Infectious Diseases. 2016;**16**(2):161-168

[36] de Oliveira DMP, Forde BM, Kidd TJ, Harris PNA, Schembri MA, Beatson SA, et al. Antimicrobial resistance in ESKAPE pathogens. Clinical Microbiology Reviews. 2020;**33**(3):e00181-19

[37] Magiorakos AP, Srinivasan A, Carey RB, Carmeli Y, Falagas ME, Giske CG, et al. Multidrug-resistant, extensively drug-resistant and pandrug-resistant bacteria: An international expert proposal for interim standard definitions for acquired resistance. Clinical Microbiology and Infection. 2012;**18**(3):268-281

[38] Hernando-Amado S, Coque TM, Baquero F, Martínez JL. Defining and combating antibiotic resistance from One Health and Global Health perspectives. Nature Microbiology. 2019;**4**:1432-1442

[39] de Kraker MEA, Stewardson AJ, Harbarth S. Will 10 million people die a year due to antimicrobial resistance by 2050? PLoS Medicine. 2016;**13**(11): e1002184

[40] Cassini A, Högberg LD, Plachouras D, Quattrocchi A, Hoxha A, Simonsen GS, et al. Attributable deaths and disability-adjusted life-years caused by infections with antibiotic-resistant bacteria in the EU and the European Economic Area in 2015: A population-level modelling analysis. The Lancet Infectious Diseases. 2019;**19**(1):56-66

[41] O'Neill J. The Review on Antimicrobial Resistance Tackling Drug-Resistant Infections Globally: Final Report and Recommendations. London, United Kingdom: Review on Antimicrobial Resistance; 2014

[42] Zhi S, Stothard P, Banting G, Scott C, Huntley K, Ryu K, et al. Characterization of water treatment-resistant and multidrug-resistant urinary pathogenic *Escherichia coli* in treated wastewater. Water Research. 2020;**182**:115827

[43] Zhen X, Lundborg CS, Sun X, Hu X, Dong H. Economic burden of antibiotic resistance in ESKAPE organisms: A systematic review. Antimicrobial Resistance and Infection Control. 2019;**8**:1-23

[44] Smith DRM, Pouwels KB, Hopkins S, Naylor NR, Smieszek T, Robotham JV. Epidemiology and health-economic burden of urinary-catheter-associated infection in English NHS hospitals: A probabilistic modelling study. Journal of Hospital Infection. 2019;**103**(1):44-54

[45] Wang M, Wang W, Niu Y, Liu T, Li L, Zhang M, et al. A clinical extensively-drug resistant (XDR) *Escherichia coli* and role of its β-lactamase genes. Frontiers in Microbiology. 2020;**11**:3178

[46] Hertz FB, Nielsen JB, Schønning K, Littauer P, Knudsen JD, Løbner-Olesen A, et al. Erratum to: Population structure of drug-susceptible,—Resistant and ESBL-producing *Escherichia coli* from community-acquired urinary tract infections. BMC Microbiology. 2016;**16**(1):1-6

[47] Hertz FB, Nielsen JB, Schønning K, Littauer P, Knudsen JD, Løbner-Olesen A, et al. Erratum: Population structure of drug-susceptible, – resistant and ESBL-producing *Escherichia coli* from community-acquired urinary tract infections. BMC Microbiology. 2016;**16**(63):1-6. DOI: 10.1186/s12866-016-0681-z

[48] Bush K, Jacoby GA. Updated functional classification of β-lactamases. Antimicrobial Agents and Chemotherapy. 2010;**54**(3):969-976

[49] Bush K, Bradford PA. Epidemiology of β-lactamase-producing pathogens. Clinical Microbiology Reviews. 2020;33(2): e00047-19.

[50] Morris S, Cerceo E. Trends, epidemiology, and management of multi-drug resistant gram-negative bacterial infections in the hospitalized setting. Antibiotics. 2020;**9**(4):196

[51] Rajapaksha P, Elbourne A, Gangadoo S, Brown R, Cozzolino D, Chapman J. A review of methods for the detection of pathogenic micro-organisms. The Analyst. 2019; **144**(2):396-411

[52] Lantz PG, Al-Soud WA, Knutsson R, Hahn-Hägerdal B, Rådström P. Biotechnical use of polymerase chain reaction for microbiological analysis of biological samples. Biotechnology Annual Review. 2000;**5**:87-130

[53] Stöckel S, Kirchhoff J, Neugebauer U, Rösch P, Popp J. The application of Raman spectroscopy for the detection and identification of microorganisms. Journal of Raman Spectroscopy 2016;**47**(1):89-109

[54] Han Q, Lau JW, Do TC, Zhang Z, Xing B. Near-infrared light brightens bacterial disinfection: Recent progress and perspectives. ACS Applied Bio Materials. 2021;**4**(5):3937-3961

[55] Frickmann H, Zautner AE, Moter A, Kikhney J, Hagen RM, Stender H, et al. Fluorescence *in situ* hybridization (FISH) in the microbiological diagnostic routine laboratory: A review. Critical Reviews in Microbiology. 2017;**43**(3): 263-293

[56] Lee WS, Lee S, Kang T, Ryu CM, Jeong J. Detection of ampicillin-resistant *E. coli* using novel nanoprobe-combined fluorescence in situ hybridization. Nanomaterials. 2019;**9**(5):750

[57] Ha SW, Camalier CE, Beck GR, Lee JK. New method to prepare very stable and biocompatible fluorescent silica nanoparticles. Chemical Communications. 2009;**20**:2881-2883

[58] Ow H, Larson DR, Srivastava M, Baird BA, Webb WW, Wiesnert U. Bright and stable core-shell fluorescent silica nanoparticles. Nano Letters. 2005;**5**(1):113-117

[59] Uehling DT, Wolf L. Enhancement of the bladder defense mechanism by immunization. Investigative Urology. 1969;**6**(5):520-526

[60] Kaijser B, Larsson P, Olling S, Schneerson R. Protection against acute, ascending pyelonephritis caused by

Escherichia coli in rats, using isolated capsular antigen conjugated to bovine serum albumin. Infection and Immunity. 1983;**39**(1):142-146

[61] O'Hanley P, Lalonde G, Ji G. Alpha-hemolysin contributes to the pathogenicity of piliated digalactoside-binding Escherichis coli in the kidney: Efficacy of an alpha-hemolysin vaccine in preventing renal injury in the BALB/c mouse model of pyelonephritis. Infection and Immunity. 1991;**59**(3):1153-1161

[62] Frankel G, Ron EZ, editors. *Escherichia Coli*, a Versatile Pathogen. Vol. 416. Springer; 2018

[63] Majhool AB, Zenad KH. Immunological study for the changes that associated with infection of mice with Shiga toxin producing *Escherichia coli* (STEC) O157: H7. Annals of the Romanian Society for Cell Biology. 2021;**25**(7):914-921

[64] Gallego CG, Salminen S. Novel probiotics and prebiotics: How can they help in human gut microbiota dysbiosis? Applied Food Biotechnology. 2016;**3**(2):72-81

[65] Hashemi A, Villa CR, Comelli EM. Probiotics in early life: A preventative and treatment approach. Food and Function. 2016;**7**(4): 1752-1768

[66] Kerr AK, Farrar AM, Waddell LA, Wilkins W, Wilhelm BJ, Bucher O, et al. A systematic review-meta-analysis and meta-regression on the effect of selected competitive exclusion products on *Salmonella* spp. prevalence and concentration in broiler chickens. Preventive Veterinary Medicine. 2013;**111**(1-2):112-125

[67] Revolledo L, Ferreira CSA, Ferreira AJP. Prevention of Salmonella Typhimurium colonization and organ invasion by combination treatment in broiler chicks. Poultry Science. 2009;**88**(4):734-743

[68] Nakamura A, Ota Y, Mizukami A, Ito T, Ngwai YB, Adachi Y. Evaluation of aviguard, a commercial competitive exclusion product for efficacy and after-effect on the antibody response of chicks to Salmonella. Poultry Science. 2002;**81**(11):1653-1660

[69] Lebeer S, Bron PA, Marco ML, van Pijkeren JP, O'Connell Motherway M, Hill C, et al. Identification of probiotic effector molecules: Present state and future perspectives. Current Opinion in Biotechnology. 2018;**49**:217-223

[70] de Vos WM, de Vos EAJ. Role of the intestinal microbiome in health and disease: From correlation to causation. Nutrition Reviews. 2012;**70**(SUPPL. 1), S45-S56.

[71] Bron PA, Kleerebezem M. Lactic acid bacteria for delivery of endogenous or engineered therapeutic molecules. Frontiers in Microbiology. 2018;**9**:1821

[72] Zhang ZF, Kim IH. Effects of multistrain probiotics on growth performance, apparent ileal nutrient digestibility, blood characteristics, cecal microbial shedding, and excreta odor contents in broilers. Poultry Science. 2014;**93**(2):364-370

[73] Roto SM, Rubinelli PM, Ricke SC. An introduction to the avian gut microbiota and the effects of yeast-based prebiotic-type compounds as potential feed additives. Frontiers in Veterinary Science. 2015;**2**:28

[74] Belizário JE, Sircili MP. Novel biotechnological approaches for monitoring and immunization against resistant to antibiotics *Escherichia coli* and other pathogenic bacteria. BMC Veterinary Research. 2020;**16**(1):1-10

[75] Serwecińska L. Antimicrobials and antibiotic-resistant bacteria: A risk to the environment and to public health. Water (Switzerland). 2020;**12**(12): 3313

[76] Fijan S, Šulc D, Steyer A. Study of the in vitro antagonistic activity of various single-strain and multi-strain probiotics against *Escherichia coli*. International Journal of Environmental Research and Public Health. 2018;**15**(7):1539

[77] Kaur J, Singh BP, Chaudhary V, Elshaghabee FMF, Singh J, Singh A, et al. Probiotics as live bio-therapeutics: Prospects and perspectives. In: Advances in Probiotics for Sustainable Food and Medicine. Singapore: Springer; 2021

[78] Schultz M. Clinical use of *E. coli* Nissle 1917 in inflammatory bowel disease. Inflammatory Bowel Diseases. 2008;**14**(7):1012-1018

[79] Balain M, Oddie S, Banait N, Embleton N, Sesham R, Clarke P. PC.99 PINC UK (Probiotics in Neonatal Collaboration in UK). Archives of Disease in Childhood - Fetal and Neonatal Edition. 2014;**99**(Suppl 1):A70

[80] Minocha A. Probiotics for preventive health. Nutrition in Clinical Practice. 2009;**24**(2):227-241

[81] Wang G, Tang H, Zhang Y, Xiao X, Xia Y, Ai L. The intervention effects of Lactobacillus casei LC2W on *Escherichia coli* O157:H7 -induced mouse colitis. Food Science and Human Wellness. 2020;**9**(3):289-294

[82] Jariwala R, Mandal H, Bagchi T. Indigenous lactobacilli strains of food and human sources reverse enteropathogenic *E. Coli* O26:H11-induced damage in intestinal epithelial cell lines: Effect on redistribution of tight junction proteins. Microbiology (United Kingdom). 2017;**163**(9):1263-1272

[83] Mirabdollah Elahi SS, Mirnejad R, Kazempoor R, Sotoodehnejadnematalahi F. Study of the histopathologic effects of probiotic Lactobacillus acidophilus in exposure to *E. coli* O157: H7 in zebrafish intestine. Iranian Red Crescent Medical Journal. 2020;**22**(4):6

[84] Zare R, Abedian Kenari A, Yazdani Sadati M. Influence of dietary acetic acid, protexin (probiotic), and their combination on growth performance, intestinal microbiota, digestive enzymes, immunological parameters, and fatty acids composition in Siberian sturgeon (Acipenser baerii, Brandt, 1869). Aquaculture International. 2021;**29**(3):891-910

[85] Elsayed MSAE, Shehata AA, Ammar AM, Allam TS, Ali AS, Ahmed RH, et al. The beneficial effects of a multistrain potential probiotic, formic, and lactic acids with different vaccination regimens on broiler chickens challenged with multidrug-resistant *Escherichia coli* and Salmonella. Saudi Journal of Biological Sciences. 2021;**28**(5):2850-2857

[86] Mathipa MG, Thantsha MS. Probiotic engineering: Towards development of robust probiotic strains with enhanced functional properties and for targeted control of enteric pathogens. Gut Pathogens. 2017;**9**:1-17

[87] Bogere P, Choi YJ, Heo J. Probiotics as alternatives to antibiotics in treating post-weaning diarrhoea in pigs: Review paper. South African Journal of Animal Sciences. 2019;49(3): 403-416.

[88] Yousefi B, Eslami M, Ghasemian A, Kokhaei P, Salek Farrokhi A, Darabi N. Probiotics importance and their immunomodulatory properties. Journal of Cellular Physiology. 2019;**234**:8008-8018

[89] Diaz-Garrido N, Fábrega MJ, Vera R, Giménez R, Badia J, Baldomà L. Membrane vesicles from the probiotic Nissle 1917 and gut resident *Escherichia coli* strains distinctly modulate human dendritic cells and subsequent T cell

responses. Journal of Functional Foods. 2019;**61**:103495

[90] Dubreuil JD. Enterotoxigenic *Escherichia coli* and probiotics in swine: What the bleep do we know? Bioscience of Microbiota. Food and Health; 2017

[91] Cañas MA, Fábrega MJ, Giménez R, Badia J, Baldomà L. Outer membrane vesicles from probiotic and commensal *Escherichia coli* activate NOD1-mediated immune responses in intestinal epithelial cells. Frontiers in Microbiology. 2018;**9**(MAR):498

[92] Yang B, Fang D, Lv Q, Wang Z, Liu Y. Targeted therapeutic strategies in the battle against pathogenic bacteria. Frontiers in Pharmacology. 2021;**12**:1090

[93] Shukla A, Jani N, Polra M, Kamath A, Patel D. CRISPR: The multidrug resistance endgame?. Molecular Biotechnology 2021;63(8): 676-685.

[94] Gholizadeh P, Köse Ş, Dao S, Ganbarov K, Tanomand A, Dal T, et al. How CRISPR-Cas system could be used to combat antimicrobial resistance. Infection and Drug Resistance. 2020;**13**:1111

[95] Parsons C, Brown P, Kathariou S. Use of bacteriophage amended with CRISPR-Cas systems to combat antimicrobial resistance in the bacterial foodborne pathogen listeria monocytogenes. Antibiotics. 2021;**10**:308

[96] Ricaño-Rodríguez J, Suárez-Medellin J, Cocoletzi Vásquez E, Ramos-Prado JM, Hipólito-Romero E. Understanding CRISPR/Cas9: A magnificent tool for plant genome editing. In: Transgenic Crops - Emerging Trends and Future Perspectives. London: IntechOpen; 2019

[97] Samai P, Pyenson N, Jiang W, Goldberg GW, Hatoum-Aslan A, Marraffini LA. Co-transcriptional DNA and RNA cleavage during type III CRISPR-Cas immunity. Cell. 2015;**161**(5):1164-1174

[98] Agarwal N, Gupta R. History, evolution and classification of CRISPR-Cas associated systems. In: Progress in Molecular Biology and Translational Science. 2021

[99] Mousavi SM, Babakhani S, Moradi L, Karami S, Shahbandeh M, Mirshekar M, et al. Bacteriophage as a novel therapeutic weapon for killing colistin-resistant multi-drug-resistant and extensively drug-resistant gram-negative bacteria. Current Microbiology. 2021;**78**:4023-4036

[100] Dufour N, Clermont O, la Combe B, Messika J, Dion S, Khanna V, et al. Bacteriophage LM33_P1, a fast-acting weapon against the pandemic ST131-O25b: H4 *Escherichia coli* clonal complex. Journal of Antimicrobial Chemotherapy. 2016;**71**(11):3072-3080

[101] Yazdi M, Bouzari M, Ghaemi EA, Shahin K. Isolation, characterization and genomic analysis of a novel bacteriophage VB_EcoS-Golestan infecting multidrug-resistant *Escherichia coli* isolated from urinary tract infection. Scientific Reports. 2020;**10**(1):1-13

[102] Huh AJ, Kwon YJ. "Nanoantibiotics": A new paradigm for treating infectious diseases using nanomaterials in the antibiotics resistant era. Journal of Controlled Release. 2011;**156**:128-145

[103] Gupta A, Mumtaz S, Li CH, Hussain I, Rotello VM. Combatting antibiotic-resistant bacteria using nanomaterials. Chemical Society Reviews. 2019;**48**:415-427

[104] Du J, Tang J, Xu S, Ge J, Dong Y, Li H, et al. A review on silver nanoparticles-induced ecotoxicity and the underlying toxicity mechanisms.

Regulatory Toxicology and Pharmacology. 2018;**98**:231-239

[105] Salem SS, El-Belely EF, Niedbała G, Alnoman MM, Hassan SED, Eid AM, et al. Bactericidal and in-vitro cytotoxic efficacy of silver nanoparticles (Ag-NPs) fabricated by endophytic actinomycetes and their use as coating for the textile fabrics. Nanomaterials. 2020;**10**(10):2082

[106] Yang P, Pageni P, Rahman MA, Bam M, Zhu T, Chen YP, et al. Gold nanoparticles with antibiotic-metallopolymers toward broad-spectrum antibacterial effects. Advanced Healthcare Materials. 2019;**8**(6):1800854

[107] Zhao X, Jia Y, Li J, Dong R, Zhang J, Ma C, et al. Indole derivative-capped gold nanoparticles as an effective bactericide in vivo. ACS Applied Materials and Interfaces. 2018;**10**(35):29398-29406

[108] Cui Y, Zhao Y, Tian Y, Zhang W, Lü X, Jiang X. The molecular mechanism of action of bactericidal gold nanoparticles on *Escherichia coli*. Biomaterials. 2012;**33**(7):2327-2333

[109] Nisar P, Ali N, Rahman L, Ali M, Shinwari ZK. Antimicrobial activities of biologically synthesized metal nanoparticles: An insight into the mechanism of action. Journal of Biological Inorganic Chemistry. 2019;**24**:924-941

Chapter 13

Understanding of Cultivability of *Escherichia coli* in Aquatic Microcosm in the Presence of Some Plant Extracts for Possible Treatment of Bacterio-Contaminated Water

Antoine Tamsa Arfao and Moïse Nola

Abstract

Escherichia coli (*E. coli*) is a bacterial indicator of sanitary and hygienic importance widely used in the evaluation of the quality of drinking water, mainly because they are easy to detect and enumerate in water. Its presence in water reflects a deterioration in water quality. The *E. coli* species is heterogenous in its biotypes, serotypes and lysotypes as in its ecology and its association with pathologies. Studies have reported several cases of infections, sometimes even fatal, caused by contact or consumption of water contaminated by pathogenic strains of *E. coli*. The detection of *E. coli* in surface water was shown in field studies to have significant information about the microbial quality of water contaminated with enteric pathogens. Studies using the properties of plant extracts for the inhibition of this bacterium have been widely carried out. Some studies show the potential exploitation of *Artemisia annua*, *Eucalyptus microcorys* and *Moringa Oleifera* extracts treatment of bacterio-contaminated water. The effect of some aqueous extracts on planktonic cells of *E. coli* in the planktonic and adhered state are summarized in this chapter.

Keywords: *Escherichia coli*, bacterial indicator, water quality, extracts plant, treatment of bacterio-contaminated water

1. Introduction

Water has a vital role in the world. In general, humans use it for their daily needs, for economic activities and recreation [1]. However, the different uses of water can become sources of pollution. The alteration of the physico-chemical and microbiological quality of water is sometimes the result of poor management by humans of waste and wastewater of domestic, agricultural and industrial origin [2]. The use of polluted water exposes populations to health risks. The medium- and long-term risks are linked to the poor chemical quality of the water, while the short-term risks are biological in origin. The poor biological quality of water is partly due to the presence of protozoa, viruses or bacteria [3]. Several microbial communities live in aquatic and soil environments, with physiologies that are sometimes similar in both types of biotope despite a significant difference in nutrient sources [4].

Bacteria are generally the most abundant microorganisms in nature and their unwanted presence in an environment can represent a health risk of varying degrees for humans. Indeed, the pathogenicity of normally harmless bacteria can occur as a result of the immunosuppression of a host. Bacteria are known to be responsible for water contamination in a community generally belong to the genera *Salmonella*, *Shigella*, *Escherichia*, *Yersinia*, *Vibrio* and *Campylobacter* [5, 6]. These bacteria can cause diarrhea, gastroenteritis and genitourinary infections in humans [7]. Their morphology and physiology in an aquatic environment vary with the general environmental conditions. The *E. coli* species belongs to the group of fecal coliforms or thermotolerant coliforms. The presence of these coliforms in water is generally a sign of the deterioration of its bacteriological quality, due to its contamination by other microorganisms that are strict pathogens or opportunistic pathogens [8–10]. For nearly a decade, numerous outbreaks attributed to pathogenic *E. coli* strains have been regularly reported worldwide [11–14]. In Ngoïla in the eastern region of Cameroon, from December 1997 to April 1998, 298 people came into contact with an epidemic of gastroenteritis caused by *E. coli* O157:H7 [15]. Contamination was by the oral route through ingestion of contaminated water or food.

In recent years, water disinfection methods using plant extracts have been proposed as a new alternative for household water treatment [16–18]. The use of plants for therapeutic purposes has been common practice for thousands of years [19]. However, little is known about the sensitivity of bacteria to these water extracts in the aquatic environment. There is still little information on the synergistic effect of the aqueous extract of *Eucalyptus microcorys* (*E. microcorys*) and light on planktonic cells of *E. coli* in the aquatic environment. Little information about the plant extracts of *Eucalyptus microcorys*, *Artemesia annua* and *Moringa oleifera* on *E. coli* bacterial cells is available. The present chapter summarizes the known effects of aqueous extract of the medicinal plants *Eucalyptus microcorys*, *Artemesia annua* and *Moringa oleifera* on the cultivability of *E. coli* in aquatic microcosm.

2. *Eucalyptus microcorys* extracts

2.1 Bioactive *Eucalyptus* compounds

Phytochemical screening showed that alkaloids, anthraquinones, flavonoïds and saponins are the major components of the aqueous extract of *E. microcorys*. Whereas anthocyanines, gallic tannins, polyphenols and triterpenes are found in negligible quantity [18].

2.2 Effect of *Eucalyptus microcorys* leaves extracts on planktonic *Escherichia coli*

A variation in the abundances of planktonic cells of *E. coli* was generally observed in the presence of the extract of *Eucalyptus microcorys* [18]. This variation is dependent not only on the concentration of the plant extract, but also on associated factors such as incubation temperature and lighting conditions. Thus, the incubation temperature affects the cultivability of *E. coli* cells with inhibition percentages varying from 3 to 100% for enteropathogenic *E. coli*, from 5 to 100% for commensal *E. coli*. *Eucalyptus microcorys* has a bactericidal property whose scope varies relatively according to the type of cell and the environmental conditions. Cultivable cells of *E. coli* happen to be relatively less abundant at temperatures 23°C and 37°C than 7°C, when grown in presence of *Eucalyptus microcorys* extract.

Planktonic cells of enteropathogenic *E. coli* have proven to be more resistant to bactericide properties of the *Eucalyptus microcorys* extract in psychrophilic conditions.

A gradual decrease in the abundance of cultivable enteropathogenic *E. coli* cells were observed during the period of exposure to light in the presence of the extract of *Eucalyptus microcorys* [20]. There is a progressive increase in the rate of cell inhibition in the cells tested in the presence of extract after exposure to light. Under dark conditions, the percentage of metabolically non-culturable enteropathogenic *E. coli* cells ranged from 17 to 99%. These inhibition rates increase under light conditions after each incubation period. Under an intensity of 1000 lx, the inhibition percentages fluctuated from 16 to 100% when considering all concentrations of *Eucalyptus microcorys* extract.

At 2000 lx, these inhibition rates fluctuated between 38 and 100%. At 3000 lx, peak inhibition rates of 100% were obtained after 12 hours of incubation at an extract concentration of 0.05%. Overall, the 3000 lx light intensity appears to result in the maximum inhibition of enteropathogenic *E. coli*.

The hourly inhibition rate of *E. coli* is very low in dark conditions and increases with increasing light intensity. In dark, cell inhibition rates are generally between 0.102 h^{-1} and 0.146 h^{-1}. These inhibition rates increase considerably in the presence of light, with values sometimes reaching 0.662 h^{-1} at 3000 lx. The more the light intensity increases, the more the percentage of cell inhibition increases, whatever the concentration of the plant extract. For a light intensity of 3000 lx, the percentages of inhibition of *E. coli* cells are greater than 80%. The combined effect of light and herbal extract *E. microcorys* influence considerably the evolution of the percentage of cellular inhibition about each concentration of extract and each light condition.

The observation of commensal *E. coli* and enteropathogenic *E. coli* abundances is different when considering all the extract concentrations and incubation temperatures. Studies showed that the number of the colony forming units (CFU) of each of the cell strains in the presence of *Eucalyptus* extract, decreased in most cases for the increase of the concentration of aqueous extract and the incubation temperature. The cell inhibition percentage varied from one strain to another and with respect to the extract concentration and temperature incubation. The enteropathogenic *E. coli* strains seem to resist the effects of the extract concentrations 1% and 1.5% at 7°C, 23°C and 37°C, in contrary to the commensal strains with which the relatively higher percentages were observed in the same conditions.

Natural or acquired resistance to antibiotics would explain the observed resistance of enteropathogenic *E. coli* strains. Indeed, it is well known that bacteria can develop protective mechanisms such as changes in cell wall permeability and structure, production of inhibitory enzymes and alteration of antibacterial molecules [21]. This could explain the difference observed between CAIRs (Cell Apparent Inhibition Rates) of both bacteria strains. Indeed, whether we consider the enteropathogenic *E. coli* strain or the commensal strain, the rate of cell inhibition per hour for each incubation temperature increases as the concentration of the extract of *Eucalyptus microcorys* increases (**Figure 1**). Extracts of crushed and dried leaves of *Eucalyptus cloeziana*, *Eucalyptus microcorys*, *Eucalyptus saligna* and *Eucalyptus grandis* exhibit inhibitory activity against *E. coli* cells. Molecules present in *Eucalyptus* leaves that provide disinfectant properties are the monoterpenes, such as 1,8-cineole, alpha and beta-terpinene, 4-terpineol and tannins. 1,8-cineole has germicidal potential against *E. coli* cells.

2.3 Effect of *Eucalyptus microcorys* leaves extracts on adhered *Escherichia coli*

The different percentages of adhered and detached enteropathogenic *E. coli* cells after contact with the *Eucalyptus microcorys* extract solution at the different

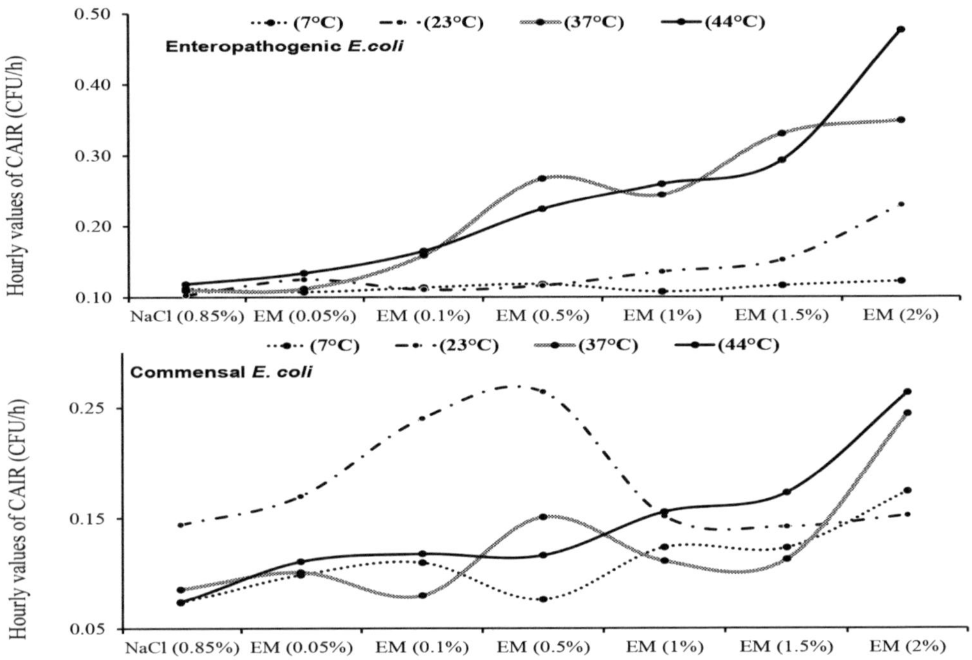

Figure 1.
Hourly inhibitory rate of Enteropathogenic and commensal E. coli *cells (determined as hourly values of cell apparent inhibition rates, CAIR) for each concentration of the* Eucalyptus microcorys *leaves extract (EM) at different incubation temperatures.*

concentrations chosen, were evaluated for the cells from each growth phase. They are presented in **Table 1**.

When the cells were from the lag phase, the percentages of cells remaining adhered after a stay in the extract solution fluctuated between 1.2 and 15%, 0.3 and 12.2%, and 0.2 and 6.4% after 1, 2 and 3 hours respectively when the concentration of the *Eucalyptus microcorys* extract was 1%. These percentages varied between 0.5 and 5.4%, between 0.3% and 3.1%, and between 0.2% and 4.7% after 1 h, 2 h and 3 h respectively in the 1.5% extract solution. In the 2% extract solution, these percentages fluctuated between 1% and 8.4%, between 0.1% and 4.9%, and between 0.9% and 4.4% after 1 h, 2 h and 3 h of residence respectively. At the same time, the percentages of detached cells varied from 1.7 to 11.0%, 0.5 to 9.2% and 2.0 to 4.5% respectively at extract concentrations of 1, 1.5 and 2% respectively (**Table 1**).

For cells from the exponential growth phase, the percentages of cells remaining adhered fluctuated between 0.4 and 10.2%, between 0.2 and 5.8%, and between 0.3 and 4.9% after 1 hour, 2 hours and 3 hours of contact with the 1% extract solution. They varied between 1.2 and 6.8%, between 0.6 and 5.4%, and between 1.0 and 5.2% after 1 hour, 2 hours and 3 hours of contact with the 1.5% extract solution. At the 2% extract concentration, these percentages fluctuated between 1.2 and 5.6%, between 0.3 and 6.5%, and between 1.0 and 1.8% after 1 hour, 2 hours and 3 hours of contact respectively. Under similar experimental conditions, the percentages of detached cells ranged from 1.6 to 11.4%, 0.9 to 68.8% and 1.0 to 12.7% at 1, 1.5 and 2% extract concentrations respectively (**Table 1**).

Solutions of *Eucalyptus microcorys* extract lead to detachment of the bacterial cells initially adhered to the polyethylene fragment. The importance of this cell detachment varies not only as a function of the concentration of the extract but also as a function of the residence time of the adhered cells in the extract solution. The bacterial adhesion to substrates involves two main steps: reversible adhesion and

Cell growth phase and preincubation time		**Percentage of detached (%D) and adhered (%A) cells of enteropathogenic Escherichia coli, after 1, 2 or 3 hours of contact with the plant extract solution at concentrations of 1%, 1.5% and 2%**																	
Growth phase	**Duration of preincubation for cell adhesion**[a]	**1%**						**1.5%**						**2%**					
		%A			**%D**			**%A**			**%D**			**%A**			**%D**		
		1 h	**2 h**	**3 h**	**1 h**	**2 h**	**3 h**	**1 h**	**2 h**	**3 h**	**1 h**	**2 h**	**3 h**	**1 h**	**2 h**	**3 h**	**1 h**	**2 h**	**3 h**
Lag	3 hours	15.0	12.2	6.4	1.7	2.8	2.0	1.2	1.3	0.2	3.9	3.5	0.4	8.4	1.7	4.4	4.4	4.5	4.0
	6 hours	1.3	0.3	0.2	2.9	0.5	2.9	0.5	0.3	0.2	0.4	3.4	2.8	1.0	0.1	0.9	1.8	0.9	4.9
	9 hours	1.2	2.2	1.3	11.0	9.2	4.5	5.4	3.1	4.7	11.1	9.9	11.2	3.6	4.9	2.7	9.3	5.7	1.5
Exponential	3 hours	10.2	5.8	3.0	10.5	3.6	9.4	6.8	5.4	2.4	13.3	9.3	62.8	5.6	6.5	1.8	12.7	1.0	7.3
	6 hours	0.4	5.3	4.9	11.4	8.8	8.6	4.5	4.9	5.2	14.6	5.2	68.8	4.4	2.9	1.0	8.1	1.4	1.3
	9 hours	0.8	0.2	0.3	5.7	1.9	1.6	1.2	0.6	1.0	4.2	0.9	12.9	1.2	0.3	1.3	8.3	1.9	2.6
Stationary	3 hours	11.3	14.0	17.1	33.3	12.4	37.6	16.1	17.2	7.0	4.6	3.7	48.5	5.6	6.8	4.8	3.7	2.8	2.9
	6 hours	4.3	9.6	7.7	16.0	24.0	33.6	3.0	5.4	10.5	3.1	2.5	33.5	1.6	8.5	7.7	2.2	1.7	1.9
	9 hours	8.8	8.8	3.3	34.0	33.4	25.6	13.5	11.1	5.4	3.0	1.9	25.1	7.6	5.9	1.3	2.9	1.3	1.8
Decline	3 hours	33.0	15.3	17.2	55.6	44.9	48.3	14.6	15.3	8.8	3.3	0.7	9.7	14.0	13.0	9.7	1.9	2.5	4.5
	6 hours	4.6	17.7	12.4	21.3	14.7	12.6	6.8	8.9	8.6	1.8	2.3	30.2	17.2	14.4	5.8	0.9	1.8	1.1
	9 hours	7.9	11.3	25.4	14.5	14.4	13.8	8.0	5.2	13.5	1.2	1.2	16.0	9.7	5.4	5.0	1.5	1.8	4.8

[a]*Incubation duration for cell adhesion process on polythene fragment in NaCl solution (0.85%) prior exposure to the Eucalyptus leaves extracts.*

Table 1.
Effect of 1%, 1.5% and 2% Eucalyptus *leaves extracts on adhered EPEC cells from different growth phases and preincubation periods in 0.85% NaCl solution.*

irreversible adhesion [22]. The reversible adhesion is governed by physico-chemical interactions of type Van der Waals and Lewis acid–base [23]. The irreversible adhesion is slower than the previous one, the irreversibility of the membership using the bacterial metabolism step.

The detachment of enteropathogenic *E. coli* cells, initially fixed on the fragments of polyethylene, would be caused by the secondary metabolites present in the plant extract, which would cause the breakdown of the hydrogen bonds within the exopolysaccharide secreted by the enteropathogenic *E. coli* cells such as a protective matrix. In bacteria, the permeabilization of membranes by these compounds is associated with a loss of ions and degradation of the ATP potential, the aromatic molecules having the highest antibacterial activity being the phenols.

The polyphenols present in the extract of *Eucalyptus microcorys* would constitute stress factors and probably deprive the bacteria of their protective glycocalyx, thus causing a disorganization of the biofilm and the dislodgement of the bacteria from the surface of the polyethylene slides. However, studies showed that the rates of detached cells remain below 15%. This low rate would be linked to the exopolymer covering the bacteria which creates a concentration gradient so that the permeabilization of the protective layer is not complete. Thus, only bacterial cells from a certain distance from the support are affected and dislodged. Some bacteria carry specific genes in their plasmids, genes that code for virulence factors (type IV fimbriae, adhesins, toxins) and which play an important role in the cell adhesion process. They allow the interconnection of bacteria in micro-colonies, promoting their stabilization, which can lead to resistance to the effect of the detachment of the extract.

The change of strains from the adhered state to the planktonic state further exposes the bacterial cells to the antibacterial effect of the flavonoids and alkaloids contained in the plant extract. Alkaloids are hydrophobic cations with antibacterial properties and targeting cellular DNA. This inhibitory effect is modulated by the adherent cell-extract contact time, the long contact times acting on targets not reached by relatively short contact times. The percentages of inhibition of enteropathogenic *E. coli*, for all the four phases of cell growth, vary between 73.56% and 99.49%, the concentration of *E. microcorys* 2% being that which results in high levels of cell inhibition.

The presence of bacterial strains still living in the planktonic state in the extract could be explained by the phenomenon of resistance such as the phenomenon of microbial resistance to antibiotics. Bacteria can synthesize enzymes capable of destroying or modifying antibacterial molecules, the enzymatic reactions leading to this destruction or this modification, although varying with the bacterial strain. The resistance mechanism observed appears to be multifactorial. Indeed two mechanisms are generally advanced to explain the resistance of biofilms to antibacterials. It can be due either to a limitation of the diffusion of the antibacterial agents in the biofilm by the polysaccharide matrix which coats the bacteria, or to the particular physiological state (low growth rate) of the bacteria of the biofilm, consequence of the nutritional limitation that undergo bacteria within the biofilm. The hydrated polyanionic matrix that coats bacteria in biofilms, limits the diffusion of molecules from the surrounding medium and more particularly of charged molecules. The hydrated polyanionic matrix that coats bacteria in biofilms, limits the diffusion of molecules from the surrounding medium and more particularly of charged molecules.

The physiological state of the cell and the extract concentrations are the first factors influencing the adhesion process of *E. coli* through the detachment or maintenance of cells after the stay of the polyethylene fragment in the extract solution of *E. microcorys*. The action of disinfectant solutions on microorganisms could depend on several factors, some of which are intrinsic to the organism and others are related to the environment. Knowledge of these factors should lead to better achievement of disinfection and sterilization. Resistance of the adhered cells of *E. coli* to the plant extract is observed after 9 hours of incubation when the cells have emerged from the exponential phase of growth. The lag phase corresponds to the adaptation of the inoculum to its new environment, while the decline phase is the period corresponding to the exhaustion of all nutritional resources. There is an accumulation of toxic metabolites. Under the action of endogenous proteolytic enzymes, cell lysis leads to a decrease in viable organisms.

3. *Artemisia annua* extracts

3.1 Bioactive *Artemisia* compounds

Phytochemical analysis reveals the presence of alkaloids, free flavonoids, tannins, triterpenes and sterols, anthocyanins, reducing compounds, mucilages and coumarins in the extracts of *A. annua* [24, 25].

3.2 Effect of *Artemisia annua* leaves extract on *Escherichia coli*

The rates of change of bacterial concentrations varied in the presence and absence of light in the different media. In the dark, the rates of change of *E. coli* cell concentrations ranged from −12 to 50 cells/h. Minimum rates were obtained in pH 4 solutions. Maximum rates were recorded in pH 9 media. Negative values were noted in the pH 4 and pH 5 solutions. These negative rates indicate a relative inhibition of cell metabolisms in the presence of the plant extracts. In the control solutions, the rates of change in cell concentrations sometimes reached −19 cells/h (**Table 2**).

In the presence of light, the evolutionary rates of cell concentrations fluctuated between −14 and − 12 cells/h. Minimum rates were obtained in solutions containing *A. annua* extract at pH 4. Maximum rates were recorded at pH 8. In the control solutions, the rates of change in cell concentrations often reached −8 cells/h (**Table 2**).

The growth of *E. coli* was favored by slightly alkaline pH in the *Artemesia annua* extract solution, when grown in dark. The secondary metabolites present in the *Artemesia annua* extract are potentially involved in the physico-chemical modifications of the medium, promoting the observed cell growth. These results suggest that the *A. annua* extract does not present a bactericidal activity in water treatment in the absence of light. The study of the impact of light on *E. coli* bacteria in the presence of *A. annua* extract leads to a significant reduction in cell densities in comparison with the evolution of cell abundances in the dark. Leaves

Experimental conditions	*Escherichia coli* growth rates (CFU/h)						
	pH 4	pH 5	pH 6	pH 7	pH 8	pH 9	Witnesses
In the dark	−12 (0,645)	−7 (0,637)	23 (0,822)	40 (0,696)	44 (0,518)	50 (0,763)	−19 (0,750)
In the presence of light	−14 (0,762)	−13 (0,836)	−13 (0,750)	−14 (0,828)	−12 (0,771)	−12 (0,767)	−8 (0,936)

Table 2.
Evolution rates of E. coli *cell concentrations (and regression coefficient) at each pH value and experimental condition [24, 25].*

of *A. annua* contain organic and inorganic substances and bioactive compounds. Molecules from extracts of *A. annua* can be a source of nutrients in the experimental conditions and allow the growth of different species. The study of the impact of light on *E. coli* bacteria in the presence of *Artemisia annua* extract leads to a significant reduction in cell densities in comparison with the evolution of cell abundances in the dark. This inhibition is explained by the effect of photosensitive compounds originating from the extract of *A. annua*. These compounds can induce a photosensitization reaction capable of inhibiting the metabolism of *E. coli* bacterial cells, since in the absence of light, no significant inhibition is observed.

4. *Moringa oleifera* extracts

4.1 Bioactive *Moringa* compounds

Phytochemical screening showed that most of the constituents obtained from aqueous and ethanoic extracts of *Moringa* spp. are alkaloids, flavonoids and phenols [26].

4.2 Effect of *Moringa oleifera* extract on *Escherichia coli*

The abundances of *E. coli* in different extract concentrations ranged from 500×10^3 to 0.92×10^3 CFU/100 mL. At 4°C, it ranged from 224.48×10^3 to 3.58×10^3 CFU/100 mL. The lowest abundance was recorded at 10 g/L and the highest at 1 g/L (**Figure 2**).

At 23°C, it ranged from 129.7×10^3 to 0.92×10^3 CFU/100 mL, with the lowest abundance recorded at 30 g/L and the highest at 1 g/L. The cell concentrations in the control (solution without seed extract) were 500×10^3 CFU/100 mL at 23°C and 4°C, respectively (**Figure 2**).

The obtained inhibition percentages and temporal variation of *E. coli* cell abundances show that *Moringa oleifera* seed extract can be used as a natural alternative for efficient water treatment. The antibacterial activity of *M. oleifera* seed extract is believed to be due to the presence of a cationic protein molecule present in the seed. This protein, commonly known as *M. oleifera* cationic protein (MOCP), is responsible for the death of bacterial cells by rapid flocculation and fusion of their inner and outer membranes. This protein would inhibit the growth of bacteria at higher concentrations, thus facilitating the antibacterial inhibition of the latter. However, this activity is dependent on the bacterial load, and an increased bacterial concentration would require a higher dose or higher concentration of the seed extract. The inhibitory effect of *M. oleifera* seed extract against bacterial cells is thought to be related to phytochemicals such as alkaloids, flavonoids and tannins, among others steroids, saponins, phenols, terpenoids and finally coumarins and anthraquinones present in the different seed extracts.

Temperature appears to be an important factor involved in cell inhibition by the aqueous extract of *M. oleifera* seeds. Incubation temperature increases the efficacy of the aqueous extract of *M. oleifera* seeds, with considerable inhibition at psychrophilic temperature. The seeds have been reported to contain calcium, magnesium, phosphorus, copper, vitamins (A, B and E) and are also rich in organic elements. These different secondary metabolites, sometimes present in large quantities in the extracts, could accumulate in the cell wall of *E. coli* and become toxic. The inhibition of bacteria could also be due to the presence of the isothiocyanate molecules α-L-rhamnosyloxy benzyl which are found in the seeds and whose antibacterial and

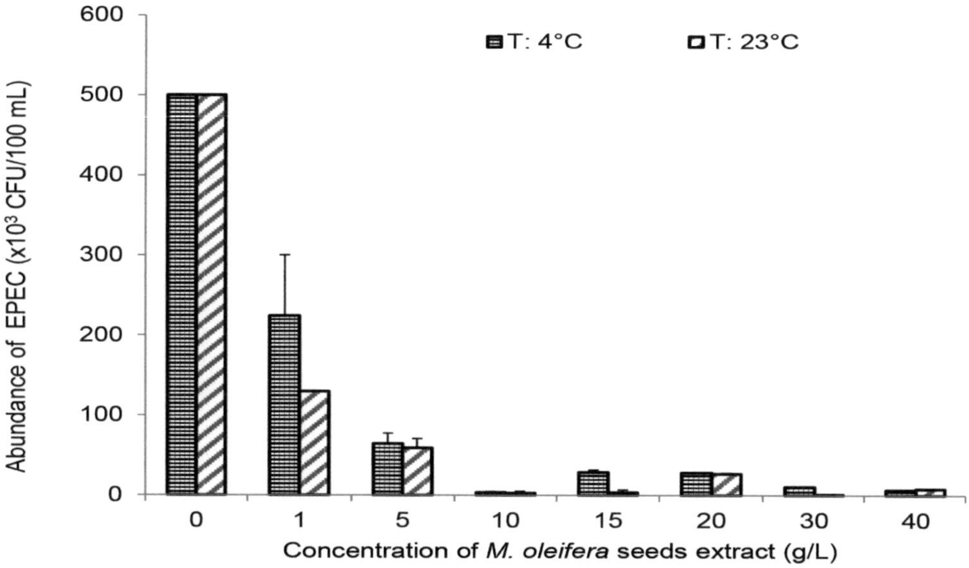

Figure 2.
Enteropathogenic E. coli *(EPEC) abundance depends on the concentration of* Moringa oleifera *seeds extracts [26].*

antifungal properties have been described. These molecules are soluble and positively charged. They can easily cross the bacterial membrane to bind to negatively charged cationic proteins on the cell membrane surface and support their inhibition. Water disinfection with Moringa seeds requires relatively high doses of 200 g/L of extract to have a germicidal effect. With a variation of extract concentrations from 1 g/L to 40 g/L, the bacterial inhibitions varied from 55.12% to 99.9% for enteropathogenic *E. coli*. This suggests that the environment, as well as the genetic characteristics of the bacteria or other abiotic properties of the water used, could affect the activity of the constituents of the seeds and other parts of the plant (root barks and stems).

5. Conclusion

Drinking water is often subject to bacteriological contamination, causing serious health problems due to diarrhoeal diseases, gastroenteritis, cholera and typhoid fever. In recent years, water disinfection methods using plant extracts have been proposed as a new alternative for household water treatment. Information on the conditions of the use of plant extracts in the treatment of bacterio-contaminated water is often not available. The chapter aimed to summarize the known effects of some plant extracts on the cultivability of *E. coli* cells.

The results show that the presence and absence of light determine the action of the plant extracts on the survival of *E. coli* bacteria in aquatic environments. In the absence of light, *A. annua* extracts can sometimes promote bacterial cell activity. This activity is influenced by the pH of the solutions, the sensitivity of the bacterial cells under monospecific conditions being observed. The impact of the pH would be linked to a variation of the assimilation coefficient of nutritive substances.

In the presence of light, the plant extract inactivates bacterial metabolism to varying degrees. This variability depends on the concentration of the extract. The rate of photo-oxidation reactions that lead to bacterial inactivation is pH dependent, and varies from one bacterial species to another. The presence of light increases the inhibitory effect of plant extracts on *E. coli* cells.

The use of medicinal plants in water disinfection offers many opportunities in a world where access to safe drinking water remains a permanent concern for public authorities, therefore it should be considered to use plant extracts as an alternative process for water disinfection.

Author details

Antoine Tamsa Arfao* and Moïse Nola
Faculty of Sciences, Hydrobiology and Environment Research Unit, University of Yaoundé 1, Yaoundé, Cameroon

*Address all correspondence to: tamsarfao@yahoo.fr

References

[1] Bleu P, Mohammed B, Gaëlle T Faire face aux crises et pénuries d'eau en Méditerranée. Les notes du Plan Bleu: environnement et développement durable en Méditerranée. Plan Bleu, Centre d'Activités Régionales; 2006. p. 4

[2] Furet JP, Firmesse O, Le Gourmelon M, Bridonneau C, Tap J, Mondot S, et al. Comparative assessment of human and farm animal faecal microbiota using real-time quantitative PCR. Microbiology and Ecology. 2009;**68**:351-362

[3] Chevallier H. L'eau, un enjeu pour demain. Etat des lieux et perspectives. Paris: Ed. Ellébore, Sang de la terre; 2007. p. 352

[4] Roszak DB, Colwell RR. Metabolic-activity of bacterial-cells enumerated by direct viable count. Applied and Environmental Microbiology. 1987; **53**(12):2889-2983

[5] Le Minor L, Richard C. Méthodes de Laboratoire pour l'Identification des Entérobactéries. Institut Pasteur; 1993. p. 217

[6] WHO. Guidelines for Drinking-Water Quality. 3rd ed. Geneva: World Health Organization; 2004

[7] Le Minor L, Véron M. Bactériologie médicale. Paris: Flammarion éd.; 1989. p. 1107

[8] CEAEQ (Centre d'Expertise En Analyse Environnementale Du Québec). Recherche et dénombrement simultanés des coliformes totaux et d'*Escherichia coli* dans l'eau potable avec le milieu de culture MI: Méthode par filtration sur membrane, MA. 700 – Ecctmi 1.0, Rév. 3, Ministère du Développement durable, de l'Environnement et de la Lutte contre les changements climatiques; 2015. p. 19

[9] GSE (Groupe Scientifique sur l'eau). *Escherichia coli*, Dans Fiches synthèses sur l'eau potable et la santé humaine. Institut national de santé publique du Québec; 2003. p. 4

[10] Haslay C, Leclerc H. Microbiologie des eaux d'alimentation. Technique et Document. Lavoisier éd., Paris. 1993

[11] Clarke SC. Diarrhoeagenic *Escherichia coli* - an emerging problem? Diagnostic Microbiology and Infectious Disease. 2001;**41**:93-98

[12] Clarke SC, Haigh RD, Freestone PPE, Williams PH. Virulence of enteropathogenic *Escherichia coli*, a global pathogen. Clinical Microbiology Reviews. 2003;**16**(3):365-378

[13] Germani Y, Le Bouguénec C. Diagnostic des *Escherichia coli* agents de diarrhée chez l'homme. Revue francophone des laboratoires. 2008;**38**(400):67-76

[14] Mainil J. Facteurs de virulence et propriétés spécifiques des souches invasives d'*Escherichia coli*: I) Les adhésines et facteurs de colonisation. Annales de Médecine Vétérinaire. 2003;**147**:105-126

[15] Cunin P, Tedjouka E, Germani Y, Ncharre C, Bercion R, Morvan J, et al. An epidemic of bloody Diarrhea: *Escherichia coli* O157 emerging in Cameroon? Emerging Infectious Disease. 1999;**5**:285-290

[16] Sunda M, Rosillon F, Taba K. Contribution to the study of water disinfection by photosensitization with plant extracts. European Journal of Water Quality. 2008;**39**(2):199-209

[17] Sunil B, Nitin W. Formulation of kinetic model to predict disinfection of water by using natural herbs. International Journal of Environmental Sciences. 2012;**2**(3):1344-1354

[18] Tamsa Arfao A, Nola M, Lontsi Djimeli C, Nougang ME, Hzounda Fokou JB. Cultivability of *Salmonella typhi, Vibrio cholerae* and enteropathogenic *Escherichia coli* in water microcosm in the presence of *Eucalyptus microcorys* leaves extract: Effect of the concentration of leaves extract and incubation temperature. International Journal of Research in BioSciences. 2013;**2**(4):32-46

[19] WHO. 2008. Traditional medicine http://www.int/medecine/areas/traditional/en/ondex.html

[20] Tamsa AA, Ewoti OVN, Onana MF, Djimeli CL, Tchakonté S, Dama NK, et al. Combined effect of *Eucalyptus microcorys* aqueous extract and light on pathogenic *Escherichia coli* survival in aquatic microcosm. Journal of Microbiology, Biotechnology and Food Sciences. 2020;**9**(5):1003-1008. DOI: 10.15414/jmbfs.2020.9.5.1003-1008

[21] Trabulsi L. and Altherthum F. 2005. Microbiologia. 5ème ed. São Paulo: Atheneu. p. 718

[22] Pouneh K. Formation de biofilm à *Pseudomonas aeruginosa*: évaluation d'inhibiteurs potentiels du Quorum Sensing. Thèse de Doctorat. Toulouse III: Université Paul Sabatier; 2009. p. 328

[23] Boutaleb N. Etude de la formation des biofilms sur les surfaces de matériaux couramment utilisés dans les canalisations d'eau potable. Thèse de Doctorat. Université de Bretagne-Sud; 2007. p. 194

[24] Mobili OB, Lontsi DC, Tamsa AA, Nougang ME, Allahdin O, Noah EOV, et al. Synergistic effect of light, pH and *Artemisia annua* extract on *enterococcus faecalis* in aquatic microcosms. International Journal of Applied Microbiology and Biotechnology Research. 2015;**3**:62-72

[25] Mobili OB, Nola M, Nougang ME, Moungang LM, Noah Ewoti OV, Ateba BH, et al. Assessment of the effect of *Artemisia annua* leave extract infusion pH under dark conditions on *Staphylococcus aureus*, *Salmonella paratyphi* and *Escherichia coli*. Journal of Applied Biosciences. 2013;**62**:4595-4609

[26] Metsopkeng CS, Lontsi C, Noah Ewoti OV, Moungang LM, Nana PA, Tamsa AA, et al. *Moringa oleifera* seeds extract activity on Enteropathogenic *Escherichia coli* and *Aeromonas hydrophila* cells in aquatic microcosm. Journal of Applied Biotechnology. 2019;**7**(2):13-30

Section 4

Useful *Escherichia coli*

Chapter 14

Potential of *Escherichia coli* Probiotics for Improved Health and Disease Management

Nareshkumar Gattupalli and Archana Gattupalli

Abstract

Although natural gut microbiota contains *Escherichia coli* as a commensal, this bacterium, along with other members of the *Enterobacteriaceae* family, are usually known for their pathogenic potential. Interestingly, *E. coli* colonizes first and remains all through life, and in fact, some strains possess beneficial properties such as antibacterial colicin secretion. Among the beneficial strains, *E. coli* Nissle, isolated in 1917, has been the most extensively explored strain. Adaptability to survive under diverse conditions coupled with facile genetic manipulations enabled the design of *E. coli* strains with properties to deliver antioxidant, anti-inflammatory, and antitumor molecules. Moreover, genetically modified *E. coli* strains secreting enzymes for converting sucrose and fructose into insulin and mannitol, respectively, were very effective in preventing the onset of metabolic disease by acting as synbiotics. Thus, *E. coli* is emerging as a very potent probiotic platform for developing strains with the potential of controlling many metabolic and multifactorial diseases, including cancer.

Keywords: *E. coli* Nissle, probiotic, prebiotic

1. Introduction

Escherichia coli resides in the gastrointestinal (GI) tract of animals along with a few hundreds and thousands of different microbiota. In humans, *E. coli* is present at less than 1% of gut microbiota and it is not among the 25 most prevalent bacteria [1, 2]. But *E. coli* is the predominant *Enterobacteriaceae* species in humans [3]. Interestingly, *E. coli* is the first to colonize the intestines and persists all through life in humans [4]. Mucin layers do not allow any direct interaction of gut microbiota with the enterocytes. However, the diversity of gut microbiota is known to influence the intestinal permeability involving LPS, peptidoglycan, lipoproteins, deoxynucleic acid (DNA), and ribonucleic acid (RNA). Most *E. coli* strains are nonpathogenic and exist as commensals, but some pathogenic strains are associated with severe diseases [5]. Additionally, some *E. coli* strains known as pathobionts do not cause any disease in healthy individuals but exacerbate chronic inflammatory diseases [1]. The *E. coli* population in the GI tract is dynamic with a turnover in months to years [3, 6]. Humans contain five different strains of *E. coli* [7]. Oxygen diffusion from intestinal epithelium is favorable for *Enterobacteriaceae* members including *E. coli* to be present in close proximity to the mucus layer [8]. *E. coli* is known to play an important role in the maintenance by decreasing oxygen content,

E. coli Probiotic	Properties	References
Mutaflor, EcN 1917	Serotype O6:K5:H1, Motile, flagella, present, Microcin M and H47	[4, 15, 16]
Symbioflor 2	20% strain G1/2 (DSM16441), 20% G3/10 (DSM16443), 20% G4/9 (DSM 16444), 10% G5 (DSM 16445), 20% G6/7 (DSM 16446), and 10% G8 (DSM 16448). Microcin S, Non-motile	[19, 35]
Colinfant (A0 34/86)	O83:K24:H3;	[16, 20, 21]
CFR16	Colicin E1/Ia1b	[9, 22]

Table 1.
Characteristics of E. coli *probiotic strains.*

by facilitating the colonization of anaerobes and vitamin K production, and protects against colonization of pathogens. *E. coli* produces microcins and colicins, which prevents the colonization of pathogens. In contrast to antibiotics, colicins act on bacteria related to the bacteria producing these antibacterial proteins. Many *E. coli* isolates from rat fecal matter were found to produce E1-, 1a/1b-, and B/D-type colicins with antimicrobial properties against enteropathogens [9].

2. *E. coli* as a probiotic

International Scientific Association for Probiotics and Prebiotics (ISAPP) defines probiotics as "live microorganisms that, when administered in adequate amounts, confer a health benefit on the host" [10]. Although fermented foods with 10^5–10^8 microorganisms per gram can be constituted probiotics, recent oral supplementations contain 10^{10-12} live cells per single dose [11, 12]. *Lactobacillus* sp., *Bifidobacterium* sp., *Streptococcus* sp., *Enterococcus* sp., and *Saccharomyces boulardii* are common probiotics [13]. Some *Bacillus*, *E. coli,* and *Clostridium* sp. were found to be probiotics.

Commensal and probiotic *E. coli* modulates innate adaptive immune responses in the intestinal epithelium by activating the secretion of defensins, cytokines, IgA, and CD4 T cells. Additionally, siderophore production and iron scavenging by probiotic *E. coli* strains prevent the survival of the pathogens [14]. Interestingly, Alfred Nissle discovered an *E. coli* strain that prevented the growth of pathogenic *Salmonella* when it was cultured with stool samples [15]. The strain was isolated from a soldier who had no diarrhea when other soldiers had suffered *Shigella* infection. Upon oral supplementation, this strain was protected from diarrhea and this strain was named as *E. coli* Nissle 1917(EcN) from the year after its use [16]. Commercially, this strain is known as Mutaflor. EcN is the most commonly used gram-negative probiotic [17]. Interestingly, EcN could get established in swine herds with variation in the colonization of individual animals [18]. Some other *E. coli* strains have also been shown to possess probiotic properties (**Table 1**).

3. Probiotic *E. coli* Nissle 1917

EcN is effective against infection by *Salmonella enterica* serovar, *Typhimurium* strain C17, *Yersinia enterocolitica*, *Shigella flexneri*, *Legionella pneumophila*, and *Listeria monocytogenes* [19]. EcN is serum sensitive, forms semi-rough colony, and

DOI: http://dx.doi.org/10.5772/intechopen.100380

has low levels of smooth lipopolysaccharide (sLPS) [23]. *EcN* prevents the colonization of pathogens by efficient adhesion with the help of fimbriae and capsule to the epitheliumt but not activating inflammation as its lipopolysaccharide (LPS) has a short O chain and weak binding to toll-like receptor 4. EcN decreases pro-inflammatory cytokine and increases anti-inflammatory cytokine formation [24]. EcN repairs leaky gut by increasing the expression and phosphorylation of tight junction protein zonula occludens-1 (ZO-1), ZO-2, and claudin 14 [25–27]. Additionally, EcN prevents disruption of epithelial tight junctions by inhibiting NF-κB-mediated activation of the MLCK-P-MLC signaling pathway [28]. EcN mediates pathogen elimination by secretion of low molecular weight microcin H47 and microcin S. Probiotic EcN, but not commensal *E. coli* MG1655, increases serotonin (5-hydroxytryptamine) secretion by enterochromaffin cells [29].

Interestingly, bacteria are known to secrete vesicles known as membrane vesicles (MVs) [30]. Gram-negative bacterial outer membrane vesicles contain LPS and the size and complexity of O-antigen, the number, and nature of fatty acid components of lipid A determining the beneficial or toxic effects on the host cells. EcN outer membrane vesicles (OMVs) prevent the inflammation and progression of dextran sodium sulfate (DSS)-induced colitis in mice [26, 31, 32]. EcN OMVs get internalized by macrophages and activate the phagocytosis, which increases pro-inflammatory cytokine secretion and killing of pathogens [33].

4. Symbioflor-2

Symbioflor-2 is a commercial product containing six *E. coli* strains, which brings about an increase in β-defensin-2 and reduces mast cell activation [19]. Symbioflor-2 is effective in reducing symptoms of irritable bowel syndrome [34]. Microcin S is produced by Symbioflor G3/10 strain. Surprisingly, virulence genes have also been detected in Symbioflor-2 genomes suggesting that the presence of virulence genes does not imply pathogenicity [2]. Transcriptomic analysis of ileum and colon upon inoculation with Symbioflor-2 strains indicated the increase in defense responses involving dual oxidase/nitric oxide pathway mediated reactive oxygen species generation along with β-defensin-2 activity [35]. Transcription profiles were distinct with EcN and Symbioflor-2.

5. Colinfant

Colinfant is an *E. coli* (A0 34/86) strain that is used as prophylactic in infants for allergy, nosocomial infection, and diarrhea [20, 21]. Additionally, it is effective in later years in preventing infections and developing allergies. Some strains of *Klebsiella oxytoca* are implicated in antibiotic-associated diarrhea, which could be reduced by the administration of Colinfant in infants. Colinfant also prevents infection of pathogenic *E. coli*.

6. Genetic modifications of probiotic *E. coli*

Probiotic *E. coli* strains have been modified to improve colonization, to secrete metabolites, proteins, and enzymes exploiting a variety of genetic manipulations (**Table 2**). EcN was tagged with a green fluorescent protein (Gfp), which facilitated monitoring the colonization and survival in stomach, ileum, colon, and Peyer's patches [36]. EcN was detected in the fecal matter at 45 days after oral inoculation.

E. coli strain	Nature of modification	Properties	References
EcN	pUC-gfp	48-h residence in stomach, cecum and rectum; the presence in Peyer's patches; detected in feces up to 45 days after oral administration in rats	[36]
Ec16 (*vgb*)	*Green Fluorescent protein-Vitreoscilla* hemoglobin (*Vgb*) gene	Improves survival and ameliorates carbon tetrachloride toxicity	[37]
Ec16::*vgb-gfp* operon	Plasmid containing *Pseudomonas fluorescens* Bf1 *pqqABCDE* gene cluster	Ameliorates dimethylhydrazine-induced colon and liver damage; improved the neurotransmitter status	[38, 39]
Ec16::*vgb-gfp (inuJ)*	pMAL-p2ΔlaclQ Inulosucrase (*inuJ*) gene	Extracellular secretion of inulosucrase	[22]
EcN::*vgb-gfp (pqqABCDE)*	Plasmid containing *Pseudomonas fluorescens* Bf1 *pqqABCDE* gene cluster	Preventing chronic alcohol-induced oxidative damage	[40]
EcN::*vgb-gfp (pqqABCDE)*	Plasmid containing *Gluconobacter suboxydans* 621 *pqqABCDE* gene cluster	Prevents rotenone-induced mitochondrial oxidative stress and improves mitochondrial biogenesis	[41]
EcN::*vgb-gfp (pqqABCDE)*	Plasmid containing *pqq pqqABCDE* gene cluster of *Gluconobacter oxydans*	EcN strain secretes PQQ, gluconic acid with citric acid supplementation decreases the Cd- and Hg-induced liver toxicity effects in rats	[42]
EcN::*vgb-gfp (pqqABCDE-gad)*	Plasmid containing *pqqABCDE* gene cluster of *A. calcoaceticus* and gluconate dehydrogenase (*gad*) operon of *P. putida* KT 2440	EcN strain secretes PQQ, gluconic and 2-ketogluconic acids decrease the Cd, Hg, and Pb toxic effects in rats	[43]
EcN(*pqqABCDE-arsM)*	*Ptac* G. oxydens pqqABCDE—Rhodopseudomonas palustris arsM* gene	EcN strain converts arsenite gets converted to non-toxic trimethyl arsenite and reduces arsenite toxicity	[44]
EcN::*vgb-gfp (pqqABCDE-glf-mtlK)*	*Ptac*-pqqABCDE* gene cluster of *G. suboxydans*-glucose facilitator (*glf*) of *Zymomonas mobilis*-mannitol dehydrogenase (*mtl*K)	EcN strain secrets PQQ and produces Glf protein and MtlK enzyme that coverts dietary fructose into mannitol	[45]
EcN::*vgb-gfp (pqqABCDE-fdh)*	*Ptac*-pqq* gene cluster of *G. suboxydans—Fructose dehydrogenase (fdh)* from *Gluconobacter frauteuri* IFO3260	EcN strain secrets PQQ and produces Fdh enzyme that converts dietary fructose to 5-keto-D-fructose	[45]
EcN::*vgb-gfp (pqqABCDE-inuJ)*	Genomic integration of *vgb*, *gfp*, *pqqABCDE*, and *inuJ* genes-	High dietary sucrose-induced oxidative damage and hyperlipidemia were decreased	[46]

E. coli strain	Nature of modification	Properties	References
EcN	Curing of Mut1 and Mut2 plasmids	Growth is similar to the wild type in Luria broth	[47]
SYNB1618	Phenylalanine ammonia lyase (*stlA*) gene	Phenyl alanine conversion to *trans*-cinnamate	[48]
	L-amino acid deaminase (*pma*) gene	Phenyl alanine conversion to phenylpyruvate	
EcN (Δ*frdA*, Δ*ldhA*, Δ*adhE*, Δ*pta*)	P_L-*atoDAEB* operon; gsA::PL-*LacO-hbd-crt-ter*	Butyric acid secretion	[49]
EcN	Trefoil factor	Curly fiber matrix restitutes intestinal epithelium effective against DSS-induced colitis	[50]
EcN	*Staphylococcus aureus* α-hemolysin	Tumor regression by forming pores	[51]
EcN	Hemolysin E (*hylE*) under *araBAD* promoter	EcN strain had regressed tumors in mice by pore formation	[52]

Gad—gluconate dehydrogenase, Gfp—green fluorescent protein, DSS—dextran sodium sulfate, Ptac—constitutive tac promoter.*

Table 2.
Characteristics of genetically modified probiotic E. coli *strains.*

EcN contains two cryptic plasmids MUT1 and MUT2, and these plasmids were cured using CRISP-Cas9-assisted double-strand breaks [47]. EcN strain cured of these plasmids had similar growth under Luria broth conditions despite differences in the DNA content. Effects of colonization and survival of the plasmid-cured strain with decreased DNA content as compared to the wild-type strain need to be investigated to determine the impact of metabolic load. Alternatively, both the cryptic plasmids of EcN have been engineered for stable maintenance and expression of recombinant proteins [53].

Vitreoscilla hemoglobin (VHb) with a high affinity for oxygen facilitates the survival and functionality of bacteria under microaerobic conditions [54] promoted colonization of genetically modified *E. coli* in the gut. *E. coli* 16 double transformants of *gfp* and *Vitreoscilla* hemoglobin (*vgb*) genes at 10^8 cfu/g were present in the rat fecal matter after 70 days of oral administration, while Ec16 *gfp* was not found after 48 days [37]. Additionally, catalase activity of VHb scavenges the reactive oxygen species, which decreased the carbon tetrachloride-induced hepatotoxicity in rats.

Pyrroloquinoline quinone (PQQ) is a water-soluble antioxidant with the highest redox cycles of 20,000, promotes mitochondrial biogenesis and cellular signaling, and provides health benefits [55]. *E. coli* 16 strain tagged with *gfp-vgb* genes and transformed with *pqqABCDE* operon from *Pseudomonas fluorescens* Bf1 prevented colon and liver damage by dimethylhydrazine (DMH) due to the combined beneficial effects of effective colonization and antioxidant properties of Vhb and PQQ, respectively [38]. DMH had systemic oxidative damage, and decreased brain serotonin and norepinephrine levels, but epinephrine levels were increased [39]. In addition to decreasing the oxidative damage, *E. coli* 16 *vgb-pqq* strain had near-normal levels of neurotransmitters in rats. These beneficial effects were not similar with treatments of Ec16, vitamin C, or PQQ alone suggesting other than its additional ability to confer antioxidant properties, probiotic *E. coli* 16 had synergistic effects related to the continuous secretion of PQQ in the gastrointestinal tract. These beneficial effects were also seen in EcN strain that was modified in a similar manner to that of Ec16 strain [40]. EcN *vgb-pqq* recombinant strain effects were monitored in rats for alcohol toxicity in chronic and acute exposure. Chronic alcohol caused extensive oxidative damage and induced hyperlipidemia and the EcN::*vgb-gfp*(*pqq*) probiotic strain prevented the deleterious effects, while EcN, PQQ, and vitamin C alone had no significant effects. These effects were also correlated with increased short-chain fatty acids (SCFA) in the colon. However, oral PQQ had better effects than recombinant EcN strain in acute alcohol damage. These studies further supported the significance of endogenous PQQ biosynthesis by probiotic *E. coli*.

Aging is associated with progressive loss of tissue functions mediated by reactive oxygen species-induced oxidative damage as a result of mitochondrial dysfunction [56–58]. EcN::*vgb-gfp* transformed with *pqq* gene cluster from *Gluconobacter suboxydans* 621 decreased the rotenone-induced mitochondrial oxidative damage in aging rats along with decreased lipogenesis and increased fatty acid oxidation genes correlated with increased colonic SCFA and PQQ in both feces and liver [41]. Additionally, an increase in mitochondrial biogenesis and metabolism indicates delaying of age-related tissue damage.

Heavy metal toxicity is mediated by reactive oxygen species [59]. Chelation of heavy metal ions and antioxidants is used to prevent the toxicity. EcN::*vgb-gfp* strain operon containing *pqq* gene cluster from *Gluconobacter oxydans* decreased the Cd and Hg toxicity upon oral supplementation citric acid due to the antioxidant effects of PQQ and chelation ability of citric acid [42]. Subsequently, EcN::*vgb-gfp* strain containing *pqq* gene cluster from *A. calcoaceticus* and gluconate dehydrogenase (*gad*) operon from *Pseudomonas putida* KT2440 secreted PQQ, gluconic and 2-ketogluconic acids, and this strain prevented toxicity caused by Cd, Hg and Pb without affecting the essential metal ions [43]. Thus, 2-ketogluconic acid produced by EcN recombinant strain is mimicking the chelating abilities of citric acid. Similarly, EcN strain containing As(III) S-adenosylmethionine (SAM) methlyltransferase (*arsM*) and *pqq* gene cluster prevented arsenite toxicity by scavenging arsenite-induced reactive oxygen species by secreted PQQ and converting arsenite into non-toxic trimethylarsenite in rats [44].

EcN recombinant strain containing *pqq* operon secretes 15 mM gluconic acid [43]. Gluconic acid was proposed for cancer therapy as cancer cells utilize citrate for growth and gluconic acid irreversibly inhibits citrate transporter, which is expressed on cancer cells [60]. Hence, EcN producing gluconic acid could prevent the progression of tumors, especially colorectal cancers. *Staphylococcus aureus* α-hemolysin expressing EcN recombinant strain forms pores in the tumor cells resulting in the regression of tumors in mice [51]. Similarly, tumor regression also occurred in mice xenografted with human colorectal cancer cells treated with EcN strain expressing hemolysin E (HlyE) a pore-forming protein [52].

SCFA such as acetate, propionate, and butyrate produced by gut microbiome is necessary for the survival of colonocytes, maintenance of intestinal integrity, mucus production, serotonin release by enterochromaffin cells, and secretion of gut hormone peptide YY in the intestine [61, 62]. Additionally, SCFA also regulates brain and liver functions while diminished SCFA signaling is associated with metabolic diseases [63]. Propionate and butyrate prevent the progression of these metabolic diseases [64]. In order to design EcN to secrete butyric acid, fumarate reductase (*frdA*), lactate dehydrogenase (*ldhA*), alcohol dehydrogenase (*adhE*), and phosphotransacetylase (*pta*) genes involved in the fermentation product formation of succinic, acetic, and lactic acids were deleted to generate EcN YF005 strain [49]. The *atoDABE* operon encodes the genes for the formation of acetoacetyl CoA and butyryl CoA to the butyric acid formation, while *hbd* and *crt* from *Clostridium acetobutylicum* and *ter* from *Treponema denticola* genes convert acetoacetyl CoA into butyryl CoA. The native promoter of *atoDABE* operon was replaced with a strong, constitutive P_L promoter from phage λ, and synthetic P_L-*LacO-hpd-crt-ter* operon was integrated at methylglyoxal synthase (*msgA*) gene to generate EcN Y2023 strain. This strain produced 0.49 g/L butyric acid on glucose. It will be interesting to determine its therapeutic potential in animal studies.

EcN deletion mutant of *dapA* gene coding for 4-hydroxytetra-hydropicolinate synthase was generated for incorporating phenylalanine degradation for the treatment of phenylketonuria [48]. Two different SYNB1618 strains were generated by incorporating phenylalanine ammonia-lyase and L-amino acid deaminase (*pma*) genes, which convert phenylalanine into *trans*-cinnamic acid (TCA) and phenylpyruvate, respectively. In humans, TCA is further transformed into hippuric acid in the liver and excreted in the urine. The oral load of 70 mg phenylalanine was reduced by 58% in the serum samples of individuals fed with the modified strain.

EcN was genetically modified for inflammatory bowel disease by probiotic-associated therapeutic curli hybrids (PATCH) approach using a fusion protein of amyloid domain for self-assembly (CsgA) linked to trefoil factor-3 with 6 His residues [50]. Oxidatively damaged inflamed regions are conducive for the growth of facultative anaerobes. Consequently, modified EcN strain numbers increased at the damaged regions and secreted curly fibers that facilitated the repair of damaged regions. The EcN-engineered strain could ameliorate the weight loss in DSS-induced colitis in mice.

EcN expressing a fusion protein of cholera toxin B domain and insulin growth factor-1 (CTB-IGF1) was proposed as a long-term therapeutic strategy for diabetes [65]. It was hypothesized that EcN expresses the fusion protein in the intestine that would cross the intestinal epithelium into blood circulation facilitated by CTB-specific interaction with GM1 ganglioside oligosaccharide and IGF will activate the insulin effects.

7. *E. coli* as synbiotic

The beneficial effects of probiotic *E. coli* strains are contributed by their functions in the small intestine as well as in the colon. However, prebiotics are nutrients for the survival and maintenance of the colonic microbiome, which secrete host-beneficial SCFA as fermentation products [66]. Synbiotics are a mixture of prebiotics and probiotics, which provide synergistic effects of both components [67]. Dietary fructose and sucrose are implicated in the onset and progression of metabolic diseases [68, 69]. EcN::*vgb-gfp* was modified with two different synthetic operons containing *Ptac-pqq-glf-mtlK* and *Ptac-pqq-fdh* that convert dietary fructose into mannitol and 5-keto-D-fructose that are prebiotics in the small intestine [45]. These prebiotics then serve as nutrients for colonic bacteria to produce SCFA. PQQ secreted by the synbiotic EcN will scavenge reactive oxygen species produced by fructose. Both mannitol and 5-keto-D-fructose producing strains demonstrated synbiotic activities by preventing dietary fructose-mediated metabolic disorders in rats. Fructose is known to improve iron status by its reductive ability compared to other sugars [70]. However, metabolic complications of fructose hindered its applicability. Since EcN synbiotics overcome the negative effects of fructose, these strains were found to also improve iron status [71].

High dietary sucrose also contributes to the metabolic disorders [68]. Inulosucrase catalyzes the conversion of sucrose into inulin [72]. Probiotic Ec16 transformed with inulosucrase of *Lactobacillus johnsonii* NCC 533 resulted in its secretion in the supernatant, while the enzyme was localized in the periplasm of *E. coli* BL21 suggested that extracellular enzyme in Ec16 could get transported using colicin E1/1a1b transport system [22]. EcN genomic integrant with *vgb-gfp-pqqABCDE-inuJ* gene cluster prevented high sucrose-induced metabolic disorders in rats by increased PQQ and SCFA [46].

8. Conclusion

The potential of probiotic *E. coli* is increasing over the years starting from maintaining the intestinal barrier in healthy individuals to the treatment of complex diseases such as colorectal cancer and inflammatory bowel disease. Since many commercial *E. coli* products were found to deviate by orders of magnitude in terms of claimed numbers, monitoring strain identity and numbers is imperative for exploiting their complete potential [73]. Distinct properties of *E. coli* probiotics coupled with the ease of developing strains with desired traits could greatly expand their applications.

Author details

Nareshkumar Gattupalli[1]* and Archana Gattupalli[2]

1 Department of Biochemistry, Faculty of Science, The Maharaja Sayajirao University of Baroda, Vadodara, Gujarat, India

2 Department of Microbiology and Biotechnology Centre, Faculty of Science, The Maharaja Sayajirao University of Baroda, Vadodara, Gujarat, India

*Address all correspondence to: gnaresh_k@yahoo.co.in

References

[1] Delmas J, Dalmasso G, Bonnet R. *Escherichia coli*: The good, the bad and the ugly. Clinical Microbiology. 2015; **4**:195. DOI: 10.4172/2327-5073.1000195

[2] Wassenaar TM, Zschüttig A, Beimfohr C, Geske T, Auerbach C, Cook H, et al. Virulence genes in a probiotic *E. coli* product with a recorded long history of safe use. European Journal of Microbiology & Immunology. 2015;**5**:81-93. DOI: 10.1556/EUJMI-D-14-00039

[3] Martinson JNV, Walk ST. *Escherichia coli* residency in the gut of healthy human adults. EcoSal Plus. 2020;**9**:10.1128/ecosalplus.ESP-0003-2020. DOI: 10.1128/ecosalplus.ESP-0003-2020

[4] Schulze J, Schiemann M, Sonnenborn U. 120 Years of *E. coli*. Germany: Alfred-Nissle-Gesellschaft; 2006

[5] Santos ACM, Santos FF, Silva RM, Gomes TAT. Diversity of hybrid and hetero-pathogenic *Escherichia coli* and their potential implication in more severe diseases. Frontiers in Cellular and Infection Microbiology. 2020;**10**:339. DOI: 10.3389/fcimb.2020.00339

[6] Martinson JNV, Pinkham NV, Peters GW, et al. Rethinking gut microbiome residency and the *Enterobacteriaceae* in healthy human adults. The ISME Journal. 2019;**13**:2306-2318. DOI: 10.1038/s41396-019-0435-7

[7] Apperloo-Renkema HZ, Van der Waaij BD, Van der Waaij D. Determination of colonization resistance of the digestive tract by biotyping of *Enterobacteriaceae*. Epidemiology and Infection. 1990;**105**:355-361. DOI: 10.1017/s0950268800047944

[8] Zeng MY, Inohara N, Nuñez G. Mechanisms of inflammation-driven bacterial dysbiosis in the gut. Mucosal Immunology. 2017;**10**:18-26. DOI: 10.1038/mi.2016.75

[9] Kumar P, Ferzin S, Chintan S, Kumar GN. Isolation and characterization of potential probiotic *Escherichia coli* strains from rat faecal samples. American Journal of Infectious Diseases. 2009;**5**:112-117. DOI: 10.3844/ajidsp.2009.112.117

[10] Hill C, Guarner F, Reid G, Gibson GR, Merenstein DJ, Pot B, et al. Expert consensus document. The International scientific association for probiotics and prebiotics consensus statement on the scope and appropriate use of the term probiotic. Nature Reviews. Gastroenterology & Hepatology. 2014;**11**:506-514. DOI: 10.1038/nrgastro.2014.66

[11] Di Cerbo A, Palmieri B. The market of probiotics. Pakistan Journal of Pharmaceutical Sciences. 2015;**28**:2199-2206

[12] Rezac S, Kok CR, Heermann M, Hutkins R. Fermented foods as a dietary source of live organisms. Frontiers in Microbiology. 2018; **9**:1785. DOI: 10.3389/fmicb.2018.01785

[13] Guarner F et al. World Gastroenterology Organisation Global Guidelines, Probiotics and Prebiotics. World Gastroenterology Organisation; 2017. Available from: https://www.worldgastroenterology.org/guidelines/global-guidelines/probiotics-and-prebiotics

[14] Deriu E, Liu JZ, Pezeshki M, Edwards RA, Ochoa RJ, Contreras H, et al. Probiotic bacteria reduce *Salmonella typhimurium* intestinal colonization by competing for iron. Cell Host & Microbe. 2013;**14**:26-37. DOI: 10.1016/j.chom.2013.06.007

[15] Sonnenborn U. *Escherichia coli* strain Nissle 1917-from bench to bedside and back: History of a special *Escherichia coli* strain with probiotic properties. FEMS Microbiology Letters. 2016;**363**:fnw212. DOI: 10.1093/femsle/fnw212

[16] Wassenaar TM. Insights from 100 years of research with probiotic *E. coli*. European Journal of Microbiology & Immunology. 2016;**6**:147-161. DOI: 10.1556/1886.2016.00029

[17] Behnsen J, Deriu E, Sassone-Corsi M, Raffatellu M. Probiotics: Properties, examples, and specific applications. Cold Spring Harbor Perspectives in Medicine. 2013;**3**:a010074. DOI: 10.1101/cshperspect.a010074

[18] Kleta S, Steinrück H, Breves G, Duncker S, Laturnus C, Wieler LH, et al. Detection and distribution of probiotic *Escherichia coli* Nissle 1917 clones in swine herds in Germany. Journal of Applied Microbiology. 2006;**101**: 1357-1366. DOI: 10.1111/j.1365-2672.2006.03019.x

[19] Beimfohr C. A review of research conducted with probiotic *E. coli* marketed as Symbioflor. International Journal of Bacteriology. 2016;**2016**:3535621. DOI: 10.1155/2016/3535621

[20] Kocourková I, Žádníková R, Žižka J, Rosová V. Effect of oral application of a probiotic *E. coli* strain on the intestinal microflora of children of allergic mothers during the first year of life. Folia Microbiologia (Praha). 2007;**52**:189-193. DOI: 10.1007/BF02932158

[21] Micenková L, Bosák J, Smatana S, Novotný A, Budinská E, Šmajs D. Administration of the probiotic *Escherichia coli* Strain A0 34/86 resulted in a stable colonization of the human intestine during the first year of life. Probiotics and Antimicrobial Proteins. 2020;**12**:343-350. DOI: 10.1007/s12602-019-09548-3

[22] Kumar P, Gopalakrishna SG, Kumar GN. In vitro comparism of the extracellular secretion of inulosucrase enzyme in potential probiotic *Escherichia coli* 16 and BL. African Journal of Biotechnology. 2013;**12**:6382-6388. DOI: 10.5897/AJB2013.12044

[23] Grozdanov L, Zahringer U, Blum-Oehler G, Brade L, Henne A, Knirel YA, et al. A single nucleotide exchange in the *wzy* gene is responsible for the semirough O6 lipopolysaccharide phenotype and serum sensitivity of *Escherichia coli* strain Nissle 1917. Journal of Bacteriology. 2002;**184**:5912-5925. DOI: 10.1128/jb.184.21.5912-5925.2002

[24] Güttsche AK, Löseke S, Zähringer U, Sonnenborn U, Enders C, Gatermann S, et al. Anti-inflammatory modulation of immune response by probiotic *Escherichia coli* Nissle 1917 in human blood mononuclear cells. Innate Immunity. 2012;**18**:204-216. DOI: 10.1177/1753425910396251

[25] Ukena SN, Singh A, Dringenberg U, et al. Probiotic *Escherichia coli* Nissle 1917 inhibits leaky gut by enhancing mucosal integrity. PLoS One. 2007;**2**:e1308. DOI: 10.1371/journal.pone.0001308

[26] Alvarez CS, Badia J, Bosch M, Giménez R, Baldomà L. Outer membrane vesicles and soluble factors released by probiotic *Escherichia coli* Nissle 1917 and commensal ECOR63 enhance barrier function by regulating expression of tight junction proteins in intestinal epithelial cells. Frontiers in Microbiology. 2016;**7**:1981. DOI: 10.3389/fmicb.2016.01981

[27] Behrouzi A, Mazaheri H, Falsafi S, Tavassol ZH, Moshiri A, Siadat SD. Intestinal effect of the probiotic *Escherichia coli* strain Nissle 1917 and its

OMV. Journal of Diabetes and Metabolic Disorders. 2020;**19**:597-604. DOI: 10.1007/s40200-020-00511-6

[28] Guo S, Chen S, Ma J, Ma Y, Zhu J, Ma Y, et al. *Escherichia coli* Nissle 1917 protects intestinal barrier function by inhibiting NF-κB-mediated activation of the MLCK-P-MLC signaling pathway. Mediators of Inflammation. 2019;**2019**:5796491. DOI: 10.1155/2019/5796491

[29] Nzakizwanayo J, Dedi C, Standen G, et al. *Escherichia coli* Nissle 1917 enhances bioavailability of serotonin in gut tissues through modulation of synthesis and clearance. Scientific Reports. 2015;**5**:17324. DOI: 10.1038/srep17324

[30] Caruana JC, Walper SA. Bacterial membrane vesicles as mediators of microbe–microbe and microbe–host community interactions. Frontiers in Microbiology. 2020;**11**:432. DOI: 10.3389/fmicb.2020.00432

[31] Fábrega MJ, Rodríguez-Nogales A, Garrido-Mesa J, Algieri F, Badía J, Giménez R, et al. Intestinal anti-inflammatory effects of outer membrane vesicles from *Escherichia coli* Nissle 1917 in DSS-Experimental Colitis in mice. Frontiers in Microbiology. 2017;**8**:1274. DOI: 10.3389/fmicb.2017.01274

[32] Cañas MA, Fábrega MJ, Giménez R, Badia J, Baldomà L. Outer membrane vesicles from probiotic and commensal *Escherichia coli* activate NOD1-mediated immune responses in intestinal epithelial cells. Frontiers in Microbiology. 2018;**9**:498. DOI: 10.3389/fmicb.2018.00498

[33] Hu R, Lin H, Li J, Zhao Y, Wang M, Sun X, et al. Probiotic *Escherichia coli* Nissle 1917-derived outer membrane vesicles enhance immunomodulation and antimicrobial activity in RAW264.7 macrophages. BMC Microbiology. 2020;**20**:268. DOI: 10.1186/s12866-020-01953-x

[34] Enck P, Zimmermann K, Menke G, Klosterhalfen S. Randomized controlled treatment trial of irritable bowel syndrome with a probiotic *E. coli* preparation (DSM17252) compared to placebo. Zeitschrift für Gastroenterologie. 2009;**47**:209-214. DOI: 10.1055/s-2008-1027702

[35] Escribano-Vazquez U, Beimfohr C, Bellet D, Thomas M, Zimmermann K, Langella P, et al. Symbioflor2® *Escherichia coli* genotypes enhance ileal and colonic gene expression associated with mucosal defense in gnotobiotic mice. Microorganisms. 2020;**8**:512. DOI: 10.3390/microorganisms8040512

[36] Schultz M, Watzl S, Oelschlaeger TA, Rath HC, Göttl C, Lehn N, et al. Green fluorescent protein for detection of the probiotic microorganism *Escherichia coli* strain Nissle 1917 (EcN) in vivo. Journal of Microbiological Methods. 2005;**61**:389-398. DOI: 10.1016/j.mimet.2005.01.007

[37] Kumar P, Ranawade AV, Kumar NG. Potential probiotic *Escherichia coli* 16 harboring the *Vitreoscilla* hemoglobin gene improves gastrointestinal tract colonization and ameliorates carbon tetrachloride induced hepatotoxicity in rats. BioMed Research International. 2014;**2014**:213574. DOI: 10.1155/2014/213574

[38] Pandey S, Singh A, Kumar P, Chaudhari A, Nareshkumar G. Probiotic *Escherichia coli* CFR 16 producing pyrroloquinoline quinone (PQQ) ameliorates 1,2-dimethylhydrazine-induced oxidative damage in colon and liver of rats. Applied Biochemistry and Biotechnology. 2014;**173**:775-786. DOI: 10.1007/s12010-014-0897-z

[39] Pandey S, Singh A, Chaudhari N, Nampoothiri LP, Kumar GN. Protection against 1,2-di-methylhydrazine-induced systemic oxidative stress and altered brain neurotransmitter status by probiotic *Escherichia coli* CFR 16 secreting pyrroloquinoline quinone. Current Microbiology. 2015;**70**:690-697. DOI: 10.1007/s00284-014-0763-9

[40] Singh AK, Pandey SK, Naresh KG. Pyrroloquinoline quinone-secreting probiotic *Escherichia coli* Nissle 1917 ameliorates ethanol-induced oxidative damage and hyperlipidemia in rats. Alcoholism, Clinical and Experimental Research. 2014;**38**:2127-2137. DOI: 10.1111/acer.12456

[41] Singh AK, Pandey SK, Saha G, Gattupalli NK. Pyrroloquinoline quinone (PQQ) producing *Escherichia coli* Nissle 1917 (EcN) alleviates age associated oxidative stress and hyperlipidemia, and improves mitochondrial function in ageing rats. Experimental Gerontology. 2015;**66**:1-9. DOI: 10.1016/j.exger.2015.04.001

[42] Raghuvanshi R, Chaudhari A, Kumar GN. Amelioration of cadmium- and mercury-induced liver and kidney damage in rats by genetically engineered probiotic *Escherichia coli* Nissle 1917 producing pyrroloquinoline quinone with oral supplementation of citric acid. Nutrition. 2016;**32**:1285-1294. DOI: 10.1016/j.nut.2016.03.009

[43] Raghuvanshi R, Archana C, Nareshkumar G. 2-Ketogluconic acid and pyrroloquinoline quinone secreting probiotic *Escherichia coli* Nissle 1917 as a dietary strategy against heavy metal induced damage in rats. Journal of Functional Foods. 2017;**37**:541-552. DOI: 10.1016/j.jff.2017.08.013

[44] Raghuvanshi R. Amelioration of heavy metal induced toxicity using probiotic *Escherichia* coli strain in rats. [thesis]. Vadodara, Gujarat, India: The Maharaja Sayajirao University of Baroda; 2018

[45] Somabhai CA, Raghuvanshi R, Nareshkumar G. Genetically engineered *Escherichia coli* Nissle 1917 synbiotics reduce metabolic effects induced by chronic consumption of dietary fructose. PLoS ONE. 2016, 2016;**11**:e0164860. DOI: 10.1371/journal.pone.0164860

[46] Pandey SK. Development of *Escherichia coli* probiotic strains for alleviation of metabolic syndrome in rats induced by dietary fructose and sucrose [thesis]. Vadodara, Gujarat, India: The Maharaja Sayajirao University of Baroda; 2014

[47] Zainuddin HS, Bai Y, Mansell TJ. CRISPR-based curing and analysis of metabolic burden of cryptic plasmids in *Escherichia coli* Nissle 1917. Engineering in Life Sciences. 2019;**19**:478-485. DOI: 10.1002/elsc.201900003

[48] Isabella VM, Ha BN, Castillo MJ, Lubkowicz DJ, Rowe SE, Millet YA, et al. Development of a synthetic live bacterial therapeutic for the human metabolic disease phenylketonuria. Nature Biotechnology. 2018;**36**:857-864. DOI: 10.1038/nbt.4222

[49] Bai Y, Mansell TJ. Production and sensing of butyrate in a probiotic *Escherichia coli* strain. International Journal of Molecular Sciences. 2020;**21**:3615. DOI: 10.3390/ijms21103615

[50] Praveschotinunt P, Duraj-Thatte AM, Gelfat I, Bahl F, Chou DB, Joshi NS. Engineered *E. coli* Nissle 1917 for the delivery of matrix-tethered therapeutic domains to the gut. Nature Communications. 2019;**10**:5580. DOI: 10.1038/s41467-019-13336-6

[51] St Jean AT, Swofford CA, Panteli JT, Brentzel ZJ, Forbes NS. Bacterial delivery of *Staphylococcus aureus* α-hemolysin causes regression and

necrosis in murine tumors. Molecular Therapy. 2014;**22**:1266-1274. DOI: 10.1038/mt.2014.36

[52] Chiang CJ, Huang PH. Metabolic engineering of probiotic *Escherichia coli* for cytolytic therapy of tumors. Scientific Reports. 2021;**11**:5853. DOI: 10.1038/s41598-021-85372-6

[53] Kan A, Gelfat I, Emani S, Praveschotinunt P, Joshi NS. Plasmid vectors for in vivo selection-free use with the probiotic *E. coli* Nissle 1917. ACS Synthetic Biology. 2021;**10**:94-106. DOI: 10.1021/acssynbio.0c00466

[54] Yu F, Zhao X, Wang Z, Liu L, Yi L, Zhou J, et al. Recent advances in the physicochemical properties and biotechnological application of *Vitreoscilla* hemoglobin. Microorganisms. 2021;**9**:1455. DOI: 10.3390/microorganisms9071455

[55] Jonscher KR, Rucker RB. Pyrroloquinoline quinone: Its profile, effects on the liver and implications for health and disease prevention. In: Watson RW, Preedy VR, editors. Dietary Interventions in Liver Disease. Massachusetts, USA: Academic Press; 2019. pp. 157-173. DOI: 10.1016/B978-0-12-814466-4.00013-6

[56] Liguori I, Russo G, Curcio F, Bulli G, Aran L, Della-Morte D, et al. Oxidative stress, aging, and diseases. Clinical Interventions in Aging. 2018;**13**:757-772. DOI: 10.2147/CIA.S158513

[57] Tan BL, Norhaizan ME, Liew WPP, Rahman SH. Antioxidant and oxidative stress: A mutual interplay in age-related diseases. Frontiers in Pharmacology. 2018;**9**:1162. DOI: 10.3389/fphar.2018.01162

[58] Luo J, Kevin Mills K, le Cessiea S, Noordam R, van Heemst D. Ageing, age-related diseases and oxidative stress: What to do next? Ageing Research Reviews. 2020;**57**:100982. DOI: 10.1016/j.arr.2019.100982

[59] Balali-Mood M, Naseri K, Tahergorabi Z, Khazdair MR, Sadeghi M. Toxic mechanisms of five heavy metals: Mercury, lead, chromium, cadmium, and arsenic. Frontiers in Pharmacology. 2021;**12**:643972. DOI: 10.3389/fphar.2021.643972

[60] Mycielska ME, Mohr MTJ, Schmidt K, Drexler K, Rümmele P, Haferkamp S, et al. Potential use of gluconate in cancer therapy. Frontiers in Oncology. 2019;**9**:522. DOI: 10.3389/fonc.2019.00522

[61] Dalile B, Van Oudenhove L, Vervliet B, Verbeke K. The role of short-chain fatty acids in microbiota-gut-brain communication. Nature Reviews. Gastroenterology & Hepatology. 2019;**16**:461-478. DOI: 10.1038/s41575-019-0157-3

[62] Silva YP, Bernardi A, Frozza RL. The role of short-chain fatty acids from gut microbiota in gut-brain communication. Frontiers in Endocrinology. 2020;**11**:25. DOI: 10.3389/fendo.2020.00025

[63] Sanna S, van Zuydam NR, Mahajan A, et al. Causal relationships among the gut microbiome, short-chain fatty acids and metabolic diseases. Nature Genetics. 2019;**51**:600-605. DOI: 10.1038/s41588-019-0350-x

[64] Coppola S, Avagliano C, Calignano A, Berni Canani R. The protective role of butyrate against obesity and obesity-related diseases. Molecules. 2021;**26**:682. DOI: 10.3390/molecules26030682

[65] Bazi Z, Jalili M, Hekmatdoost A. The long term oral regulation of blood glucose in diabetic patients by using of *Escherichia coli* Nissle 1917 expressing CTB-IGF-1 hybrid protein. Medical Hypotheses. 2013;**81**:961-962. DOI: 10.1016/j.mehy.2013.08.035

[66] Peredo-Lovillo A, Romero-Luna HE, Jiménez-Fernández M. Health promoting microbial metabolites produced by gut microbiota after prebiotics metabolism. Food Research International. 2020;**136**:109473. DOI: 10.1016/j.foodres.2020.109473

[67] Malik JK, Raina R, Ahmad AH, Kalpana S, Prakash A, Gupta RC. Synbiotics: Safety and toxicity considerations. In: Gupta RC, Lall R, Srivastava A, editors. Nutraceuticals Efficacy, Safety and Toxicity. 2nd ed. Massachusetts, USA: Academic Press; 2021. pp. 1107-1123. DOI: 10.1016/B978-0-12-821038-3.00066-5

[68] Stanhope KL. Sugar consumption, metabolic disease and obesity: The state of the controversy. Critical Reviews in Clinical Laboratory Sciences. 2016;**53**:52-67. DOI: 10.3109/10408363.2015.1084990

[69] Taskinen MR, Packard CJ, Borén J. Dietary fructose and the metabolic syndrome. Nutrients. 2019;**11**:1987. DOI: 10.3390/nu11091987

[70] Tsuchiya H, Ebata Y, Sakabe T, Hama S, Kogure K, Shiota G. High-fat, high-fructose diet induces hepatic iron overload via a hepcidin-independent mechanism prior to the onset of liver steatosis and insulin resistance in mice. Metabolism. 2013;**62**:62-69. DOI: 10.1016/j.metabol.2012.06.008

[71] Chaudhari AS, Raghuvanshi R, Kumar GN. Genetically engineered *Escherichia coli* Nissle 1917 synbiotic counters fructose-induced metabolic syndrome and iron deficiency. Applied Microbiology and Biotechnology. 2017;**101**:4713-4723. DOI: 10.1007/s00253-017-8207-7

[72] Anwar MA, Kralj S, van der Maarel MJEC, Dijkhuizen L. The probiotic *Lactobacillus johnsonii* NCC 533 produces high-molecular-mass inulin from sucrose by using an inulosucrase enzyme. Applied and Environmental Microbiology. 2008;**74**:3426-3433. DOI: 10.1128/AEM.00377-08

[73] Zimmer C, Dorea C. Enumeration of *Escherichia coli* in probiotic products. Microorganisms. 2019;**7**:437. DOI: 10.3390/microorganisms7100437

Chapter 15

DNA-FACE™ - An *Escherichia coli*-based DNA Amplification-Expression Technology for Automatic Assembly of Concatemeric ORFs and Proteins

Piotr M. Skowron and Agnieszka Zylicz-Stachula

Abstract

DNA-FACE™ (DNA Fragment Amplification & Concatemeric Expressed Nucleic Acids and Proteins) is a universal biotechnological platform, developed as *Escherichia coli* (*E. coli*) system. It is based on the ordered, head-to-tail directional ligation of the amplified DNA fragments. The technology enables the construction of targeted biomolecules - genetically programmed, concatemeric DNA, RNA, and proteins, designed to fit a particular task. The constructed, "artificial" (never seen in Nature) tandem repeat macromolecules, with specialized functions, may contain up to 500 copies of monomeric units. The technology greatly exceeds the current capabilities of chemical gene synthesis. The vector-enzymatic DNA fragment amplification assembles the DNA segments, forming continuous Open Reading Frames (ORFs). The obtained ORFs are ready for high-level expression in *E. coli* without a need for subcloning. The presented method has potential applications in pharmaceutical industry and tissue engineering, including vaccines, biological drugs, drug delivery systems, mass-production of peptide-derived biomaterials, industrial and environmental processes. The technology has been patented worldwide and used successfully in the construction of anti-HBV vaccines, pro-regenerative biological drugs and, recently, the anti-SARS-CoV-2 vaccine. The anti-SARS-CoV-2 vaccine, developed using the DNA-FACE™ technology, is non-toxic and induces strong immunological response to recombinant human spike and nucleocapsid proteins, as shown in animal studies.

Keywords: DNA-FACE™; DNA amplification, concatemer, gene expression, amplification vector, expression vector, protein overproduction, biotechnology, HBV, SARS-CoV-2, vaccine, biological drug, tissue regeneration

1. Introduction

The recombinant DNA technology, born in early 1970 ties, has nowadays become mature and highly sophisticated. It plays an indispensable role in medicine, industry, and scientific research. Its rapid proliferation into a wide variety of techniques and molecular solutions for genetic engineering is largely owed to its historically first and still dominating recombinant DNA "vehicle" – *E. coli*. These bacteria were used for the construction of the most robust genetic expression systems, such as those based on: T7-*lac* hybrid promoter, bacteriophage lambda *pL* and *pR* promoters or arabinose operon promoter. Other advantages of the systems include fast bacterial growth, cost-effectiveness and deep understanding of the molecular basis of recombinant proteins biosynthesis, as *E. coli* is the most thoroughly studied microorganism. Bioactive molecules, including DNA, RNA, and proteins, designed *de novo*, are being utilized in increasingly diverse scientific, biomedical, industrial and environmental applications. The rapidly developing synthetic biology generates growing demands for synthetic genes of various types, from native to those with a highly modified sequence. The capability of generating DNA molecules of any sequence or size is especially important in biomedical research.

DNA sequences of the designed, synthetic genes may require long repetitive DNA fragments. However, the chemical synthesis of such fragments still poses a great challenge. Thus, there is a current need for the development of new technologies, enabling modular construction of complex synthetic genes.

Vast majority of synthetic recombinant DNA constructs are intended for recombinant protein production. Besides the need for recombinant proteins, the protein fragments, polypeptides, functional domains, protein-derived or natural peptides are being increasingly used, not only for functional studies but also in a wide variety of applied sciences, like material engineering and medicine. The intensive research on peptide-based biomaterials as biologically active tools has resulted in the development of a wide range of peptides and polypeptides with extended functionalities. Both natural and engineered versions of the bioactive peptides have found applications in the construction of biosensors, drug-delivery systems, and medicine [1–4].

To allow detailed investigation and scaled-up production of new macromolecular biomaterials (built from the repetitive DNA, RNA, or protein segments), a reliable method is needed to obtain a high yield of complex recombinant genes and their expression products. Some of the methods enabling construction of such sophisticated recombinant DNA molecules are based on the targeted plasmid vectors, capable of carrying of the multiple, joined (concatemeric) genes. Such concatemerization of the designed DNA, RNA and peptides may improve their stability and boost the biosynthesis level. What is even more interesting, concatemerization can also bring novel and sometimes unexpected features, such as an increase of bioactivity, a formation of bionanoparticles or more precise targeting by means of a gradual release of the bioactive molecule monomers or oligomers at the intended destination.

In order to solve the problems encountered in the techniques existing so far, a new genetic engineering method was developed. The method is dedicated to the formation of DNA concatemers necessary to produce "artificial", repetitive genes, encoding concatemeric RNAs and proteins (of any nucleotide (nt)/amino acid sequence), in a format suitable for a high-level genetic expression [5–7]. The technology is world-wide protected by patents: Polish no. 228341, (2018; first filed in 2015 [8]), United States no US 10,874,735 B2 (2020 [9]), European no EP 3 134 426 B1 (2020 [10]), Japanese no P6692796 (2020 [11]); Israeli – temporary no 248011 (2021 [12]) and patent applications: Chinese no CN 106488983 A (2016 [13]) and Indian no 201647039411 (2020 [14]).

2. DNA-FACE™ - a DNA fragment amplification-expression technology

2.1 The concept of the DNA-FACE™ method for directional DNA fragment amplification and protein concatemers construction

The rapid development of synthetic biology has generated a high demand for synthetic "artificial" genes that do not exist in Nature. Whatever their application, the construction of such synthetic genes may require the use of repetitive DNA fragments. However, one of the major limitations of the chemical synthesis of DNA is the difficulty in assembling repeated segments into longer DNA sequences. The ability to construct DNA molecules of any sequence or size is crucial for numerous applications, especially in the areas of biomedical and biotechnological research.

The biosynthesis-based strategies, that can ensure control over joining repeated DNA segments (multimerization or concatemerization), which would enable head-to-tail arrangement of the monomers of DNA, RNA, and peptides, require the development of special DNA manipulation methods. Otherwise, the obtained arrangements of monomeric units would be random and result in a mixture of head-to-head, tail-to-tail orientations of DNA fragments within the assembled multimeric DNA. Such randomized monomer arrangement could render any DNA construct useless for any rational applications, as it would disable a constructed DNA molecule from performing its desired function of encoding the genetic information about specific RNA and protein. For example, even a single undesired tail-to-tail segment within a constructed DNA multimer causes a nonsense amino acid sequence translation within this segment or even further downstream, an appearance of stop codons with inverted DNA segment, or a prematurely terminated translation of the constructed gene. Thus, controlling the mode of multimerization is pivotal in downstream processes after DNA multimerization. Furthermore, the controlled head-to tail-arrangement of the multimerized DNA segments provides stability of the recombinant DNA plasmid-vector and allows for a constructed operon expression.

Several alternative strategies for the construction of concatemeric genes have been developed so far [15–29]. However, most of the established technologies suffered from several problems, such as (*i*) limitations on the sequence of the DNA segment serving as the monomer; (ii) technical difficulties in joining of the DNA segments; (*iii*) an excessively complicated reaction, leading to the necessity of tedious DNA manipulations (*iv*) an inadequate copy number of the monomers within the formed multimer; (*v*) inability to repeat another round of the DNA fragment multimerization, if a desired monomer copy number within the resulting concatemeric DNA was not obtained. There were also two critically important problems; (*vi*) completion of a DNA concatemer formation without the ability to express coded RNA and proteins; (vii) codon discontinuity in the newly created ORF, which would prevent its expression and the production of the final result (a polypeptide/protein, containing multiple, linked together, bioactive peptides with programmed functions, without off-frame segments) [15–26].

A simple and efficient method was developed by us to make concatemeric "artificial" proteins or to emulate the old novel – "Frankenstein" proteins, composed of multiple functional parts, dedicated to suit a particular task. The DNA-FACE™ technology enables both homoconcatemers and heteroconcatemers formation, which highly enhances the pre-programmed functionality of the resulting "artificial" proteins (**Figure 1**). The technology allows for insertion of the synergistically acting bioactive

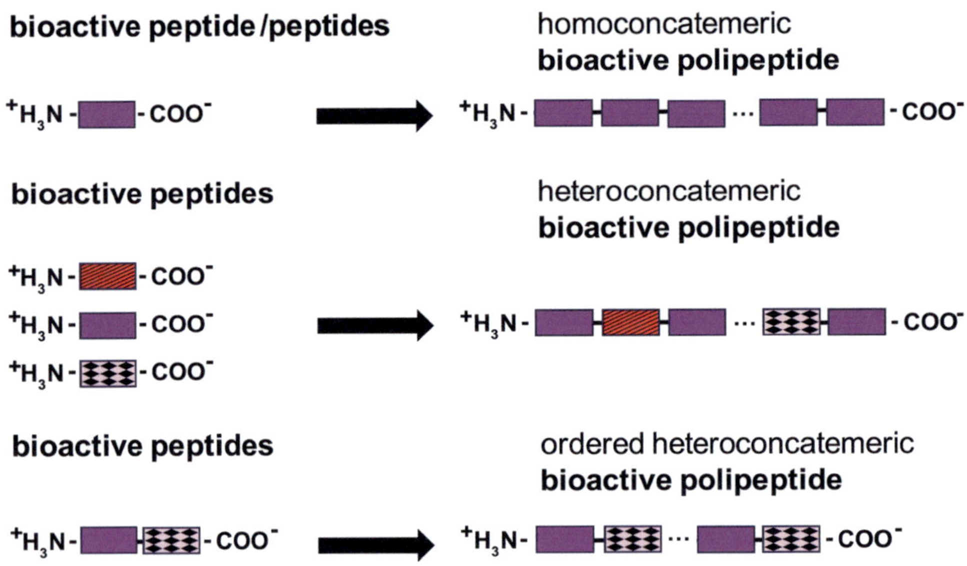

Figure 1.
Potential protein products are obtainable with the DNA-FACE™ technology.

peptides into the nascent concatemeric "Frankenstein" protein. The examples of such bioactive peptides are: different epitopes or antigen domains incorporated during a vaccine construction or combinations of various pro-regenerative peptides/protein segments. The technology is based on the custom vector-enzymatic system, which employs: (*i*) atypical Type IIS restriction endonucleases (REases). These Type IIS REases possess unique features: the ability to recognize 4-7 base pairs (bp) DNA sequence and to cleave at a fixed distance outside this sequence. Out of the known Type IIS REases, the DNA-FACE™ uses SapI (or its isoschizomers), which generate 3-nt protruding DNA ends; (*ii*) DNA ligase and (*iii*) dedicated amplification-expression vectors. DNA-FACE™ offers a significant improvement from earlier strategies [17]. It highly improves the construction of DNA concatemers, additionally allowing for the formation of continuous, multimeric ORFs as well as concatemeric proteins, with the desired monomer copy number and polymer/co-polymer length. **Figure 2** shows schematically the DNA fragment amplification reaction and its potential for employing multiple amplification cycles.

2.2 Molecular components and mechanism of the directional DNA fragment amplification reaction

2.2.1 DNA-interacting enzymes used and the mechanism of amplification reaction

All the DNA-FACE™ amplification-expression DNA vectors share variants of a universal DNA amplification module. The module may be custom modified and transferred into other DNA vectors, either prokaryotic or eukaryotic, containing alternative antibiotic resistance genes, origins of replication, transcriptional promoters, and translation initiation signals, among others. The amplification module contains two convergently oriented recognition sequences of the Type IIS REase - SapI, able to recognize asymmetric 7-bp 5′-GCTCTTC-3′ and cleave the upper DNA strand to the 3′ direction, at a distance of 1-nt and the bottom strand at a distance of 4-nt, thus leaving 3-nt 5′ cohesive ends. The SapI (and its isoschizomers) are unique among the discovered, atypical Type IIS REases. The SapI DNA recognition sequence is long and the protruding ends of the SapI cleaved substrate form a codon

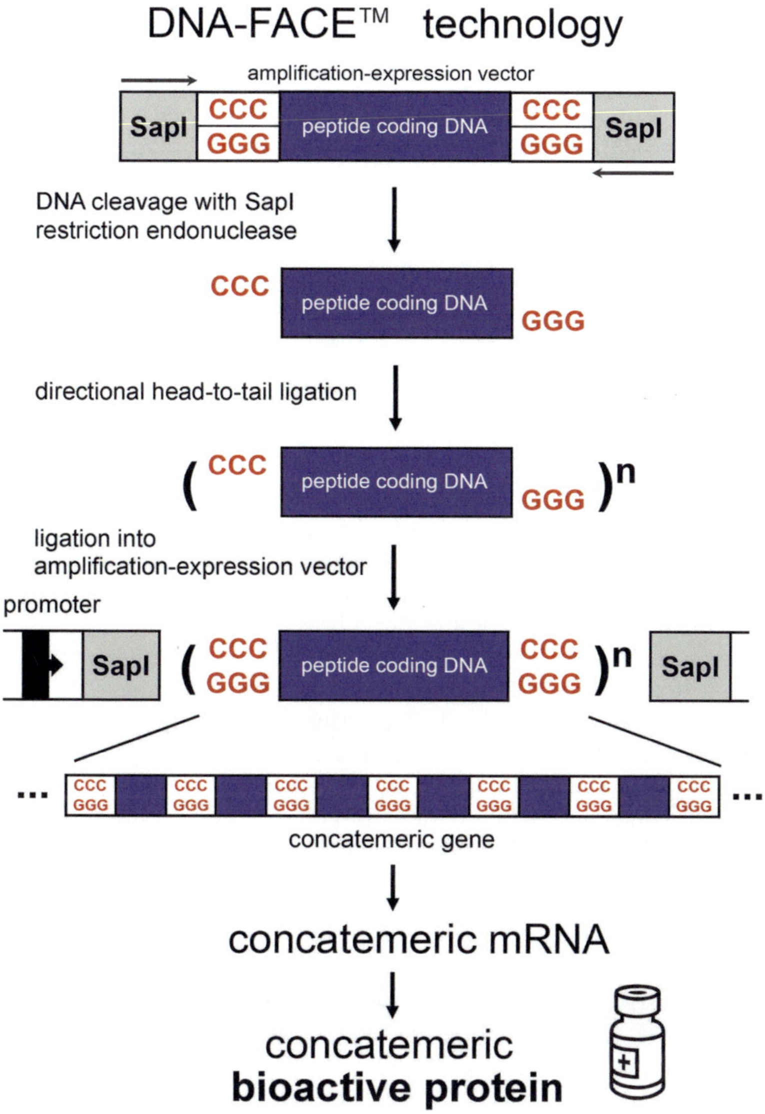

Figure 2.
Principles of the DNA-FACE™ technology.

length upon ligation. The long DNA recognition sequence highly decreases the probability of its accidental and undesired appearance with both the amplification-expression vector used and the DNA fragment to be amplified. Furthermore, the key feature of the SapI is a very rare occurrence of the 3-nt long protruding ends, that enable codon length formation in-between linked the amplified coding DNA segments, thus ensuring continuity of the ascending ORF. The SapI sites are separated by the orthodox Type II SmaI REase recognition sequence (5′-CCC|GGG-3′). SmaI cleaves its recognition sequence, leaving blunt ends. This is a convenient setup for cloning of any synthetic DNA fragments, as typically they are synthesized/delivered as double-stranded (ds) forms. The amplification-expression module provides three cloning options for a DNA fragment to be amplified (*i*) cohesive end cloning of the SapI generated 5′-CCC/5′-GGG sticky ends, (*ii*) blunt-end cloning of the SapI cohesive ends, previously filled in by T4 DNA polymerase/deoxyribonucleotides triphosphates; (*iii*) blunt-end cloning into SmaI site.

Whatever cloning option is used, the general protocol needs to be followed for all the amplification-expression vectors: (1) selection of bioactive peptides from a natural source or design of the monomeric DNA fragment to be amplified; (2) generation of the monomer by chemical synthesis of DNA, PCR amplification or REases excision from natural DNA; (3) cloning of the DNA fragment to the selected amplification-expression vector.

The cloning process is preferentially conducted using the cohesive end approach (mentioned above). The asymmetric 5′ -CCC/5′-GGG cohesive ends can be generated (a) through their addition at 5′- and 3′- termini of the dsDNA monomer during chemical synthesis or (b) *in vitro*, from SapI recognition sequences, added during chemical synthesis of the monomer and further clipped-off with SapI REase or (c) by PCR amplification with primers, containing SapI sites at their 5′- overhangs and clipping-off with SapI or (d) by excision of a SmaI-cloned monomer from an amplification-expression vector.

Subsequent stages of the DNA-FACE™ procedure include: (*i*) purification of the ds DNA fragment equipped with SapI-compatible 5′ -CCC/5′-GGG cohesive ends, ordered self-ligation of DNA monomers in directional, head-to-tail orientation, driven by asymmetric cohesive ends, (*ii*) ligation of generated concatemers mixture or of a selected gel-purified concatemer into the SapI-cleaved amplification vector, (*iii*) transformation into a suitable *E. coli* host strain, tolerant to atypical DNA sequences, such as DH5alfa, Top10, JM109, Endura™; (*iv*) selection of *E. coli* clones containing a concatemeric ORF segment with the desired number of monomers; (*v*) expression of the concatemeric ORF directly from an amplification-expression vector, containing strong transcription promoter, resulting in concatemeric protein biosynthesis or (*vi*) excision of the concatemer with SapI from the vector, which results in a DNA concatemeric segment equipped with SapI-cohesive ends and repeating steps (*i–iv*), until a desired number of monomeric DNA segments within a concatemer is obtained (**Figure 2**).

2.2.2 Amplification-expression vectors

Four categories of DNA amplification-expression vectors were designed for the purpose of the DNA-FACE™ technology: (I) pAMP series of six vectors for concatemeric protein biosynthesis in *E. coli* cytoplasm, (II) DNA amplification-expression pET21AMP-HisA vector for IPTG-controlled concatemeric protein biosynthesis in *E. coli* cytoplasm, (III) pET28AMP_SapI-Ubq vector for cytoplasmic biosynthesis of concatemeric proteins fused with ubiquitin at N-terminus and (IV) pET28AMP_PhoA or pET28AMP_MalE vectors for secretion of concatemeric proteins, produced under IPTG control, to the *E. coli* periplasm.

The pAMP series ((I); **Figure 3**) was constructed on the basis of the vector pACYC184 and its derivative pRZ4737 (W. S. Reznikoff) [5, 6, 30]. For the pAMP DNA vectors, six versions of the amplification modules were used, which differed by the presence/absence of His6_tag for metal affinity chromatography in three different reading frames (**Figure 3**). The pAMP DNA vectors are compatible with the *colE1* origin vectors, such as pET-series, and can be maintained in the same *E. coli* cell, if needed [5–14].

The pET21AMP-HisA vector ((II); **Figure 4**) is based on the pET-21d(+) expression vector (Novagen, EMD Millipore Corporation).

The pET28AMP_SapI-Ubq vector ((III); **Figure 5**) was designed as a modification of the pET-28d(+) expression vector (Novagen, EMD Millipore Corporation) and enables the fusion of a concatemeric protein with ubiquitin [5–7]. As concatemeric proteins contain repeated peptide modules, depending on their sequence, they may not form typical, natural protein structures, with hydrophobic amino acids residues forming a “core” surrounded by exposed polar and charged amino acids residues. This may affect their solubility, thus for some applications a fusion

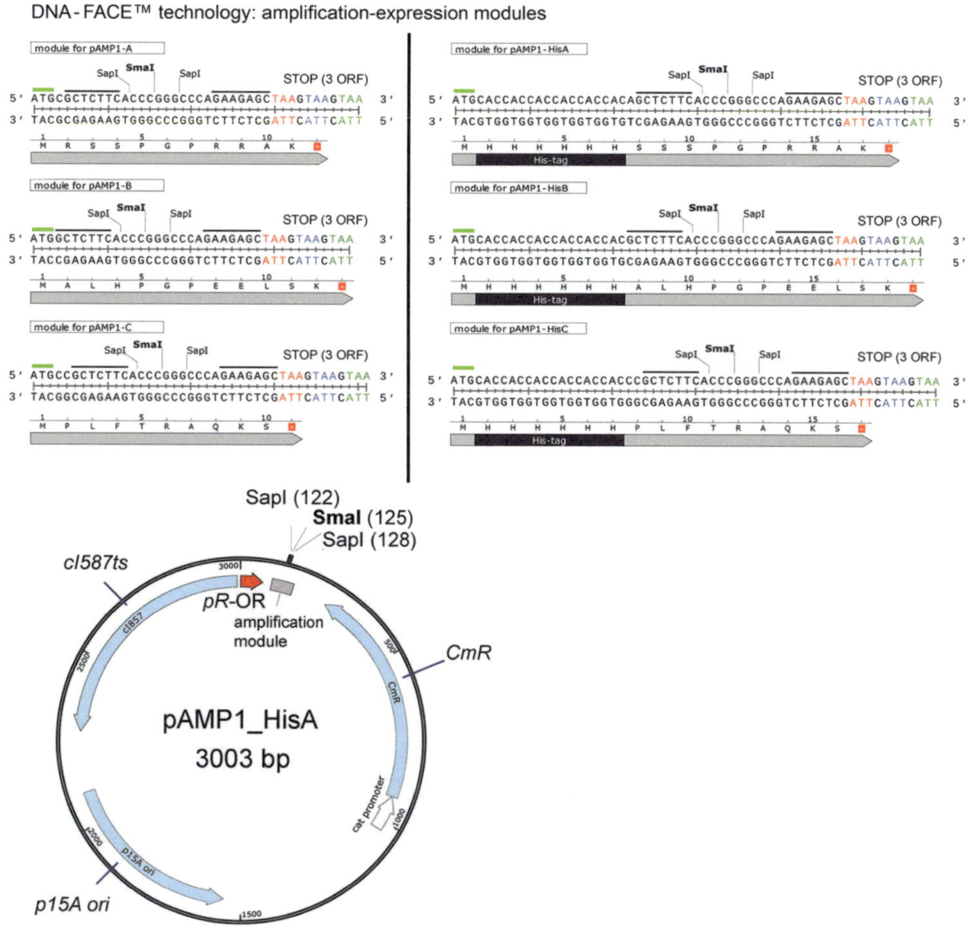

Figure 3.
The first set of amplification-expression modules designed for the DNA-FACE™ technology, used for the construction of the series of six pAMP DNA vectors (GenBank: MK606505, MK606506, MK606507, MK606519, MK606520, MK651654). All pAMP vectors are composed of: (a) p15A origin, (b) strong, temperature-regulated bacteriophage lambda pR transcription promoter, (c) bacteriophage lambda cI857ts repressor gene for control of pR promoter and host-independence, (d) an amplification module, containing two convergent SapI sites and a SmaI site and (e) chloramphenicol resistance gene.

with ubiquitin may be beneficial. The ubiquitin domain can be removed from a fusion protein by deubiquitinating proteases [7, 31].

The pET28AMP_PhoA or pET28AMP_MalE vectors ((IV); **Figure 6**) contain two alternative DNA segments, coding for *E. coli* secretion leaders MalE or PhoA. The MalE/PhoA encoding DNA segments are located at the 5′ end of the fused ORFs.

Detailed protocols, maps, sequences of all the amplification-expression vectors series have been published elsewhere [5–7].

2.3 Proof of the DNA-FACE™ concept: biosynthesis of a polyepitopic proteins, containing a model HBV antigen S epitope, pro-regenerative concatemeric proteins, and environment remediation/monitoring proteins

2.3.1 Testing the DNA-FACE™ with model HBV antigen S epitope and construction of functional concatemeric proteins

To evaluate the theoretical assumptions made during DNA-FACE™ biotechnology design, a 7-aa HBV epitope derived from S protein was selected [5]. The *E. coli*

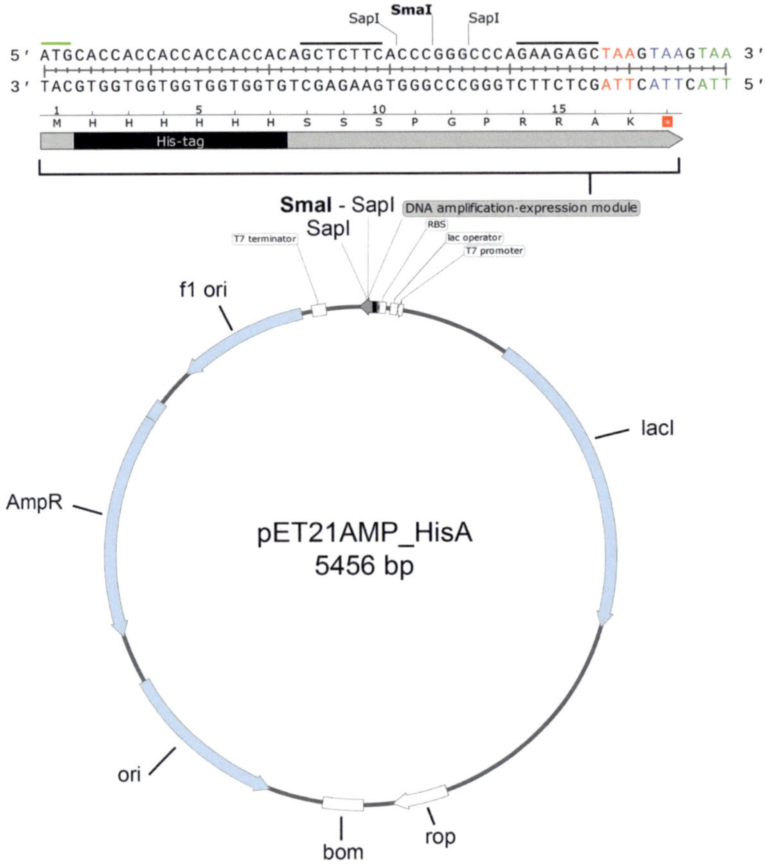

Figure 4.
The pET-derived, amplification-expression DNA vector, designed and constructed for the DNA-FACE™ technology. The pET21AMP-HisA DNA vector (GenBank MK606521) contains (a) colE1 origin, (b) f1/M13 origin, (c) very strong T7-lac transcription promoter, inducible by lactose or IPTG, (d) amplification module HisA and (e) ampicillin resistance gene.

expression optimized synthetic 21-bp DNA fragment, encoding the epitope TKPTDGN was cloned into the amplification-expression pAMP1-HisA vector. The detailed procedure of cloning of the synthetic 21-bp DNA fragment and its further amplification (based on the DNA-FACE™ technology) was described by Skowron et al. [5]. The results are shown in **Figure 7** (see the first round of amplification).

Further, the selected 5-mer was subjected to the second round of amplification (**Figure 7**; see the second round of amplification). The amplification-expression vectors were designed in such a way, that no SapI recognition sites were left within the amplified DNA segment. Such a vector design makes it possible to use a multimeric DNA fragment, obtained in the first round, as a "monomer".

Subsequently, an alternative or hybrid route of a DNA fragment amplification was tested. A possibility of a combination of chemical synthesis of the pre-formed HBV epitope-coding DNA, pushed to its technical limits with DNA-FACE™ method, was investigated. The limits of such chemical synthesis strongly depend on the DNA sequence and the size of the DNA fragment to be concatemerized. The model HBV epitope coding DNA turned out to be a rather "friendly" case, as compared to the other designed DNA sequences. Testing several commercial services, chemically synthesizing the designed DNA molecules, a maximum of 25-mers within a single synthetic gene was obtained. The 25-mer was then used as a "monomer" in DNA-FACE™ biotechnology amplification. As a result, bacterial clones

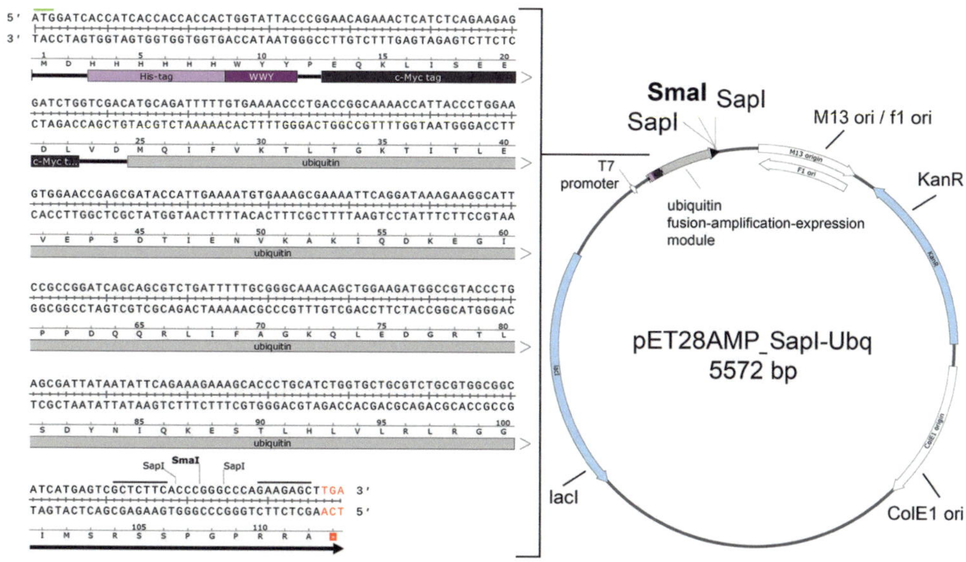

Figure 5.
The pET-derived pET28AMP_SapI-Ubq DNA vector, designed and constructed for the DNA-FACE™ technology. The pET28AMP_SapI-Ubq DNA vector (GenBank MK606527) is composed of: (a) colE1 origin, (b) f1/M13 origin, (c) T7-lac transcription promoter, inducible by lactose or IPTG, (d)) the DNA fragment amplification module His6_c-Myc_WYY_ubiquitin_SapI-Sma-SapI, enabling ubiquitin gene fusion and (e) kanamycin resistance gene.

containing up to 500-copies of the 21-bp HBV epitope were obtained (**Figure 7**; see the alternative round of amplification [5, 7–14].

During the next stage of the technology testing, the ability of the amplification-expression vector to yield an efficient translation of a highly atypical concatemeric gene, was investigated. The selected pAMP-HisA constructs, exemplified by 10-mer, 13-mer, 15-mer, 20-mer and 30-mer, were expressed (**Figure 8**). The expression of the recombinant constructs with up to 450 repeats, (composed of the 7-aa monomers), was tested.

It is known that the upper limit of molecular weight of a single polypeptide, biosynthesized by *E. coli*, is app. 150–200 kDa. However, due to a potential "slippage" of the translation machinery on mRNA repeat and a possible premature translation termination, a "smear" on SDS-PAGE gels was typically observed. The "smear" was located near the expected position of the recombinant protein, with the size corresponding to its molecular weight (**Figure 8**) [5]. Nevertheless, even a mixture of translation products is expected to be fully functional in planned applications, as each monomeric unit is semi-independent in genetically programmed functions, such as comprising an immunologically condensed "artificial" antigen, built from immunoactive epitopes only.

Afterwards, the DNA-FACE™ biotechnology was validated in the construction of prospective, pro-regenerative drugs and in the concatemeric proteins designed for remediation of the environment and new generation biosensors. Taken together, over 50 concatemeric ORFs and the resulting concatemeric proteins were constructed. Among those, a series of prospective pro-regenerative drugs was developed [5, 6]. For this purpose, the amplification (concatemerization) of four types of the designed DNA fragments was performed. The selected DNA fragments encoded either the laboratory-developed/predicted peptides or the peptides originally derived from wound healing stimulatory proteins.

The first selected peptide -TSRGDHELLGGGAAPVGG, which originated from the angiopoietin-related growth factor (AGF), was used for the construction of a poly-signal protein [5]. The peptide was linked to the elastase recognition sequence

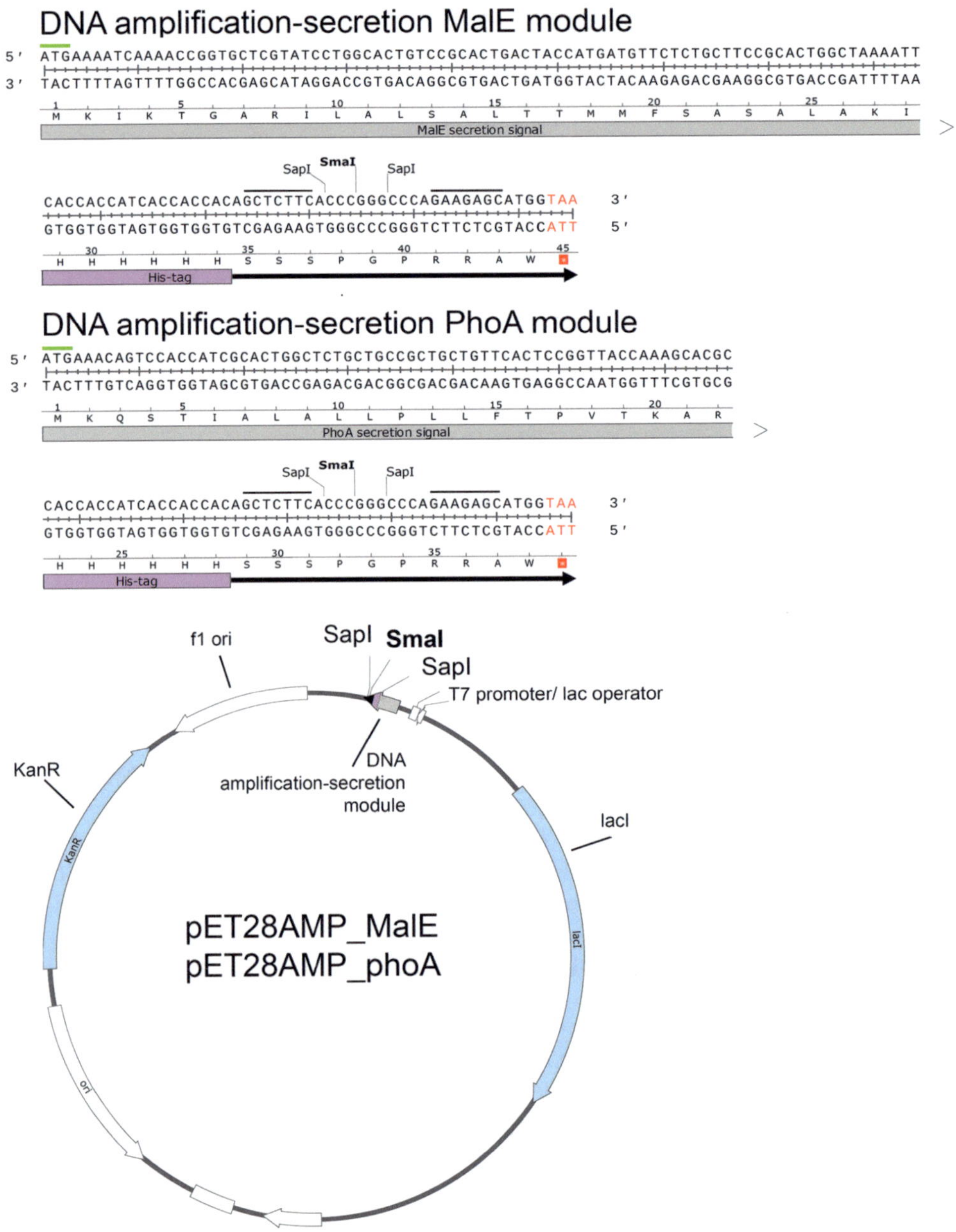

Figure 6.
The pET-derived, amplification-expression-secretion DNA vectors: pET28AMP_PhoA and pET28AMP_MalE, designed and constructed for the DNA-FACE™ technology. The pET28AMP_PhoA DNA vector (GenBank MK606526) and the pET28AMP_MalE vector (GenBank MK606522) contain: (a) colE1 origin, (b) f1/ M13 origin, (c) T7-lac transcription promoter, inducible by lactose or IPTG, (d) the amplification module His6_PhoA_SapI-Sma-SapI or His6_MalE_SapI-Sma-SapI, (e) kanamycin resistance gene.

to facilitate a gradual enzymatic release of the monomers/oligomers, cleaved by the elastase present in human serum [5, 6]. The AGF is known to promote epidermal proliferation, new blood vessel formation, and wound healing in the skin.

The second type of the selected peptides – the RGD and RGDGG peptides – originated from fibronectin. These motifs function as crucial cell-binding factors.

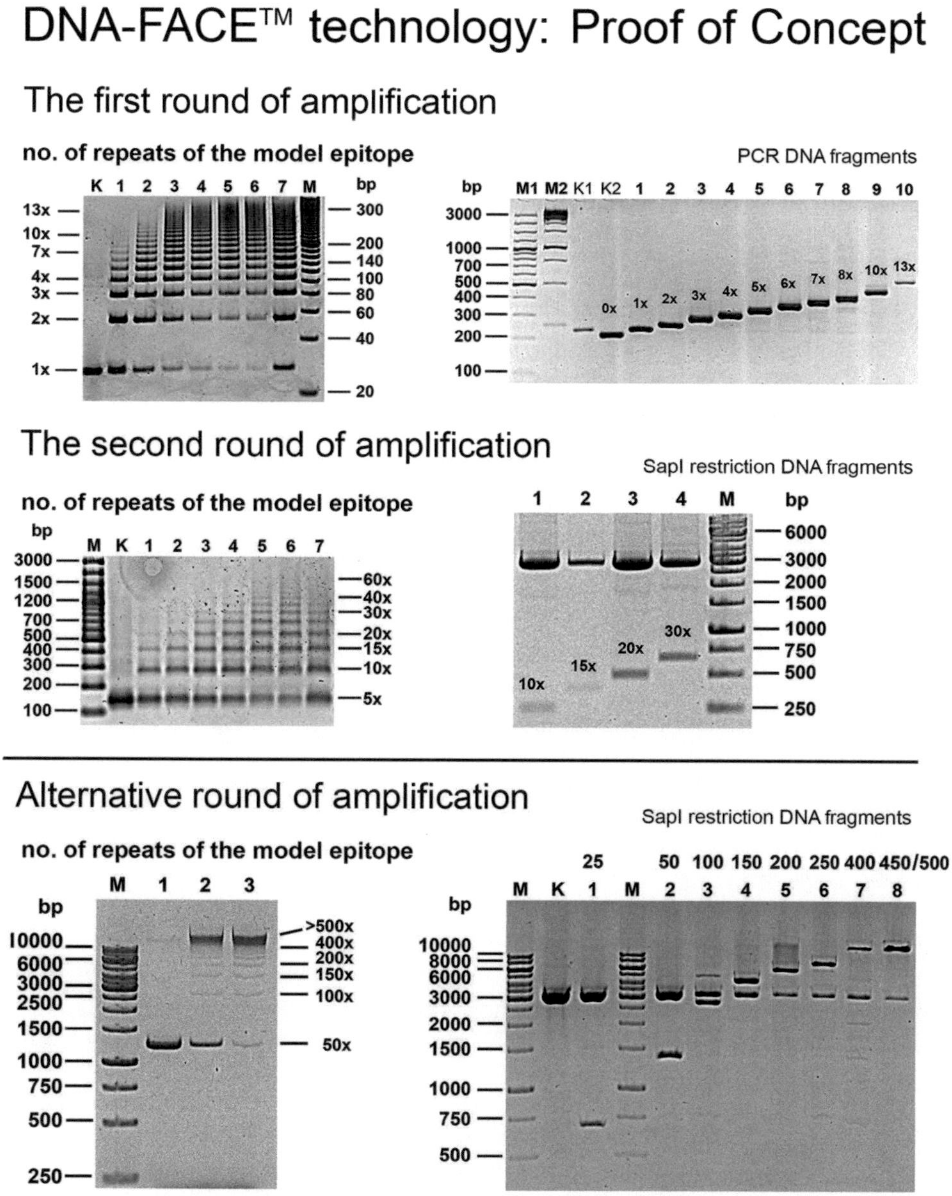

Figure 7.
The DNA-FACE™ technology proof of concept – amplification of the model epitope encoding DNA fragment. The first round of amplification: the synthetic DNA fragment, encoding the model epitope, was cloned into the amplification-expression pAMP1-HisA vector as described by Skowron et al. [5]. Then, the PCR amplification of the appropriate DNA segment was performed. The PCR product was cleaved with SapI and subjected to autoligation at 16°C using T4 DNA ligase and aliquots were taken at intervals of 5, 10, 20, 40, 80, and 160 min. A series of DNA segments of increasing length was obtained. The resulting concatemers were pooled and cloned in pAMP1-HisA, cleaved with SapI. The obtained bacterial clones were analyzed by colony PCR [5]. The second round of amplification: the selected 5-mer was subjected to the second round of amplification [5]. The appropriate DNA fragment was excised from the E. coli clone plasmid and subjected to autoligation. The reaction products were pooled, ligated back to the pAMP1-HisA. The plasmid DNAs from the positive bacterial clones were cleaved by SapI and the obtained restriction fragments were analyzed electrophoretically. Alternative round of amplification: the synthetic 25-mer was subjected to the amplification as described by Skowron et al. [5]. The autoligation products were ligated to the pAMP1-HisA. The plasmid DNAs from the positive bacterial clones were cleaved by SapI and the obtained restriction fragments were analyzed electrophoretically.

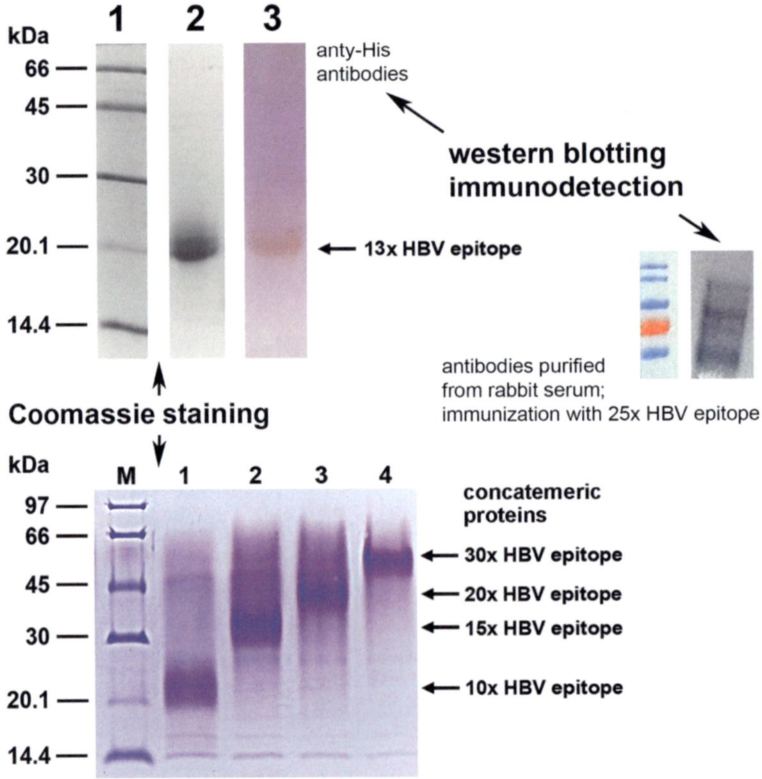

Figure 8.
Concatemeric, recombinant proteins consisting of multiple repeats of a model HBV epitope, obtained with the DNA-FACE™ technology.

The RGD sequence is present in several extracellular matrix (ECM) molecules and is responsible for the mediation of cell attachment. It is known to promote cell/tissue interaction with artificial biomaterials and shows a pro-regenerative effect [32, 33].

The third designed peptide - RLIDRTNANFLGGGAAPVGGG originated from the platelet-derived growth factor (PDGF B). PDGF B functions as a mitogen for fibroblasts and smooth muscles cells and regulates embryonic development. The peptide was extended by GG helical breakers and an AAPV peptide, known to be effectively cleaved by human elastase [34–36].

The fourth peptide series: GHK, GHKGG, GHKGGGAAPVGG, KGHKGGGAAPVGG was designed on the basis of the GHK peptide, which naturally occurs in human plasma and can be released by the injured tissues. The peptide is responsible for diverse protective and healing actions. For example, it is known to improve tissue repair, stimulate blood vessel and nerve outgrowth, boost collagen, elastin, and glycosaminoglycan synthesis [37, 38].

Another explored application of the DNA-FACE™ biotechnology includes construction and evaluation of concatemeric proteins for the purpose of environment remediation and biosensors development. These proteins target toxic heavy metal ions: Pb^{2+}, Hg^{2+} Ag^{+}, As^{3+}, Ni^{2+}, Cd^{2+}, and the uranyl ion (UO_2^{2+}). The details will be released to public domain following submission of the patent application.

2.3.2 Final note concerning concatemeric proteins constructions

It should be noted that the maximum possible monomer copy number within a constructed DNA concatemer could be lower or higher than in the case of the model

HBV epitope [5], as it strongly depends on the DNA sequence and the length of the DNA fragment to be concatemerized. Further precautions concern the downstream applications of the DNA concatemers. Namely, some of the constructed concatemeric genes/ORFs may not be efficiently or error-free transcribed or expressed in *E. coli*. This is again highly dependent on the nt sequence of the resulting mRNA as well as on its resistance to form stable secondary and tertiary structures, which may hide the translation initiation/termination signals or stall the translating ribosomes. Moreover, certain translated amino acids sequences, especially these appearing in ascending concatemeric proteins as repeated segments, may yield low expression levels due to the depletion of highly used aminoacyl-tRNAs, as well as cause the insolubility or toxicity to the recombinant host. Although these potential problems generally concern recombinant genes expression, they may be more pronounced due to the "artificial" nature of the concatemeric proteins. If needed, an implementation of additional strategies can be helpful, such as testing various cultivation/expression conditions, fusions with non-concatemeric proteins, *E. coli* strains, alternative prokaryotic or eukaryotic expression systems, among others. It is worth noting that the DNA amplification-expression modules can be easily transferred to other DNA vectors, including eucaryotic, if necessary.

2.4 Application of the DNA-FACE™: development of a novel type of anti-SARS-CoV-2 vaccine

The DNA-FACE™ concept of construction of "artificial", concatemeric protein with a vaccine functionality has been used to construct an anti-SARS-CoV-2 vaccine.

The co-polymerization type construct contains multiple copies of various epitopes clusters, derived from Spike and Nucleoprotein of the SARS-CoV-2 virus.

The clusters were selected using a proprietary method developed in BioVentures Institute Ltd. (Poland), which has also developed the DNA-FACE™ biotechnology.

The vaccine is the most "Frankenstein", as it is composed of: (*i*) N-terminal polyepitopic clusters of various amino acid sequences and (*ii*) protein adjuvant to enhance the immune response. The entire protein construct has a molecular weight of approximately 70 kDa only, as compared to Spike protein (140.3 kDa kDa) and Nucleoprotein (45.6 kDa). Nevertheless, immunogenicity and toxicity studies in animals (rabbits) have shown that the protein is not toxic and induces a strong, specific immune response (**Figure 9**). Additional advantages of such composite vaccine are: (*i*) no need for refrigeration, as the antigen does not contain native conformation protein, which has to be protected from denaturation, and (*ii*) potential for rapid module exchange if a new virus variant contains changes epitopes. Panel (a) shows a high level of biosynthesis of the recombinant anti-SARS-CoV-2 vaccine protein in *E. coli*. The recombinant gene was cloned into pETAMP1-A amplification-expression vector. Upon induction with IPTG (**Figure 9A, lane 2**), a dominant band of approximately 70 kDa becomes evident, as compared to control, uninduced recombinant *E. coli* culture (**Figure 9A, lane 1**). The induced sample was subjected to western blotting, using primary anti-His6-tag antibodies (**Figure 9B, lane 1**) or rabbit anti-anti-SARS-CoV-2 vaccine (**Figure 9C,** lanes 1 and 2). In both lines, multiple bands appear, which is expected, as rabbits carry also their own anti-*E. coli* proteins antibodies. The band pattern in the lane 2 is different - the dominant protein band of approximately 70 kDa comprises the anti-SARS-CoV-2 vaccine, the remaining bands are most probably a mixture of the vaccine degradation products, reacting *E. coli* proteins, including those, which are co-induced with IPTG. Further assay included western blotting using native Spike and Nucleoprotein proteins (purchased from an independent, foreign company), expressed in human cells,

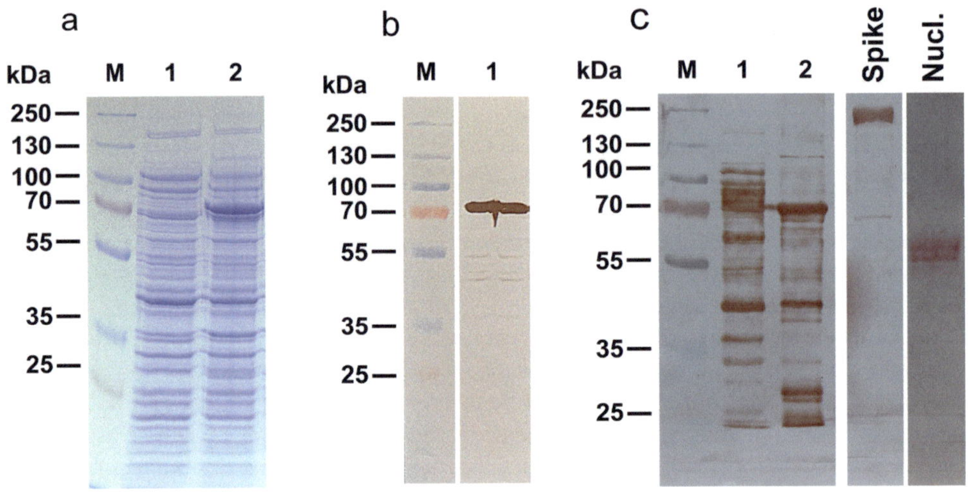

Figure 9.
Biosynthesis and immunogenicity of the recombinant anti-SARS-CoV-2 vaccine. Lane M, Page Ruler™ Plus Prestained Protein Ladder (Thermo Scientific) (a) SDS-PAGE analysis of the recombinant anti-SARS-CoV-2 vaccine protein biosynthesis in E. coli.; lane 1, control E. coli lysate, without induction of the recombinant gene expression; lane 2, E. coli lysate, 3 hours after induction of the recombinant gene expression. (b) Western blotting and immunodetection of the recombinant anti-SARS-CoV-2 vaccine protein using anti-His-tag antibodies. (c) Western blotting and immunodetection of the recombinant anti-SARS-CoV-2 vaccine protein, using antibodies purified from the immunized rabbit serum. Lane 1, E. coli lysate, 3 hours after induction; lane 2, purified anti-SARS-CoV-2 vaccine protein; lane Spike, immunodetection of human, recombinant SARS-CoV-2 Spike protein; lane Nucl., immunodetection of the human, recombinant SARS-CoV-2 Nucleocapsid protein.

thus identical to those present in the virus, including posttranslational modification, absent in *E. coli*. Nevertheless, strong, specific immunological reaction was obtained. Currently, this novel type of anti-SARS-CoV-2 vaccine undergoes further full-scale evaluation, the regulated pre-clinical animal and *in vitro* tests and its clinical tests will follow shortly. Important aspect of this vaccine design is that it does not contain intact Spike and Nucleoprotein, which are known to be toxic to human immunological system, among other negative effects. Thus, it is expected, that the vaccine will have much reduced, if any, side effects upon vaccination. It needs to be emphasized that the DNA-FACE™ concept used to develop anti-SARS-CoV-2 vaccine also applies to essentially all types of microbial pathogens as well as to cancer cells.

3. Conclusions

The DNA-FACE™ biotechnology was developed for the construction of "artificial", repetitive genes, encoding concatemeric RNAs and proteins of any nt and aa sequence. The DNA-FACE™ is capable of formatting of ordered polymers in a controlled process, containing 500 or more copies of DNA, RNA, or peptide repeats within a concatemer.

The constructed concatemeric genes yield efficient genetic expression of concatemeric proteins, which were tested in the development of:

i. New generation of vaccines with an enhanced stimulation of the immune system, including anti-SARS-CoV-2 vaccine.

ii. Concatemeric proteins contain modules for toxic and/or rare metal ions chelation for their industrial obtainment, environmental remediation, or organism detoxification.

iii. Reservoirs for bioactive peptides, either developed or derived from the signaling proteins, which can stimulate tissue regeneration.

iv. Protective, therapeutic concatemeric proteins, containing peptide activators or inhibitors of biological functions, for a new type of biological drugs development, aiming at the treatment of molecular, viral, and bacterial diseases.

The DNA-FACE™ technology is uniquely suited for wide applications in the scientific research, biotechnology, pharmaceutical, and medical industries. The method goes far beyond current chemical genes synthesis capabilities. It allows for gene and protein design solutions, which were impossible before the development of this technology. This opens new research avenues not only for studying biological systems but also for practical solutions, such as novel types of cancer inhibitors, currently under development, using DNA-FACE™. We believe that the scientific and industrial community will recognize the potential of the DNA-FACE™ technology, and several new applications of the technology will soon appear.

Acknowledgements

The authors thank Joanna Żebrowska and Edyta Raczuk for providing the unpublished data, concerning SARS-CoV-2 vaccine. The authors also thank Olga Żołnierkiewicz, Natalia Krawczun, Daria Krefft, Joanna Żebrowska, Beata Łubkowska and all the others who have contributed to the development and validation the DNA-FACE™ technology. Jakub Zaleski is appreciated for English corrections. This work was supported by the National Center for Research and Development (Poland) [grant numbers: POIG.01.04.00-22-140/12 to BioVentures Institute Ltd. (Poznan, Poland), STRATEGMED1/235077/9/NCBR/2014 to University of Gdansk, Faculty of Chemistry, Molecular Biotechnology Department and BioVentures Institute Ltd., and TECHMATSTRATEG2/410747/11/NCBR/2019.

Conflict of interest

The authors declare no conflict of interest.

Author details

Piotr M. Skowron* and Agnieszka Zylicz-Stachula*
Department of Molecular Biotechnology, Faculty of Chemistry, University of Gdansk, Gdansk, Poland

*Address all correspondence to: piotr.skowron@ug.edu.pl
and a.zylicz-stachula@ug.edu.pl

References

[1] Fosgerau K, Hoffmann T. Peptide therapeutics: Current status and future directions. Drug Discovery Today. 2015;**20**(1):122-128. DOI: 10.1016/j.drudis.2014.10.003

[2] Lau J, Dunn M. Therapeutic peptides: Historical perspectives, current development trends, and future directions. Bioorganic and Medicinal Chemistry. 2018;**26**(10):2700-2707. DOI: 10.1016/j.bmc.2017.06.052

[3] Chow D, Nunalee ML, Lim DW, Simnick AJ, Chilkoti A. Peptide-based biopolymers in biomedicine and biotechnology. Materials Science and Engineering R: Reports. 2008;**62**(4): 125-155. DOI: 10.1016/j.mser.2008.04.004

[4] Lee H, Piao D, Lee J, Choi J, Bok J, Cho C, et al. Artificially designed recombinant protein composed of multiple epitopes of foot-and-mouth disease virus as a vaccine candidate. Microbial Cell Factories. 2017;**16**(1):33. DOI: 10.1186/s12934-017-0648-2

[5] Skowron PM, Krawczun N, Zebrowska J, Krefft D, Zolnierkiewicz O, Bielawa M, et al. A vector-enzymatic DNA fragment amplification-expression technology for construction of artificial, concatemeric DNA, RNA and proteins for novel biomaterials, biomedical and industrial applications. Materials Science and Engineering C: Materials for Biological Applications. 2020;**108**:110426. DOI: 10.1016/j.msec.2019.110426

[6] Skowron PM, Krawczun N, Żebrowska J, Krefft D, Żołnierkiewicz O, Bielawa M, et al. Data regarding a new, vector-enzymatic DNA fragment amplification-expression technology for the construction of artificial, concatemeric DNA, RNA and proteins, as well as biological effects of selected polypeptides obtained using this method. Data in Brief. 2019;**28**:105069. DOI: 10.1016/j.dib.2019.105069

[7] Skowron PM, Krawczun N, Żebrowska J, Krefft D, Żołnierkiewicz O, Bielawa M, et al. An efficient method for the construction of artificial, concatemeric DNA, RNA and proteins with genetically programmed functions, using a novel, vector-enzymatic DNA fragment amplification-expression technology. MethodsX. 2020;**7**:101070. DOI: 10.1016/j.mex.2020.101070

[8] Skowron P, Zylicz-Stachula A, Zołnierkiewicz O, Skowron M, Janus Ł, Jezewska-Frąckowiak J, Szymańska A. Sposób uzyskiwania białka poliepitopowego oraz wektor DNA do realizacji tego sposobu (A method of obtaining a polyepitopic protein as well as a DNA vector for embodying this method), Republic of Poland patent office no 228341, 2018

[9] Skowron P, Zylicz-Stachula A, Zolnierkiewicz O, Skowron M, Janus L, Jezewska-Frackowiak J, Nidzworski D., Szemiako K, Maciejewska N, Nowak M, Szymanska A. Method of obtaining a polyepitopic protein and polyepitopic DNA vector. United States Patent US 10,874,735 B2, 2020

[10] Skowron P, Zylicz-Stachula A, Zolnierkiewicz O, Skowron M, Janus L, Jezewska-Frackowiak J, Nidzworski D, Szemiako K, Maciejewska N, Nowak M, Szymanska A. Method of obtaining a polyepitopic protein and polyepitopic DNA vector. European Patent Office EP 3 134 426 B1, 2020

[11] Skowron P, Zylicz-Stachula A, Zolnierkiewicz O, Skowron M, Janus L, Jezewska-Frackowiak J, Krefft D, Nidzworski D, Szemiako K, Maciejewska N, Nowak M, Szymanska A. A method of obtaining a polyepitopic protein as well as a DNA vector for embodying this method. Japanese Patent Office no P6692796, 2020

[12] Skowron P, Zylicz-Stachula A, Zolnierkiewicz O, Skowron M, Janus L, Jezewska-Frackowiak J, Krefft D, Nidzworski D, Szemiako K, Maciejewska N, Nowak M, Szymanska A: A method of obtaining a polyepitopic protein as well as a DNA vector for embodying this method. Israel Patent Office no 248011, 2021

[13] Skowron P, Zylicz-Stachula A, Zolnierkiewicz O, Skowron M, Janus L, Jezewska-Frackowiak J, Krefft D, Nidzworski D, Szemiako K, Maciejewska N, Nowak M, Szymanska A. A method of obtaining a polyepitopic protein as well as a DNA vector for embodying this method. China Patent Office no CN 106488983 A, 2016

[14] Skowron P, Zylicz-Stachula A, Zolnierkiewicz O, Skowron M, Janus L, Jezewska-Frackowiak J, Krefft D, Nidzworski D, Szemiako K, Maciejewska N, Nowak M, Szymanska A. A method of obtaining a polyepitopic protein as well as a DNA vector for embodying this method. Indian Patent Office no 201647039411, 2016

[15] Kempe T, Kent SB, Chow F, Peterson SM, Sundquist WI, L'Italien JJ, et al. Multiple-copy genes: Production and modification of monomeric peptides from large multimeric fusion proteins. Gene. 1985;**39**(2-3):239-245. DOI: 10.1016/0378-1119(85)90318-x

[16] Kim SC, Szybalski W. Amplification of cloned DNA as tandem multimers using BspMI-generated asymmetric cohesive ends. Gene. 1988;**71**(1):1-8. DOI: 10.1016/0378-1119(88)90071-6

[17] Lee JH, Skowron PM, Rutkowska SM, Hong SS, Kim SC. Sequential amplification of cloned DNA as tandem multimers using class-IIS restriction enzymes. Genetic Analysis. 1996;**13**(6):139-145. DOI: 10.1016/s1050-3862(96)00164-7

[18] Lee JH, Minn I, Park CB, Kim SC. Acidic peptide-mediated expression of the antimicrobial peptide buforin II as tandem repeats in *Escherichia coli*. Protein Expression Purification. 1998;**12**(1):53-60. DOI: 10.1006/prep.1997.0814

[19] Wang YQ, Cai JY. High-level expression of acidic partner-mediated antimicrobial peptide from tandem genes in *Escherichia coli*. Applied Biochemistry and Biotechnology. 2007;**141**(2-3):203-213. DOI: 10.1007/BF02729062

[20] Lennick M, Haynes JR, Shen SH. High-level expression of alpha-human atrial natriuretic peptide from multiple joined genes in *Escherichia coli*. Gene. 1987;**61**(1):103-112. DOI: 10.1016/0378-1119(87)90369-6

[21] McDaniel JR, Mackay JA, Quiroz FG, Chilkoti A. Recursive directional ligation by plasmid reconstruction allows rapid and seamless cloning of oligomeric genes. Biomacromolecules. 2010;**11**(4):944-952. DOI: 10.1021/bm901387t

[22] Won J, Barron A. A new cloning method for the preparation of long repetitive polypeptides without a sequence requirement. Macromolecules. 2002;**35**(22):8281-8287. DOI: 10.1021/ma020892i

[23] Shen SH. Multiple joined genes prevent product degradation in *Escherichia coli*. Proceedings of the National Academy of Sciences of the United States of America. 1984;**81**(15):4627-4631. DOI: 10.1073/pnas.81.15.4627

[24] Li MV, Shukla D, Rhodes BH, Lall A, Shu J, Moriarity BS, et al. HomeRun vector assembly system: A flexible and standardized cloning system for assembly of multi-modular DNA constructs. PLoS One. 2014;**9**(6):e100948. DOI: 10.1371/journal.pone.0100948

[25] Meyer DE, Chilkoti A. Genetically encoded synthesis of protein-based polymers with precisely specified molecular weight and sequence by recursive directional ligation: Examples from the elastin-like polypeptide system. Biomacromolecules. 2002;**3**(2):357-367. DOI: 10.1021/bm015630n

[26] Chu HS, Ryum J, Park SY, Kim BG, Kim DM, Won JI. A new cloning strategy for generating multiple repeats of a repetitive polypeptide based on non-template PCR. Biotechnology Letters. 2011;**33**(5):977-983. DOI: 10.1007/s10529-010-0510-7

[27] Mi L. Molecular cloning of protein-based polymers. Biomacromolecules. 2006;**7**(7):2099-2107. DOI: 10.1021/bm050158h

[28] Ott W, Nicolaus T, Gaub HE, Nash MA. Sequence-independent cloning and post-translational modification of repetitive protein polymers through Sortase and Sfp-mediated enzymatic ligation. Biomacromolecules. 2016;**17**(4):1330-1338. DOI: 10.1021/acs.biomac.5b01726

[29] Rodríguez-Cabello JC, Girotti A, Ribeiro A, Arias FJ. Synthesis of genetically engineered protein polymers (recombinamers) as an example of advanced self-assembled smart materials. Methods in Molecular Biology. 2012;**811**:17-38. DOI: 10.1007/978-1-61779-388-2_2

[30] Chang AC, Cohen SN. Construction and characterization of amplifiable multicopy DNA cloning vehicles derived from the P15A cryptic miniplasmid. Journal of Bacteriology. 1978;**134**(3):1141-1156. DOI: 10.1128/jb.134.3.1141-1156.1978

[31] Wojtowicz-Krawiec A, Sokolowska I, Smorawinska M, Chojnacka-Puchta L, Mikiewicz D, Lukasiewicz N, et al. Use of Ubp1 protease analog to produce recombinant human growth hormone in *Escherichia coli*. Microbial Cell Factories. 2014;**13**(1):113. DOI: 10.1186/s12934-014-0113-4

[32] Schaffner P, Dard MM. Structure and function of RGD peptides involved in bone biology. Cellular and Molecular Life Sciences. 2003;**60**(1):119-132. DOI: 10.1007/s000180300008

[33] Czupryn A., Mucha P., Skowron PM, Mazuryk J, Wiśniewska M, Koziński K, Beresewicz M, Krupska O, Załuska I, Palczewska-Groves M, Żylicz-Stachula A, Żebrowska J, Piotrowski P, Pikuła M, Sachadyn P, Janus Ł, Rodziewicz-Motowidło S. PolyRGD peptides obtained by chemical and biotechnological method with antidemarcation and/or neuroprotective effect and a set for determination of their activity. European Patent Office no EP 18000319.6 (2018) and Republic of Poland Patent Office no P.425131 (2018). Patent cloud applications pending

[34] Skowron PM, Żylicz-Stachula A, Żebrowska J, Palczewska-Groves M, Maciejewska N, Czupryn A, Janus Ł, Mucha M, Pikuła M, Piotrowski A, Sachadyn P, Rodziewicz-Motowidło S, Sawicka J, Dzierżyńska M, Karpowicz P. Modified amplification and expression vectors, fusion proteins, composition and their application. Republic of Poland Patent Office no P.427146 (2018). Patent cloud applications pending.

[35] Deptuła M, Karpowicz P, Wardowska A, Sass P, Sosnowski P, Mieczkowska A, et al. Development of a peptide derived from platelet-derived growth factor (PDGF-BB) into a potential drug candidate for the treatment of wounds. Advances in Wound Care (New Rochelle). 2020;**9**(12):657-675. DOI: 10.1089/wound.2019.1051

[36] Shim AH, Liu H, Focia PJ, Chen X, Lin PC, He X. Structures of a platelet-derived growth factor/propeptide complex and a platelet-derived growth

factor/receptor complex. Proceedings of the National Academy of Sciences of the United States of America. 2010;**107**(25): 11307-11312. DOI: 10.1073/pnas. 1000806107

[37] Pickart L. The human tri-peptide GHK and tissue remodeling. Journal of Biomaterials Science, Polymer Edition. 2008;**19**(8):969-988. DOI: 10.1163/ 156856208784909435

[38] Pickart L, Margolina A. Regenerative and protective actions of the GHK-Cu peptide in the light of the new gene data. International Journal of Molecular Sciences. 2018;**19**(7):1987. DOI: 10.3390/ijms19071987